Broadband Dielectric Spectroscopy: A Modern Analytical Technique

ACS SYMPOSIUM SERIES **1375**

Broadband Dielectric Spectroscopy: A Modern Analytical Technique

William Henry Hunter Woodward, Editor
Core R&D, The Dow Chemical Company
Midland, Michigan, United States

Sponsored by the
ACS Division of Analytical Chemistry

American Chemical Society, Washington, DC

Library of Congress Cataloging-in-Publication Data

Names: Woodward, William Henry Hunter, editor. | American Chemical Society. Division of Analytical Chemistry, sponsoring body.
Title: Broadband dielectric spectroscopy: a modern analytical technique / William Henry Hunter Woodward, editor, Core R&D, The Dow Chemical Company Midland, Michigan, United States ; sponsored by the ACS Division of Analytical Chemistry.
Description: Washington, DC : American Chemical Society, [2021] | Series: ACS symposium series; 1375 | Includes bibliographical references and index.
Identifiers: LCCN 2021003378 (print) | LCCN 2021003379 (ebook) | ISBN 9780841298484 (hardcover) | ISBN 9780841298477 (ebook other)
Subjects: LCSH: Broadband dielectric spectroscopy.
Classification: LCC QC454.B74 B763 2021 (print) | LCC QC454.B74 (ebook) | DDC 543/.6--dc23
LC record available at https://lccn.loc.gov/2021003378
LC ebook record available at https://lccn.loc.gov/2021003379

The paper used in this publication meets the minimum requirements of American National Standard for Information Sciences—Permanence of Paper for Printed Library Materials, ANSI Z39.48n1984.

Foreword

The ACS Symposium Series is an established program that publishes high-quality volumes of thematic manuscripts. For over 40 years, the ACS Symposium Series has been delivering essential research from world leading scientists, including 36 Chemistry Nobel Laureates, to audiences spanning disciplines and applications.

Books are developed from successful symposia sponsored by the ACS or other organizations. Topics span the entirety of chemistry, including applications, basic research, and interdisciplinary reviews.

Before agreeing to publish a book, prospective editors submit a proposal, including a table of contents. The proposal is reviewed for originality, coverage, and interest to the audience. Some manuscripts may be excluded to better focus the book; others may be added to aid comprehensiveness. All chapters are peer reviewed prior to final acceptance or rejection.

As a rule, only original research papers and original review papers are included in the volumes. Verbatim reproductions of previous published papers are not accepted.

ACS Books

Contents

Applied Dielectric Spectroscopy of Aqueous Systems

Indexes

Preface

The public discussion of early scientific work with colleagues and peers has been an essential part of the scientific process since the birth of modern conferences in the nineteenth century. New research and new ideas can be presented and debated in an amicable and professional atmosphere. This exchange often results in interlaboratory collaborations and cross pollination with outside fields. Also existing for two centuries has been the immediate documentation of presentations at said meetings. Making discussion topics rapidly available to those unable to attend or who would like to reference the work is an important part of the scientific process, and American Chemical Society Meetings are no exception. Replacing the decades-old Advances in Chemistry Series, the peer-reviewed Symposium Series was born in 1974 following the 167th Spring Meeting in Los Angeles, with the first volume "Approaches to Automotive Emissions Control" (edited by Richard Hurn). These books are summaries of symposia sponsored by divisions of the ACS with the intention of capturing the topics of discussion. Chapters are individually contributed by invited authors/ presenters, and the expansion of the eBook format across all academic fields now allows interested readers to access and reference single chapters or the complete text instantly.

This Symposium Series book, sponsored by the Analytical Chemistry Division of the ACS, captures the discussion that would have occurred at the cancelled Spring 2020 Meeting in Philadelphia at a Symposium dedicated to the Analytical Chemistry aspects of Broadband Dielectric Spectroscopy (BDS). This book is organized into three sections. The first section discusses the science and methods by which to perform and analyze BDS experiments, beginning with motivation for this Symposium and this book. This introduces the technique by outlining the growth of BDS over the recent past, describing experiments in the modern analytical chemistry laboratory, and briefly reviewing recent complementary studies utilizing BDS. The next chapter expands upon the experimental nuances and provides some historical references to the development of modern equipment. The remainder of this section is devoted to pertinent experimental details, including specific experimental examples on the measurement of a 3-D printed polymer over a wide frequency range, the effect of electric field on the analytical aspects of BDS, and the use of probe molecules in order to improve experimental resolution when studying the dynamics of nonpolar systems.

The second section focuses on polymer dynamics—a huge field of BDS research. The use of BDS to probe the physical aging of polymers under various conditions is discussed with a detailed review of polymer dynamics experiments and the limitations of measuring a changing system. The study of polymers on the nanoscale via BDS is then introduced; a topic that would have occupied much of the Symposium in Philadelphia. The authors' myriad contributions to this field through years of interlaboratory collaborations is clear, and four chapters follow this trend of investigating the dynamics of nanoscale polymers; first in the study of polymer nanocomposites, then in the measurement of thin polymer films under various topologies to examine the effects of the surroundings/experimental conditions, then under 2-D confinement as polymers are filled into nanoscale inorganic tubes, and finally as the shape and size of nanostructures are altered. The section is concluded by providing a study of how sample preparation conditions affect the electrical properties of crosslinked polymers for high-voltage applications.

The third and final section provides two studies on aqueous systems, which recent conference discussions affirm is an extremely important and active topic in BDS today. First is a study on how the dynamics of water in hydrophilic solutions is affected by isotope substitutions, and second is a review of the nuanced experimental considerations in the study of water at microwave frequencies. Both topics reflect the importance in studying the behavior of water via BDS and provide detailed experimental descriptions for the interested analytical chemist.

I am very grateful to the ACS and to the authors for supporting and contributing to this book. The 33 contributors of the 13 chapters hail from 18 institutions and 6 countries, which supports the precept that conference symposia are an excellent catalyst for interlaboratory scientific discussion. I am also thankful to the more than 60 peer reviewers who contributed to the quality of this work, the Analytical Chemistry Division and its officers, and especially to Amanda Koenig and Tracey Glazener for their support from the Publications Division. I hope that readers find this book a useful reference for years to come. The topics of discussion here are valuable not just for the newer scientist looking to develop these capabilities in their own laboratory but also as a snapshot in time of the state-of-the-art research in BDS today. In Chapter 1 several references are given to thirty years ago and predictions for thirty years in the future. This number is arbitrary: it simply represents the working span that career scientists can expect to experience. However, it is worth mentioning that the publication of this book marks the thirtieth anniversary of the passing of Robert Cole, a ubiquitous name in the field who is referenced heavily herein. It therefore seems appropriate to dedicate this book to his memory. For an excellent perspective on the relationship between chemistry and dielectric spectroscopy, Robert Cole's "Dielectrics in Physical Chemistry" (Annu. Rev. Phys. Chem. 40, **1989,** 1-28) is highly recommended. Although I never had the pleasure of meeting Professor Cole, I am given to understand that he was an excellent scientist and collaborator. I am fortunate enough to have worked with his contemporaries—including the authors of this book—and am grateful for the cooperative, inclusive, and friendly atmosphere that has prevailed at scientific discussions on BDS all over the world.

William Henry Hunter Woodward
Dow Core Research and Development
Midland, Michigan 48667, United States

Applied Dielectric Spectroscopy: Experimental Strategies

Chapter 1

Broadband Dielectric Spectroscopy—A Practical Guide

W. H. Hunter Woodward*

Core R&D, The Dow Chemical Company, Midland, Michigan 48667, United States

***Email: wwoodward@dow.com**

The purpose of this chapter is to provide a practical overview of broadband dielectric spectroscopy (BDS) as it applies to the modern analytical chemistry laboratory. The word "practical" is carefully chosen here—this chapter will focus on summarizing the everyday practice of this rapidly growing technique for the unfamiliar but interested analytical chemist. There are more-comprehensive reviews, perspectives, and books on BDS which will be suggested here, but over the next several pages it is hoped that the reader who is almost entirely new to this technique will gain an appreciation for its applicability toward solving unique problems in cooperation with several other ubiquitous analytical techniques—and will gain a confident understanding of the capabilities needed to duplicate these efforts. Much of this chapter can be ignored by the experienced dielectric spectroscopist, although it is the desire of this author that the perspective of an industrial chemist provides at least some degree of education for all. The chapter will first provide a brief history of the technique and justification for this book, followed by a simple review of data analysis, then an overview of state-of-the-art commercial instrumentation, and finally a review of applications of the technique from the recent literature.

Introduction

Broadband dielectric spectroscopy (BDS) is the study of the movement of electronic dipoles (e.g. dipolar molecules) in the presence of an electric field. It was invented by the Dutch-American physicist Peter Debye in 1912 and so may appear to be a technique suited to the field of physics (*1–3*), but the polarization of a system by the reorientation of molecular dipoles is very much within the borders of chemistry. Philip Ball (former editor of Nature) once said "chemistry is more about electrons than elements" (*4*). Take, for example, the curing of an epoxy resin. Small polar reactant molecules rotating in the presence of an external oscillating field are "tied up" as covalent bonds are formed, leading to slower rotations and reduced dielectric strength. By studying the speed and activation energy of dipole rotation, reaction cure kinetics and molecular dynamics can be concomitantly gleaned—a feat not possible using calorimetry or Fourier transform infrared spectroscopy alone (Figure 1) (5). Although phenomena such as molecular mobility, dynamics, and phase transitions (all represented by the movement of dipoles) may bridge the gap between chemistry and physics, their measurement is a valuable part of analytical chemistry.

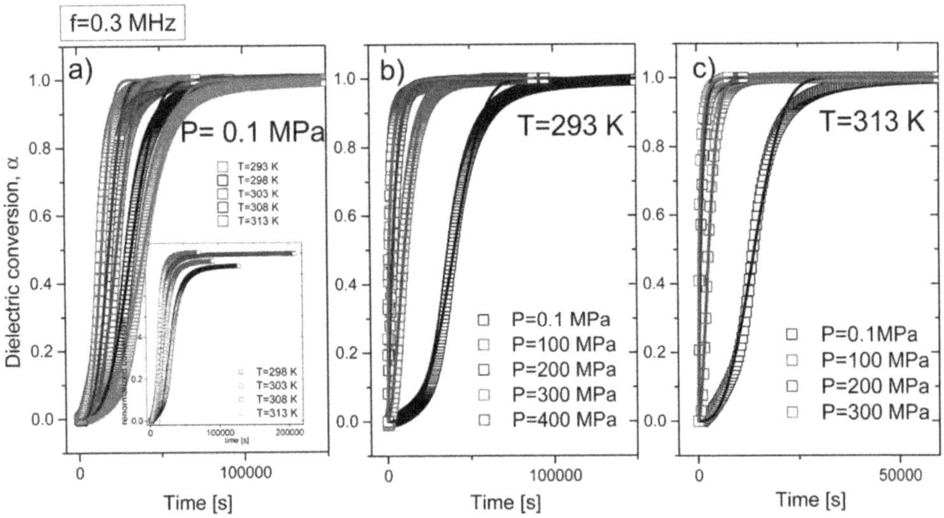

Figure 1. Reaction conversion of bisphenol A diglycidyl ether and 2-ethylhexylamine at a) different temperatures and b) and c) pressures as determined via broadband dielectric spectroscopy (BDS). Lines are fits to the Avrami equation. Inset is the conversion when using a modified Avrami equation that delivers good agreement with Fourier transform infrared spectroscopy (FTIR) at 0.1 MPa. At elevated pressures FTIR was not possible due to technical constraints, but BDS measurements observed that the activation volume was not constant with pressure. Reproduced with permission from reference (5). Copyright 2014 American Chemical Society.

As will be discussed in this chapter, BDS is a spectroscopic technique that produces experimental results directly comparable to other analytical methods such as optical spectroscopy, NMR, rheology, calorimetry, and scattering methods. It is a near benchtop analytical tool that uniquely covers eighteen decades of time scale/frequency and with commercially-available fully-automated systems can probe a wide range of temperatures, pressures, and electric field strengths with minimal supervision. The purpose of this chapter is to provide a brief overview of BDS with common experimental practices and direct comparisons with other analytical techniques in the recent literature. First, however, the growth of this century-old technique will be briefly reviewed.

The Growth of Broadband Dielectric Spectroscopy

Although dielectric spectroscopy has been practiced in a largely unchanged form since the 1940's, the concept of BDS as a modern analytical chemistry method has enjoyed a rapid growth of this technique in the literature. To emphasize this point, Figure 2 demonstrates the number of studies that mention infrared (IR) and nuclear magnetic resonance (NMR) spectroscopies over the past thirty years. One would have a difficult time locating an analytical chemist who does not place these techniques at the top of their toolbox: they are ubiquitous. It is said that knowledge doubles every seventeen years (6), and from this graph it can be seen that IR and NMR studies are doubling every eleven and nineteen years, respectively. Also on this graph is dielectric spectroscopy, which although having far fewer studies is currently doubling every five years. At this pace dielectric spectroscopy is set to match IR and NMR in the next thirty years, although it will surely slow down as it reaches the same saturation limit. One should be critical of any technique that plateaus in this fashion and make every effort to ignite an innovation revolution.

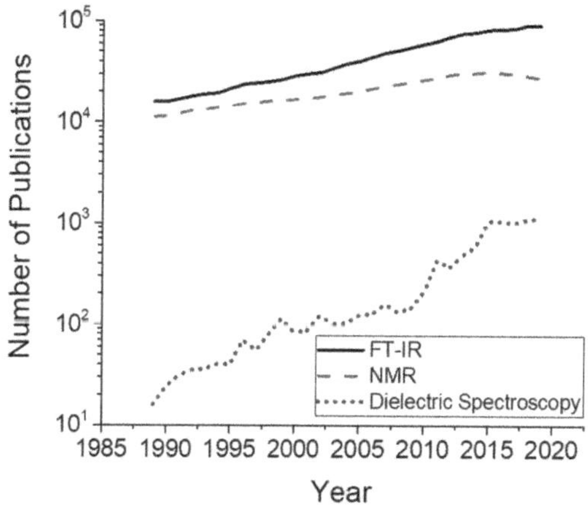

Figure 2. Number of studies that mention (solid line) infrared spectroscopy, (dashed line) nuclear magnetic resonance, and (dotted line) dielectric spectroscopy between 1989 and 2019. Data was collected from SciFinder^n on August 1, 2020. Results were fit to the line $ln(y) = A + bx$, where $ln(2){\cdot}b^{-1}$ is the doubling rate.

This book is specifically sponsored by the Analytical Chemistry Division of the American Chemical Society—the first of its kind dedicated to BDS by this Division—and so it is worthwhile to recount to interested ACS associates how the dedicated organizations and geographic centers of the technique have evolved. This content is, to put it mildly, brief. Many important contributors to the field are not mentioned by name. Nevertheless the history of the technique helps to explain the motivations for the data found in the decades of literature.

Prior to the 1950's many physicists aimed to model and measure the dielectric response of small molecules in the gas and liquid phase to consistently progressing success (notably R. Cole, C. Smyth, and H. Fröhlich from Brown, Princeton, and Liverpool Universities, respectively) (*7–11*). The disagreements between experiment and early theories led to a sub-field of physics that continues today (*12–14*). It is impossible to overstate the value of this work to analytical chemistry studies. Although principally dedicated to non-reactive systems, the study of how the shape, size, and intermolecular dynamics of molecules affected dipolar relaxations supported the continuous development of instrument best-practices as experimentalists began to study complicated liquids and solids to test the latest theoretical models (e.g. liquid crystals, semicrystaline polymers, nanoscale systems, etc.). Additionally, industrial chemists entered the field and began to apply the techniques to their own applications (*15, 16*). A steady expansion of BDS into laboratories all over the world had begun. In the 1950's the burgeoning field of macromolecules welcomed the use of dielectric spectroscopy in the study of polymer dynamics and polymerization kinetics (*17, 18*). In the 1960's and 1970's C. Smyth, M. Davies (University College Wales, Aberystwyth), L. Sutton FRS (Oxford University), and G. Mikhailov (Russian Academy of Sciences) were notable to the field, the latter explaining why more than one quarter of all studies of that time were published in Russian. During this time G. Williams (University of Wales) became an important member of the dielectric and polymer communities (*19*). In the 1980's contributions by the Russian Academy of Sciences continued with a strong presence also emerging from several universities in Germany, Poland, and Japan. Today (1990 to 2019) the field of dielectric spectroscopy enjoys top contributors to the field from every direction, which is perhaps why modern scientific meetings dedicated to dielectric spectroscopy are not centralized to the United States and are most commonly held in Europe.

On this topic: as this is a Symposium Series book it is worthwhile to summarize sources of information from past and future scientific meetings. Since the 1960's there have been professional organizations that meet regularly to discuss recent advancements in the field, and the wide variety of physics, chemistries, and materials backgrounds made topics of discussion very diverse. The formation of these discussion groups was originally out of a need for communication within the field, but today their reach has decidedly grown broader. The proceedings from these meetings are well documented and are perhaps the best source of information to appreciate the development of BDS throughout the 20th century. Of particular value is the history section of the website for the International Dielectric Society and a pair of papers by J. Tanaka (*20–22*). The timeline for the evolution of these groups are presented in Figure 3.

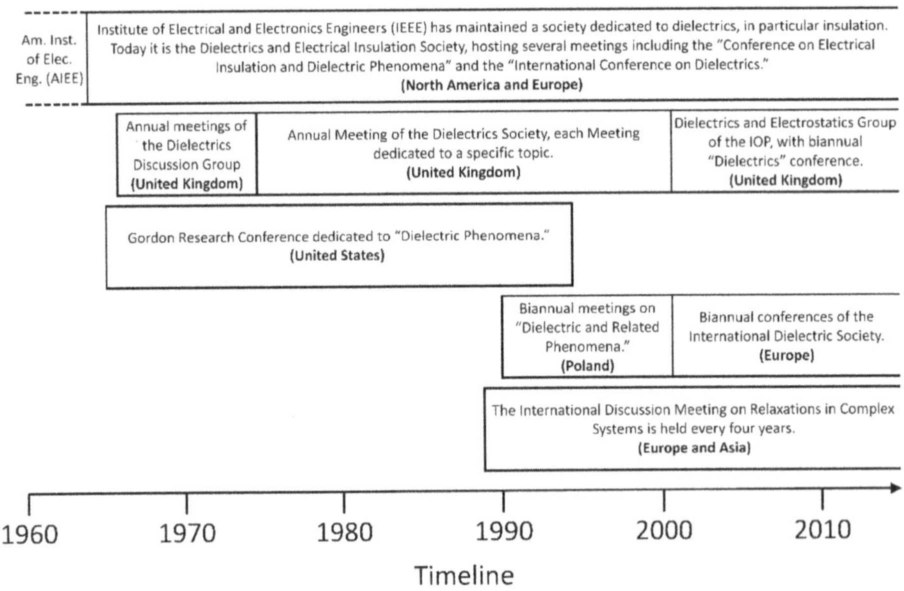

Figure 3. Timeline highlighting the evolution of various discussion groups in different geographies.

Despite these consistent gatherings the American Chemical Society hosted its first Symposium dedicated to the technique in 1997 (*23*). This is perhaps due to the preference of publications in physics journals rather than chemistry journals (*24*), but this chapter will hopefully convince the interested chemist that this technique has historically been extremely relevant to the field of physical chemistry and should be considered a valuable tool.

Still, due to the complex geographical and disciplinary history of this technique, information dedicated to dielectric spectroscopy from the standpoint of analytical chemistry is understandably dispersed through the literature.

It is for this reason that this book was compiled: a collection of modern dielectric spectroscopy studies applicable to the field of analytical chemistry. This chapter serves as an introduction to this technique so that the following chapters can be appreciated by those less familiar with BDS. It is therefore necessary to first describe the basic assumptions that underlie the measurement of a dipolar relaxation.

"Broadband Dielectric Spectroscopy"

Spectroscopy is the study of the interactions between matter and electromagnetic radiation; typically the absorption or emission of specific types of radiation described by frequency, wavelength, or energy (Figure 4). Beginning at the highest frequencies/energies, gamma-ray spectroscopy measures emitted radiation to identify traces of radioactive decay in the range of ~10^{19}-10^{21} Hz. X-ray and ultraviolet (UV) photoelectron spectroscopies use the photoelectric effect to eject core and valence electrons from atomic nuclei to elucidate elemental makeup as well as oxidation state (~10^{15}-10^{17} Hz). Below this energy level, UV-visible spectroscopy uses light to excite electrons into unoccupied orbitals; the measured absorption or emission of radiation thus illuminating the electronic energy structure of a system (~10^{13}-10^{15} Hz). Finally, infrared spectroscopy measures the vibration, rotation, and stretching of chemical bonds and bond angles in molecules and functional groups (~10^{11}-10^{14} Hz). These techniques are incredibly valuable to the modern chemical laboratory for the purposes of determining the elemental and chemical makeup of a system. However, below the infrared an escarpment is reached, both because light sources below 10^{12} Hz are uncommon but also because electrons and atoms are less interested in these sluggish speeds.

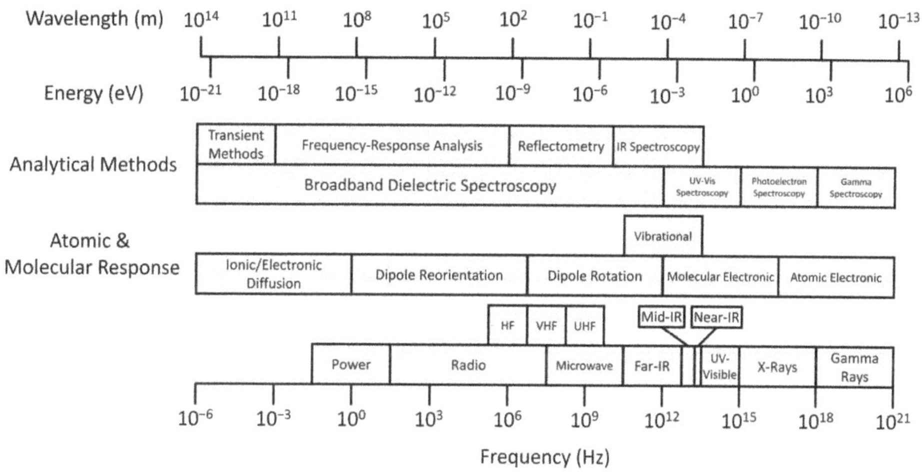

Figure 4. Electromagnetic spectrum with wavelength, frequency, and energy with common spectroscopic techniques listed.

As follows, below ~10^{12} Hz there is a vast field of spectroscopy known as broadband dielectric spectroscopy which studies the rotation and relaxation of dipoles in the presence of an electric field. In BDS the light source is replaced with an oscillating voltage supplied by modern solid-state electronics and the interaction of matter is determined by the perturbation of the field. Since the late 1980's several instrumentational advances have made commercially available dielectric spectrometers more turnkey (25), and modern lock-in amplifiers and frequency-response analyzers (FRAs) can programmatically scan from 10^{-6}-10^{12} Hz with a simple graphical user interface (26). Eighteen decades of frequency (27), compared with the two or three decades from each of the other spectroscopies, is quite the feat. Hence, traditional "dielectric spectroscopy" has become known as "broadband dielectric spectroscopy" since the development of the current instrumentation. The term was first used in the late 1980's and is prevalent today (*28–32*). It should be noted that BDS is often defined in the region between 10^{-6} and 10^{12} Hz. Although modern FRAs can reach frequencies up to 10^7 Hz, eighteen decades is not possible with a single measurement technique (Figure 4). It is therefore summarized that BDS is a field that incorporates all frequencies below infrared spectroscopy, with sub-fields for some specific frequency ranges (e.g. THz spectroscopy and transient spectroscopy methods). Regardless of the timescale, the purpose of BDS is to study of the movement of electronic dipoles in the presence of an electric field. The frequency at which a particular dipole responds is related to the "relaxation time" of that system.

"Relaxation Time"

The fundamental mathematics behind BDS are readily available to the responsive reader and need not be repeated here (*3*, *14*, *19*, *33–36*). A solid understanding of these concepts as well as experience with the equivalent circuits shown in the following section are valuable when examining new systems and are especially helpful in the recognition of flawed data. Nevertheless, a greatly simplified introduction to dielectric spectroscopy to the chemist unfamiliar with these fundamental mathematics does not yet exist. Here a brief foray into the basics of dielectric relaxations will be taken, but it should be understood that the leaps between conclusions will be large. Any reader who wishes to bridge the gaps with a better description of the math is encouraged to do so via the provided references.

The ability of a system to store an electric charge was studied by the 19th century English scientist Michael Faraday and hence he is honored with the unit "farad" (symbol F). The capacitance of a system of two charged parallel plates with electrodes of area A separated by a distance d and filled with a material of relative permittivity ε is (*37*):

$$C = \varepsilon\varepsilon_0 \frac{A}{d} \tag{1}$$

where ε_0 is the permittivity of vacuum: an experimentally determined physical constant approximately equal to 8.854×10^{-12} F·m^{-1}. A predominant amount of dielectric spectroscopy consists of determining the permittivity of a system under different conditions by measuring the capacitance. This is because the permittivity is related to the dielectric displacement D of a material interacting with a small electric field E as predicted by Maxwell's equations:

$$D = \varepsilon\varepsilon_0 E \tag{2}$$

which in essence means that the measured capacitance in the presence of an electric field estimates the susceptibility of a system to polarization just as light or magnetic fields for optical or magnetic spectroscopies, respectively. In particular, BDS measures polarization due to the number and strength of dipoles.

The theory of a molecular dipole was first proposed by P. Debye in 1912 as it applied to charge distributions in molecules (*1*). This work in part earned him the 1936 Nobel Prize in Chemistry (*38*). A dipole is any system such that a plane can be used to dissect it into unequal charges. Dipoles typically consist of electron clouds, atoms, molecules, or parts of molecules. In the presence of an external electric field the electrons will orient on the order of picoseconds (or faster) in a process called “induced polarization” or “deformation.” This can be the electrons of an atom or the electron cloud of a molecule or part of a molecule. On a much slower timescale the entire dipole will orient with the field in a process called “orientational polarization” or simply “orientation.” This process involves the movement of atoms and is appreciable: consider that a water molecule and a neon atom have the same number of electrons, and yet the water molecule exhibits a relative permittivity forty times higher than neon. Clearly the orientation of the dipole is a huge contributor to the amount of charge that a system can store. Advancing through the theory, there will be thermal energy that encourages each dipole to reorient randomly, and many dipoles in proximity (e.g. liquids) will shield interior dipoles from the outer applied electric field and also interact with their neighbors. The result is a complicated amalgam of Maxwell-Boltzmann statistics and Maxwell’s equations that have led to a century of theoretical modelling attempting to predict empirical measurements on a myriad of systems. This work is extensive and broadly successful, with many sources of information available (*3, 7, 8, 10, 11, 14, 19, 39–43*). It will simply be assumed here that at a long enough time to reach steady-state the system will be polarized. The total amount of charge stored in this process is related to the relative permittivity and will consist of both the induced polarization and the orientational polarization (Figure 5).

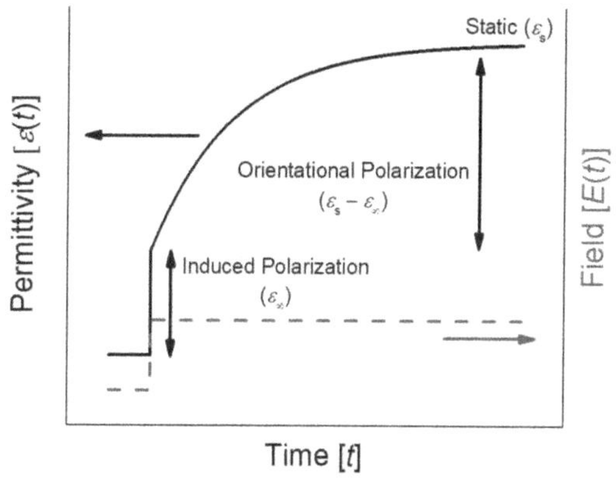

Figure 5. Time-dependent permittivity response (solid line) of a molecular dipole in the presence of an electric field (dashed line).

Once the system has fully charged (i.e. is “static”) the field is removed and the dipoles will relax back to a random configuration. Represented by the reverse of Figure 5, this process is not instantaneous, and so there is a “relaxation time,” designated by the letter τ (*44–46*). As is common and convenient with naturally decaying phenomena, the relaxation time is not the time to completely discharge but is rather the time when the charge reaches e^{-1} of its initial value (~37%). The term relaxation time is used extensively throughout the literature and so it is important to remember what it refers to; an implicit value partway through the relaxation that is seldom explicitly measured.

Equivalent Circuits and Complex Equations

In a classical BDS experiment the sample is placed between two metal electrodes, an electric field is applied, and the signal is measured via connecting electronics. Thus far the system has been discussed pedagogically as an ideal capacitor. Any real-world system will allow some electrical conductance, and so Debye-like dipole behavior can be depicted by an idealized equivalent circuit comprised of resistors (*R*) and capacitors (*C*) (Figure 6). The relaxation time of such a circuit is $\tau = (C_s - C_\infty)\cdot R_s$ where C_∞ is the capacitance of the system due to induced polarization and C_s is the capacitance of the static system (i.e. when fully charged).

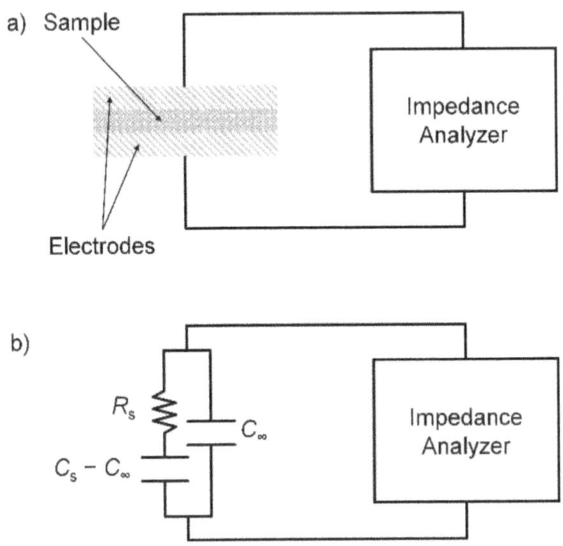

Figure 6. "Ideal" Debye circuit with a conductor and two capacitors, used to emulate a Debye relaxation.

In practice BDS is a frequency-dependent technique, and the response of this idealized circuit to an oscillating field is highly dependent on the frequency. Circuits are commonly described by their complex impedance (Z^*) or admittance ($Y^* = Z^{*-1}$) (47), which accounts for the frequency-dependent electrical resistance of the system. For convenience the formula for the impedance of this idealized Debye circuit is given here:

$$Z^*(\omega) = Z'(\omega) + jZ''(\omega) \tag{3}$$

$$Z'(\omega) = \frac{\omega^2 R_s (C_s - C_\infty)^2}{[\omega^2 R_s C_\infty (C_s - C_\infty)]^2 + \omega^2 {C_s}^2} \tag{4}$$

$$Z''(\omega) = \frac{\omega^3 {R_s}^2 (C_s - C_\infty)^2 C_\infty + \omega C_s}{[\omega^2 R_s C_\infty (C_s - C_\infty)]^2 + \omega^2 {C_s}^2} \tag{5}$$

where ω is the angular frequency (i.e. $2\pi f$), j is $\sqrt{-1}$ (48), and Z' and Z'' are the real and imaginary parts of the impedance, respectively. The real and imaginary parts of the complex impedance are commonly represented via a Nyquist plot (Figure 7a) or a Bode plot (Figure 7b). The former allows for a quick view of the circuit (represented by a depressed semicircle), while the latter depicts the frequency-dependent response.

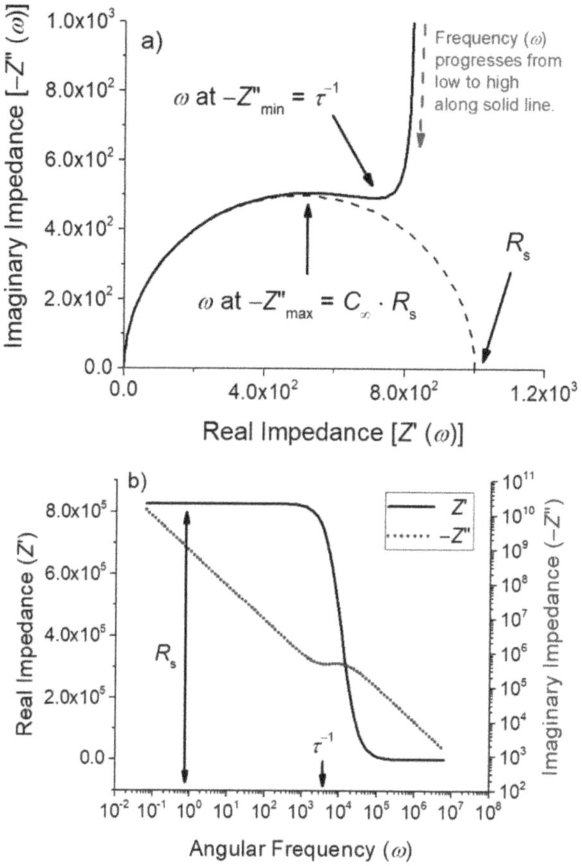

Figure 7. a) Frequency-dependent impedance represented by a Nyquist plot, and b) frequency-dependent impedance represented by a Bode plot using the ideal Debye circuit from Figure 6 ($R_s = 10^6\,\Omega$; $C_s - C_\infty = 10^{-9}$ F; $C_\infty = 10^{-10}$ F). τ is the ideal Debye relaxation time equal to $(C_s - C_\infty)\cdot R_s$.

Complex numbers are often depicted by an Argand diagram and this is somewhat helpful for this discussion (Figure 8).

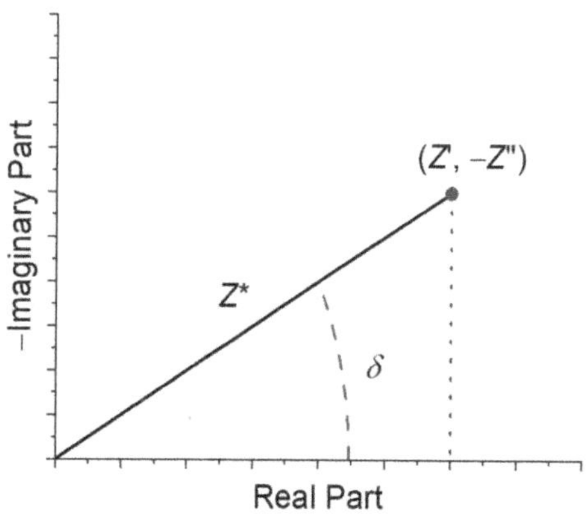

Figure 8. Argand diagram of the complex impedance Z^* at a single frequency. δ is the argument. Consider that the blue data point can be any point along the spectrum (e.g. the Nyquist plot in Figure 7a).

It can be seen that the magnitude of the vector Z* is related to the real and imaginary parts as well as the argument δ via simple trigonometry (49):

$$Z^* = \sqrt{Z'^2 + Z''^2} \tag{6}$$

$$Z' = Z^* \cos\delta \tag{7}$$

$$-Z'' = Z^* \sin\delta \tag{8}$$

$$\frac{-Z''}{Z'} = \tan\delta \tag{9}$$

When $\tan\delta < 1$, $Z' > Z''$ and thus the system is considered more insulating than conducting. This is the definition of a dielectric (50), although to be sure dielectric spectroscopy often studies systems where $\tan\delta > 1$.

Referring back to the permittivity (Equation 1), if the real and imaginary parts of the impedance are measured, then the frequency-dependent permittivity can be calculated:

$$\varepsilon^*(\omega) = \varepsilon'(\omega) - j\varepsilon''(\omega) \tag{10}$$

$$\varepsilon'(\omega) = \frac{d}{\omega\varepsilon_0 A}\frac{-Z''(\omega)}{Z^{*2}} \tag{11}$$

$$\varepsilon''(\omega) = \frac{d}{\omega\varepsilon_0 A}\frac{Z'(\omega)}{Z^{*2}} \tag{12}$$

$$\tan\delta = \frac{\varepsilon''}{\varepsilon'} \tag{13}$$

where the real part ε' is the relative permittivity, the imaginary part ε'' is the dielectric loss or loss factor, and $\tan\delta$ is the dissipation factor (sometimes abbreviated Df, tanD, or D). For the ideal Debye circuit shown in Figure 6:

$$\varepsilon'(\omega) = \frac{d}{\omega\varepsilon_0 A}\omega\left[C_\infty + \frac{(C_s - C_\infty)R_s^{-2}}{R_s^{-2} + \omega^2(C_s - C_\infty)^2}\right] = \varepsilon_\infty + \frac{(\varepsilon_s - \varepsilon_\infty)}{1 + (\omega\tau)^2} \tag{14}$$

$$\varepsilon''(\omega) = \frac{d}{\omega\varepsilon_0 A}\left[\frac{\omega^2(C_s - C_\infty)^2 R_s^{-1}}{R_s^{-2} + \omega^2(C_s - C_\infty)^2}\right] = \frac{(\varepsilon_s - \varepsilon_\infty)\cdot\omega\tau}{1 + (\omega\tau)^2} \tag{15}$$

Again considering an Argand diagram, it is clear that the complex permittivity will be related to both the real and imaginary parts, and that larger values of ε'' or $\tan\delta$ suggest a stronger deviation from an ideal (lossless) dielectric. In essence the dielectric loss determines the amount of energy "lost," or absorbed by the system:

$$P(\omega) = \pi\omega E(\omega)^2\varepsilon''(\omega)\varepsilon_0 A d \tag{16}$$

where E is the amplitude of the AC field in $V{\cdot}m^{-1}$ and P is the power in W. The use of dielectric loss to heat materials through dipole rotation is a commonly utilized phenomenon, for example in radio-frequency welding or microwave ovens.

The negative sign in Equation 10 is because the current lags behind in phase to the electric field. Another helpful way of visualizing the system is by the current response to an AC field. If an AC field is applied to an ideal capacitor the AC current ($I(\omega) = d{\cdot}E(\omega){\cdot}Z^{*}(\omega)^{-1}$) would be phase shifted by exactly $-90°$. But again, the system is not ideal and thus the actual phase shift is $-90° + \delta$ (Figure 9).

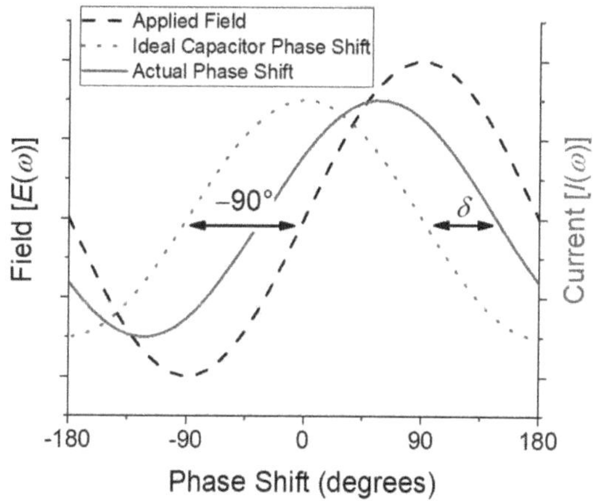

Figure 9. Time-dependence of the current in the presence of an oscillating electric field demonstrating phase shift. Applied field is represented by the dashed line. Phase shift of $-90°$ is represented by the dotted line. Hypothetical nonideal phase shift is represented by the solid line.

The practice of data acquisition will be discussed further in this chapter, but it is worth mentioning that most dielectrics have small values of δ. If there is noise in the signal, the instrumentation is inadequately precise, or an inductance element L exists in the equivalent circuit it is possible to erroneously measure phase shift values less than $-90°$ and hence $\tan\delta$ and ε'' will be negative. For practical purposes the complex permittivity is defined by physical non-negative quantities, and so too must ε', ε'', and $\tan\delta$ be non-negative.

One final note: BDS is intimately related to a technique known as electrochemical impedance spectroscopy (EIS) (*51, 52*). Instead of studying relaxation processes EIS has foundations firmly implanted in electrochemistry and traditionally presents data using impedance instead of permittivity. EIS also commonly interprets data in the form of equivalent circuit representations as opposed to refined molecular interpretations. A DC bias can be applied to the system which allows for electrochemical measurements in addition to the impedance measurements described here. Commonly EIS is used to study the interface of a system, for example the double layer at an electrode or the outer layers of colloidal particles. In these studies the interface under study acts as the equivalent circuit, and the bulk is electrically conductive as to remove its effect on the frequency-dependent impedance spectra. To be sure, EIS has many valuable analytical applications many of which can be achieved via a commercial dielectric spectrometer (many have DC-bias capabilities). Despite the similarities for decades both techniques did not interact or overlap substantially, and this is true to a large extent even today. There are experimental considerations when attempting electrochemistry such as three- or four-electrode setups which will not be addressed here, but there are excellent literature sources available for the interested reader (*51, 53–55*).

Data Representation and Analysis

And so the permittivity of an ideal Debye relaxation (i.e. an RC equivalent circuit) will follow the frequency-dependent complex equation:

$$\varepsilon^*(\omega) = \varepsilon_\infty + \frac{\Delta\varepsilon}{(1 + (j\omega\tau))} \tag{17}$$

where ε_∞ is the permittivity due to induced polarization (i.e. Figure 5), and $\Delta\varepsilon$ is the permittivity due to orientational polarization of a dipole, commonly called the dielectric strength (*56*), and τ is the orientational relaxation time of the dipole.

The Havriliak-Negami Equation

Most commonly BDS data is represented as the real and imaginary part of the permittivity across several decades of frequency akin to a Bode plot (*57*). In this form relaxation times manifest themselves as sigmoid-like steps in the real part and peaks in the imaginary part. Unfortunately, most real-world systems do not exhibit Debye relaxations. In reality most systems exhibit broadened, asymmetric relaxations (Figure 10). Over the past several decades there have been many attempts to describe this effect, with several book chapters dedicated to the topic of empirically fitting dielectric relaxations using models or model-free analyses (*58, 59*). J. Berberian and R. Cole made the pacifying argument that for most models the similarities are more important than the differences (*60*). The discussion will therefore be substantially condensed here: most often in the literature today dielectric relaxations are fit to the Havriliak-Negami (H-N) equation which was proposed by S. Havriliak and S. Negami (Rohm and Haas Co.) in 1967 (*61*):

$$\varepsilon^*(\omega) = \varepsilon_\infty + \frac{\Delta\varepsilon}{(1 + (j\omega\tau)^a)^b} \tag{18}$$

where a and b are empirical values between zero and one (sometimes represented by α and β, respectively) (*62*). When proposed this equation was not accompanied by a phenomenological description of dielectric response but was instead a combination of previous models (*63, 64*), resulting in an equation that fit most studied systems (*65*).

For convenience the real and imaginary parts of the H-N equation are reproduced here. These can be used to fit experimental results using data analysis and graphics software or by one of several recommended methods (*66–68*). Software dedicated to the analysis of dielectric or impedance spectra is also available (WinFIT, ZPlot, RelaxIS, DAK, Grafity, etc.).

$$\varepsilon' - \varepsilon_\infty = \Delta\varepsilon \cdot r(\omega)^{-b/2} \cos[b\Theta(\omega)] \tag{19}$$

$$\varepsilon'' = \Delta\varepsilon \cdot r(\omega)^{-b/2} \sin[b\Theta(\omega)] \tag{20}$$

$$r(\omega) = 1 + 2(\omega\tau)^a \cos\frac{\pi a}{2} + (\omega\tau)^{2a} \tag{21}$$

$$\Theta(\omega) = \tan^{-1}\left(\frac{(\omega\tau)^a \sin\frac{\pi a}{2}}{1 + (\omega\tau)^a \cos\frac{\pi a}{2}}\right) \tag{22}$$

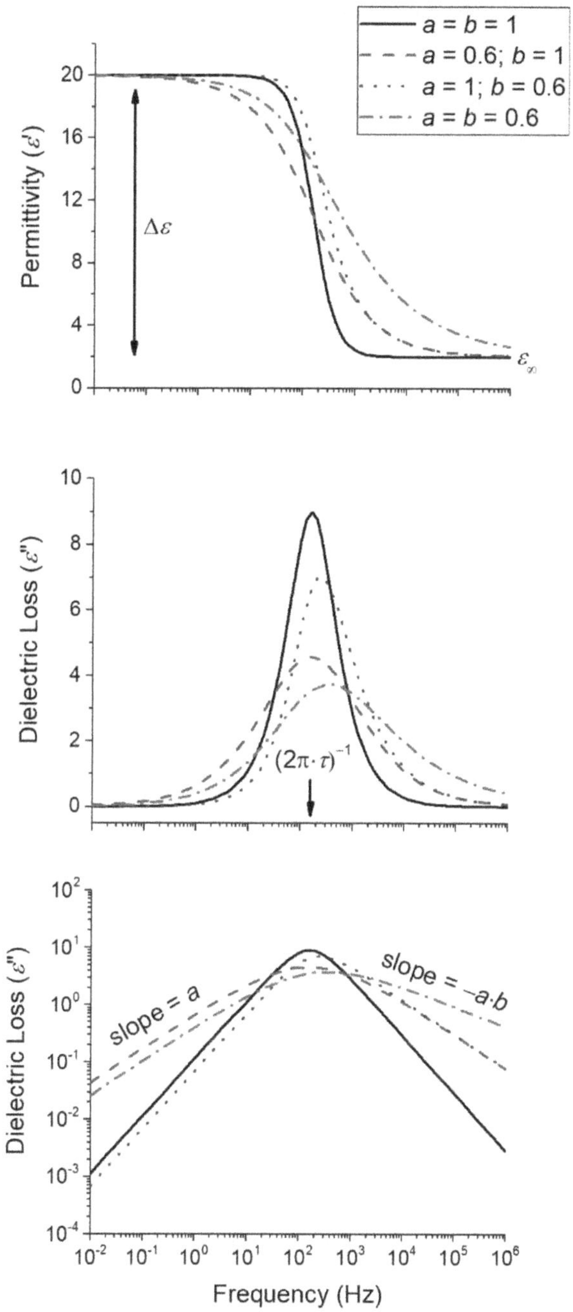

Figure 10. Dielectric spectra demonstrating how a and b from the H-N equation shift the relaxation. All three plots share the same x-axis. Note how the peak in the imaginary spectrum does not correspond to the relaxation time unless b = 1. $\Delta\varepsilon = 18$; $\tau = 10^{-3}$ s; $\varepsilon_\infty = 2$. When $\varepsilon_0 \cdot A \cdot d^{-1} = 2\times10^{-10}$ F this relaxation is similar to the ideal Debye relaxation from Figure 7. Bottom graph is equivalent to the middle graph but uses a logarithmic y-axis, demonstrating the linear wings of a relaxation on a log-log plot.

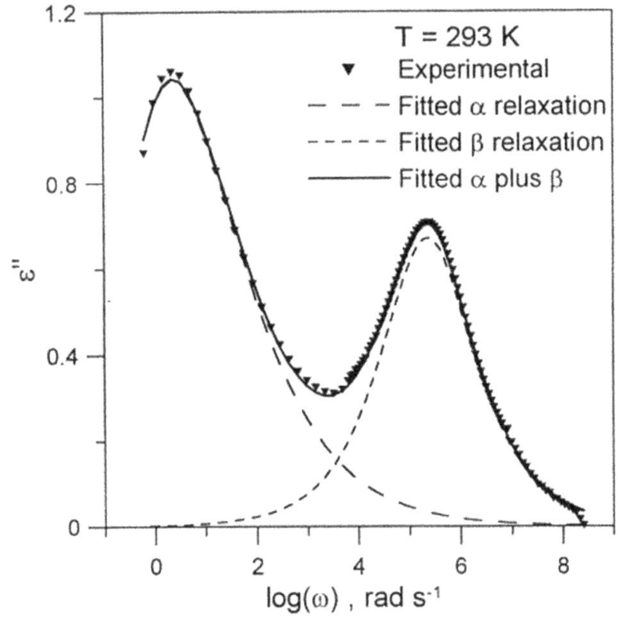

Figure 11. Imaginary part of the permittivity of poly(2-phenyl-5-acryloxymethyl-5-ethyl-1,3-dioxacyclohexane) at 293 K demonstrating two relaxations, each fit using the KWW function. Reproduced with permission from reference (94). Copyright 2002 American Chemical Society.

It has been experimentally demonstrated (*69*) for a myriad of material types that relaxation broadening is due to dynamic heterogeneity—a distribution of discrete relaxation times that summarily create the broadened result (*70–79*). It is possible (and tempting) to computationally fit the data to a set of ideal Debye relaxations or a complicated computer-determined equivalent circuit and then analyze the distribution of relaxation times, but this is not always prudent (*80*). In fact with only a few H-N relaxations most spectra will have an infinite number of solutions with sufficiently good fits. It was shown by Havriliak and Havriliak that most scientists will have a bias toward one fit over others, and thus simply adding more relaxations to a system to achieve a better fit is ill-advised (*81*). The true value of applying BDS toward chemistry comes not from fitting the data such that a minimal residual analysis is found, but from studying the change in an identified dipole relaxation as a function of chemical and physical effects (*82*). When analyzing BDS data, use of the scientific method is of the utmost importance. What dipoles and relaxation times are expected based on past studies? How would experimental variables such as temperature or pressure affect τ or $\Delta\varepsilon$ for these dipoles? It is also valuable to consider how the H-N parameters a and b change with respect to experimental variables—and to be careful that they do not change suddenly due to error in the data analysis (*83–85*).

Complicating things further, for some systems there may be more than one relaxation and these relaxations may overlap (Figure 11). This true, for example, for polymers with segmental and normal mode relaxations, side branching, and/or functional groups (*18*), as each dipole type will exhibit a different relaxation on time scales applicable to BDS. Because of their complicated dielectric behavior, no less than four books are dedicated to BDS of polymers and the many relaxations that they exhibit (*18, 19, 81, 86*). It is common to label each relaxation with a lowercase Greek letter, beginning with the slowest relaxation. For many systems this is the glass transition, and thus the "alpha relaxation" and the glass transition are often used interchangeably. Depending on historical studies of a system, sometimes relaxations are given specific Greek or Arabic letters, are subscripted, or are represented by roman numerals (e.g. α_1, α_2, β_1, β_{JG} (*87–89*), D (*90–93*), I, II, etc.). Again, a thorough literature search will afford consistencies in future studies.

The Lower Frequencies: Polarization Effects

It is pedagogically convenient to describe samples as homogeneous materials of microscopic relaxing dipoles. In truth, systems are rarely homogeneous. One special type of relaxation that manifests itself due to heterogeneities is Maxwell-Wagner-Sillars (MWS) polarization (*95–97*). This effect was originally hypothesized by Maxwell in 1873 when he asserted that heterogeneous materials would demonstrate polarization effects even though neither component would exhibit such behavior individually. A pedagogical (equivalent circuit) example of this is two parallel *RC* circuits in series which do not exhibit relaxations individually but together they create a frequency-dependent permittivity, resulting in an apparent relaxation at lower frequencies. This can be conceptually visualized as a bilayer sample comprised of two layers of dielectric material (Figure 12). If one or both of the layers has a legitimate dielectric relaxation, the MWS polarization can obscure or misalign the measured relaxation parameters.

Figure 12 provides an exaggerated pedagogical warning to the duplicitous effects of MWS via equivalent circuits, but in practice MWS polarization often manifests itself as a broad, subtle shift of the entire spectrum (i.e. not a clear peak) when there is heterogeneity in the electrical conductance or permittivity of a sample, for example in stratified materials or colloids. This is especially important in systems where a relaxation of interest exists, as the MWS polarization effect can increase permittivity at lower frequencies, thereby shifting the studied relaxation (*98–101*). The effect is most prominent when one of three conditions are met: 1) there is an insulating layer that covers one or both electrodes, 2) there are conductive filler particles in a dielectric medium (or vice versa), and 3) there are two domains in the material which have drastically different ε' values. All of these conditions can be taken into account and used to an analytical advantage if conditions are known (*35*), but the results are undoubtedly complicated and so a simple mathematical model is not sufficient (e.g. the shape and size of heterogeneities, the frequency-dependent value of ε', the non-binary transition from one layer to another, etc.). Realistically MWS polarization is not to be underestimated as an unpredictable effect on heterogeneous samples and their resulting spectra (*102*).

Additionally, if interfaces exist such that ions or electrons cannot pass through (a common example is ions collecting at an impermeable metal electrode at low frequencies), one or more layers of charge will form. This is known as electrode polarization and has been the enemy of many a dielectric spectroscopist for over a century (*103, 104*). There are strategies to mitigate or control electrode polarization, but it is often persistent (*105–107*). A strong disadvantage of electrode polarization is that the collection of charge will shield the sample from the electric field, giving the impression of increased ε'. The effects of macroscopic electrode polarization and microscopic MWS polarization are similar and the two are often discussed together, but to be sure they are different (*108*). In some experiments (for example, certain electrochemical impedance measurements), electrode polarization is an important part of the experiment (*53*). This is yet another argument for familiarizing oneself with the resulting equivalent circuit of a particular system under study.

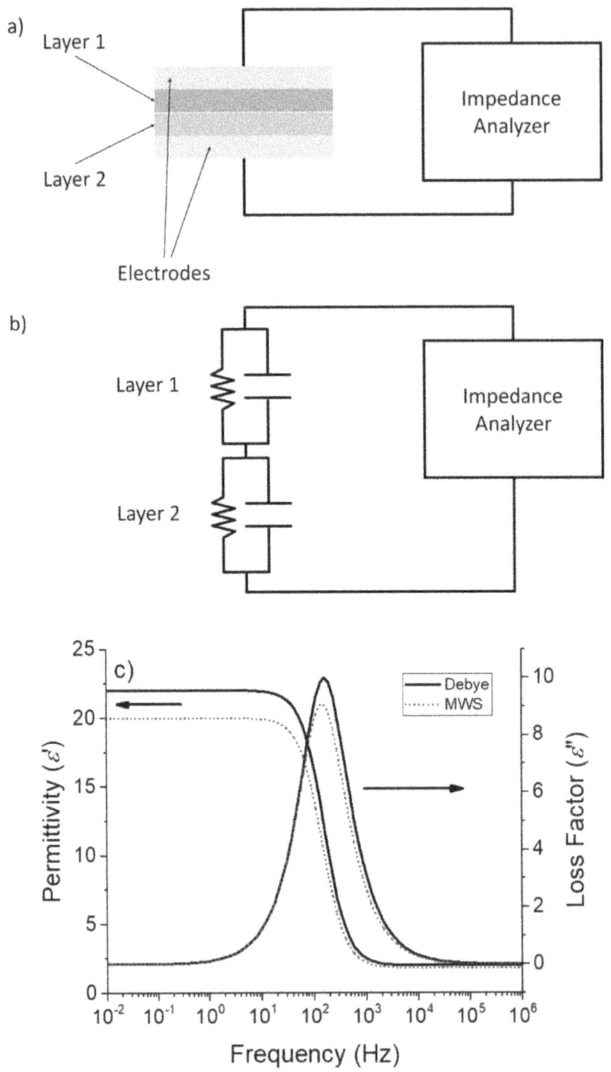

Figure 12. a) Cartoon of a sample in which two dielectric materials b) without Debye relaxation circuits are layered between the electrodes. c) The resulting permittivity spectrum, where a relaxation is seen. Black line is the ideal Debye relaxation from Figure 7. Red dashed line is the Maxwell-Wagner-Sillars circuit, where the first parallel RC circuit is ($R = 10^6\ \Omega$; $C = 10^{-10}$ F) and the second parallel RC circuit is ($R = 10^{14}\ \Omega$; $C = 10^{-9}$ F).

Any action that dissipates the electric field will contribute to the dielectric loss. This is also true for conductivity such as electronic and ionic conduction. The relationship between dissipation factor and conductivity has been known for some time (*109*), and it was proposed by J. Dyre in 1988 that low-frequency dielectric loss is directly related to the conductivity and is empirically supported by the random free-energy barrier model (*110*):

$$\varepsilon^*(\omega) = -j\left(\frac{\sigma_{DC}}{\varepsilon_0 \omega}\right)^N \tag{23}$$

where σ_{DC} is the conductivity in the limit of low frequency and N is an empirical constant. Note that this equation only affects the imaginary part of the permittivity. Because of the inverse dependence on frequency, this effect is seen in the ε'' spectrum at low frequencies and is more pronounced when σ_{DC} is high. When slow relaxation processes are being studied, the conductivity component can overlap with the relaxation peak making analysis difficult. It is possible to fit the low-frequency data to Equation 23 and subtract away the conductivity from the imaginary part of the spectrum, but depending on instrument resolution and the extent of the overlap these efforts are not always fruitful. One final note on conductivity and polarization effects: when ε'' is viewed as a log-log plot all of these effects can exhibit linear behavior. It can therefore be difficult to discern the high-frequency wing of a slow dipole relaxation and conductivity at first glance. If the former is suspected, it behooves one to collect data at lower frequencies—values for $\Delta\varepsilon$ tend to be orders of magnitude lower than ε'' when MWS or σ_{DC} are present (*99*).

Kramers-Kronig Transformations

It has been demonstrated that the real and imaginary parts of a complex number are intimately related. For example, the integrated area under a relaxation peak in the imaginary part of the permittivity is approximately equal to the real part of the permittivity (e.g. Figure 10). Kramers-Kronig (KK) relations, developed by H. Kramers and R. Kronig in the 1920's (*111, 112*), are mathematical tools used to calculate the real part from the imaginary part or vice versa. This can be seen as a hugely advantageous tool, since the imaginary part of the permittivity may be plagued by undesirable effects (e.g. conductivity) and thus transforming the "cleaner" real part into the imaginary part can remove these effects (Figure 13) (*113*). In reality, the mathematics behind KK transformations only hold for ideal sets of data which are available at frequencies from $-\infty$ to $+\infty$ and so the result is not rigorously perfect. Studies have shown that for simple H-N fits the relaxation peak is broadened but results can be acceptable for the purposes of the study (*83, 114*). Additionally, electrode polarization effects which appear in the real part of the permittivity will be erroneously translated as relaxations in the imaginary part unless the transformation is performed on the complete model (Equation 18).

In truth the static dielectric constant and low-frequency polarization effects are inseparable. One must always be clear when and how KK transformations are used, but the following equation is nevertheless highly useful for broadened relaxations:

$$\varepsilon'' \approx -\frac{\pi}{2}\frac{\partial \varepsilon'(\omega)}{\partial \ln \omega} \tag{24}$$

Note that sometimes in the literature loss spectra are simply represented as the derivative of ε' in order to give an estimation of the frequency and shape of relaxations.

And thus a comprehensive equation for describing a single dielectric spectrum with multiple relaxations and DC conductivity can be represented as:

$$\varepsilon^*(\omega) = \varepsilon_\infty + \sum_{i=1}^{n} \frac{\Delta\varepsilon_i}{(1 + (j\omega\tau_i)^{a_i})^{b_i}} - j\frac{\sigma_{DC}}{\varepsilon_0 \omega} \tag{25}$$

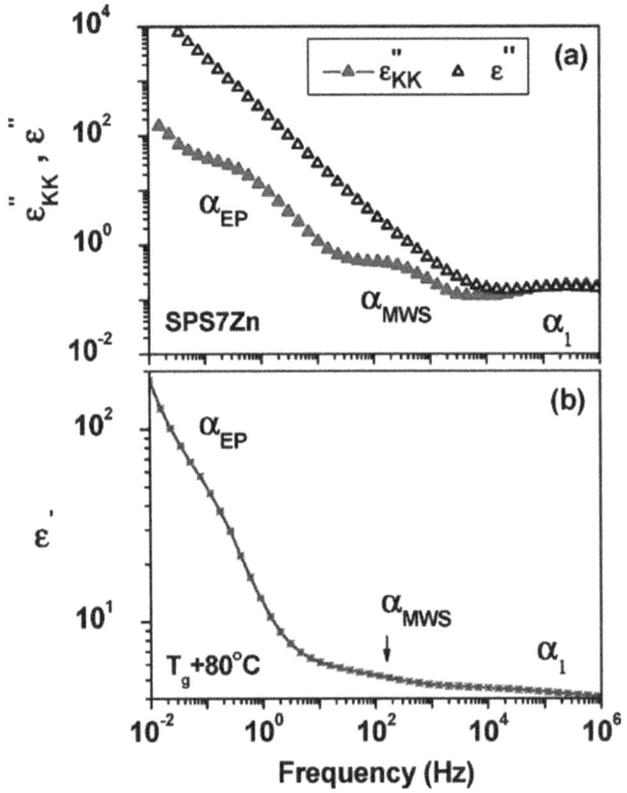

Figure 13. Real and imaginary part of the permittivity for a sulfonated polystyrene ionomer (6.8 mol% sulfonation, 89% neutralization). As seen with the black triangles, the conductivity completely dominates the imaginary part of the spectrum. As seen with the red asterisks electrode and MWS polarizations both contribute to the real part of the permittivity. When the real part is treated via a Kramers-Kronig transformation (see section further on in this chapter) the conductivity can be removed from the imaginary part (blue triangles), but the polarization effects remain. This is an example of how conductivity and polarization effects are an intimate part of many low-frequency BDS spectra. Reproduced with permission from reference (113). Copyright 2007 American Chemical Society.

An example of such a complete spectrum is shown in Figure 14 (*115*). The purpose for this figure is to demonstrate that for a complicated spectrum, the peak of the dielectric loss does not always correspond to the relaxation time. Overlapping effects and the consequences of the empirical H-N relaxation do not allow a convenient means of identifying τ: a computational fit is required. Additionally, temperature will affect relaxations in different ways and therefore relaxations may shift towards or away from each other (the same is true for all fitting parameters). If two relaxations exhibit a similar dependence on temperature then the underlying dynamics can be considered coupled—a valuable piece of information toward understanding a complex system.

In the literature the frequency at which ε'' is at a maximum is sometimes used, but it is important to remember that a) this is not the relaxation time as defined earlier and b) the frequency must be multiplied by 2π to equal the angular frequency ω. If the measurements are made carefully and the experimental data is reproducible, complicated systems unconducive to traditional computational fits can be studied in this manner provided the discussion is clear on which method was used to derive any conclusions.

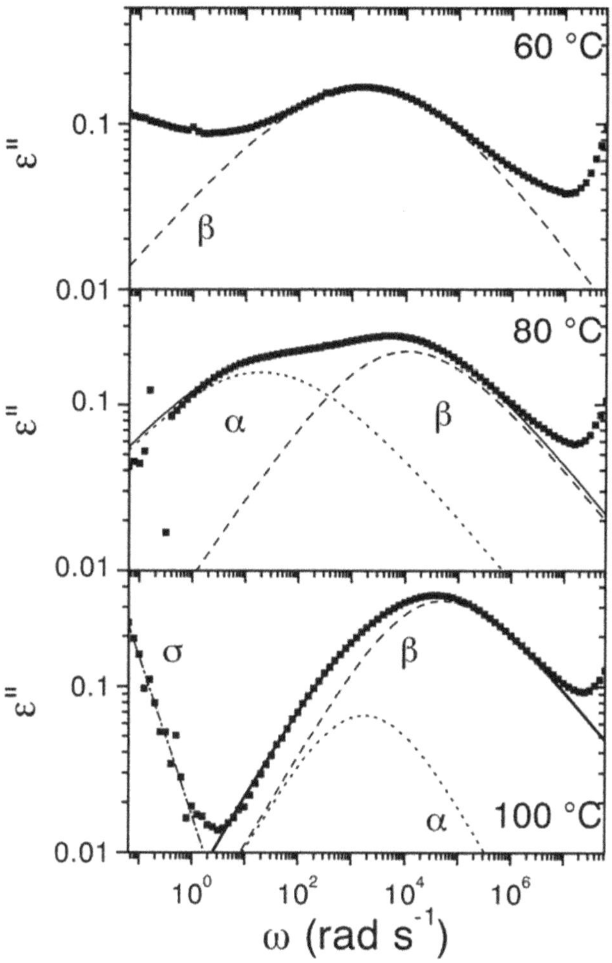

Figure 14. Imaginary part of the permittivity of poly(ethyl methacrylate) at different temperatures. Individual relaxations are represented by dashed lines. Complete spectrum is fit using Equation 25 (solid lines). Reproduced with permission from reference (115). Copyright 1998 American Chemical Society.

Alternative Methods of Data Analysis and Scaling Effects

Because of conductivity and polarization effects it has been said that dielectric spectroscopy is a technique suited for higher frequencies. Most practitioners of the technique will take exception to this statement! The conductivity components of a dielectric spectrum—both electronic and ionic—can contain valuable information. Again, a solid understanding of the system under study as experimental variables are controlled can reveal much about the chemistry and physics at play as the low-frequency signal changes. The use of electrode polarization to study mass transport in ionic systems is yet another advantage of BDS. This technique has been especially helpful in the field of ionic liquids and polymer electrolytes where dipolar relaxations and electrode polarization can be measured concomitantly (*116–121*).

If the complex permittivity spectrum appears featureless due to conductivity and polarization effects, the data can always be analyzed via the raw impedance data (Figure 7). It is also sometimes convenient to plot the data in terms of electric modulus (M^*) or AC conductivity (σ^*). Although they are related mathematically, the resulting plots are considerably different and can provide some clues as to the conduction mechanism.

$$M^*(\omega) = \varepsilon^*(\omega)^{-1} \tag{26}$$

$$\sigma^*(\omega) = j\omega\varepsilon_0\varepsilon^*(\omega) \tag{27}$$

One example demonstrating the advantage of using these methods is shown in Figure 15 (*122*). When plotting the imaginary part of the electric modulus, the DC conductivity value manifests itself as a relaxation peak $\tau_{DC} \approx \varepsilon_0\varepsilon_\infty\sigma_{DC}^{-1}$. When plotting the conductivity, plateaus and transitions can elucidate different regions of conductivity such as electrode polarization at low frequencies. Further examples of various analysis strategies are given in Chapter 3 of Reference (*14*).

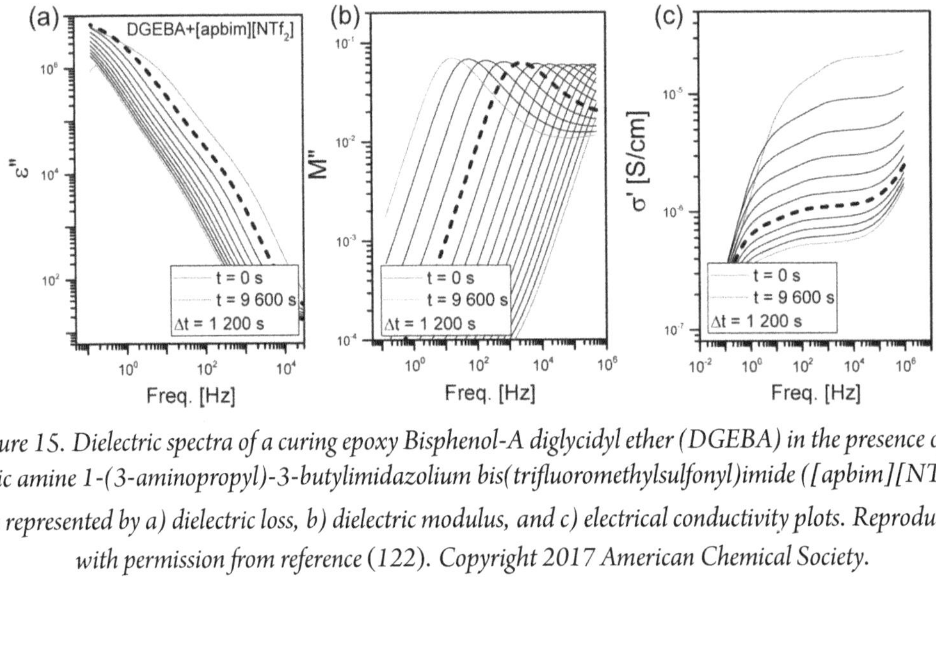

Figure 15. Dielectric spectra of a curing epoxy Bisphenol-A diglycidyl ether (DGEBA) in the presence of an ionic amine 1-(3-aminopropyl)-3-butylimidazolium bis(trifluoromethylsulfonyl)imide ([apbim][NTf$_2$]) as represented by a) dielectric loss, b) dielectric modulus, and c) electrical conductivity plots. Reproduced with permission from reference (122). Copyright 2017 American Chemical Society.

In addition to Bode-like plots shown earlier in this chapter, it is also sometimes valuable to plot the real part of the permittivity versus the imaginary part in a manner similar to Nyquist plots for impedance (Figure 7). These are called Cole-Cole plots and are often seen in historical literature sources. Their use today is less common due to the analytical advantages of an automated computational fit to a complex permittivity spectrum. Cole-Cole plots are convenient in specific instances (Figure 16) (*123*) as they dramatically demonstrate two-dimensional deviation from Debye behavior (skewed arc, suppressed arc, H-N, etc.).

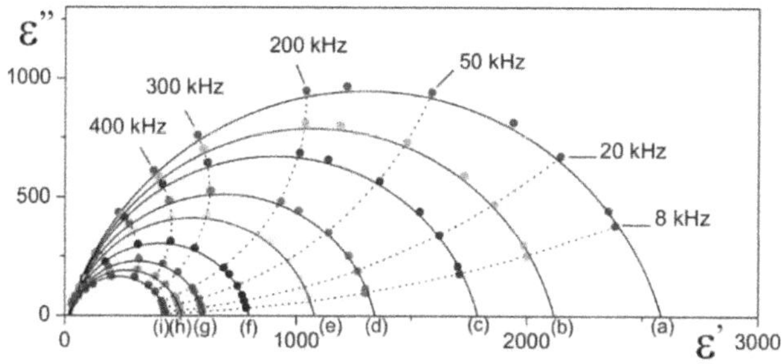

Figure 16. Cole-Cole plot at selected temperatures showing the relaxation of the ferroelectric bis(imidazolium) pentachloroantimonate(III) as the Curie temperature is approached from the paraelectric phase (II) side. a) is 0.2 K above the Curie temperature and i) is 1.2 K above the Curie temperature. Note the extremely high permittivities which are a characteristic of ferroelectrics. Reproduced with permission from reference (123). Copyright 2014 American Chemical Society.

Once a dielectric spectrum has been collected and relaxations have been identified, variables are used to change the relaxation processes in order to understand how the dipoles respond to different conditions. The most common variable employed is temperature, since all aspects of permittivity are sensitive to Maxwell-Boltzmann statistics (including conductivity and electrode polarization effects). These experiments can be performed isothermally (temperature is held while the frequency is swept) or isochronally (the frequency is held while the temperature is swept). Fitted parameters from Equation 25 are then examined, for example via an Arrhenius (or activation) plot where the activation energy of a dipole relaxation can be elucidated. Similar to the Havriliak-Negami equation, many alternate equations exist for the fitting of nonlinear relaxation times on an activation plot. One common example worth mentioning here is the Vogel-Fulcher-Tammann-Hesse or VFT equation (*14, 124–128*):

$$\tau(T) = \tau_0 \exp\left(\frac{B}{T - T_0}\right) \tag{28}$$

where B is an empirical constant, τ_0 is the "infinite-temperature limit" of the relaxation time, assuming that the material structure remains unchanged, and T_0 is the empirical Vogel temperature at which the relaxation time diverges. For glassy organic systems T_0 is often a few tens of kelvin below the glass transition temperature.

An example of a comprehensively studied system is shown in Figure 17 (*129*). In this study on amorphous ibuprofen (*129*) multiple relaxations were measured across a wide frequency and temperature range and the functions used to fit their frequency- and temperature-dependence are discussed in detail (*87, 124–126, 130–132*). Many variations of such activation plots exist, and so it is important to pay attention to the *y*-axis when reviewing literature data. The shifting of dielectric relaxations at various temperatures is a huge field of BDS and is referred to as the scaling of relaxation processes. Because of the broad frequency range available to modern experimentalists, entire relaxations can be studied across a large temperature range without the need for master plots as in similar techniques (e.g. dynamic mechanical analysis) and it has been found that in most cases the α-relaxation maintains its shape over a wide range of times, temperatures, and pressures (*133*). Master plots are still a valuable means of visualizing the consistency of a dielectric relaxation under variable conditions, especially how neighboring relaxations shift at different rates (Figure 18) (*129*).

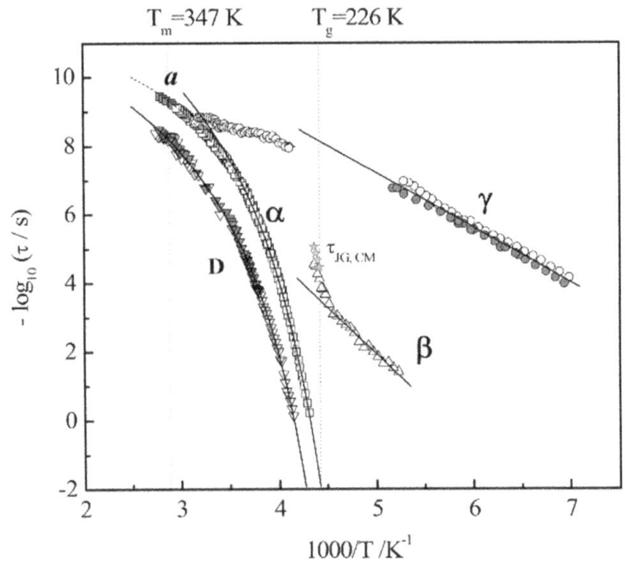

Figure 17. Activation plot of the relaxations measured in ibuprofen via BDS. Relaxation times were found using the H-N equation. Open symbols represent data collected isothermally via stepwise cooling from above the glass transition temperature (T_g) and the melting temperature (T_m). Filled symbols represent data collected isochronally by cooling at 1 K·min^{-1}. Lines are fits to the data using Arrhenius and VFT equations. Stars are the Johari-Goldstein relaxation times calculated using the coupling model. Note the negative sign on the y-axis: this is not uncommon with activation plots. Reproduced with permission from reference (129). Copyright 2014 American Chemical Society.

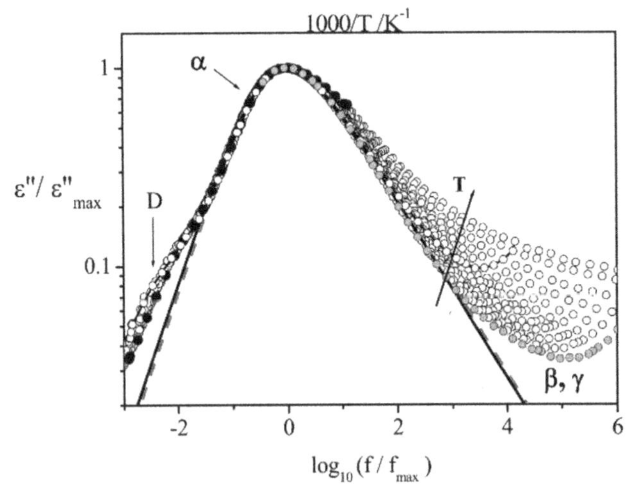

Figure 18. Master plot of the alpha relaxation of ibuprofen between 232 K (grey circles) and 262 K (black circles) with data in between (open circles). The relaxation maintains its overall shape across this temperature range, with H-N parameters remaining constant (a = 0.81, b = 0.56). Neighboring relaxations (D, β, and γ) can be seen on either side of the alpha relaxation. The black line is the fit to the H-N equation. The dashed line is the fit to the KWW function. Note that the D relaxation appears to be coupled to the α relaxation, while the β and γ relaxations do not. Reproduced with permission from reference (129). Copyright 2014 American Chemical Society.

It was suggested earlier in this chapter that BDS data is directly comparable to other techniques, and this is demonstrated in Figure 19 (*134*). Here NMR data and BDS data for the α- and β-relaxations of poly(ethylene-*alt*-propylene) at different molecular weights are plotted on the same activation plot where excellent agreement can be seen. Further examples of this via several techniques (rheology, calorimetry, scattering, etc.) are given at the end of this chapter.

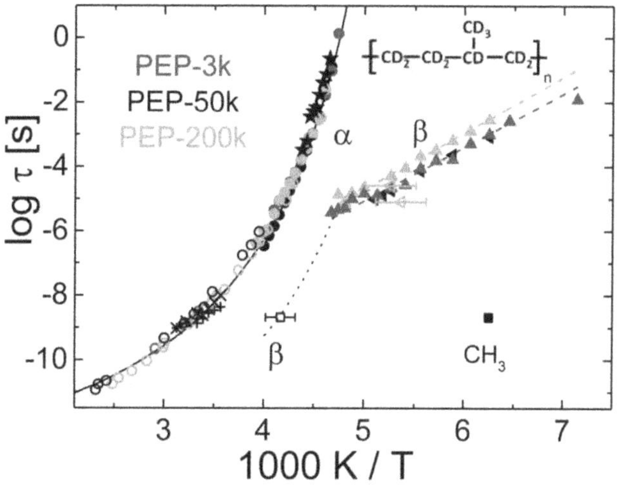

Figure 19. Activation plot of fully deuterated poly(ethylene-alt-propylene) as measured via BDS and NMR. Filled circles are the BDS data for the α-relaxation. Open circles are field-cycling ¹NMR data for the α-relaxation. Stars, crosses, pluses, and squares are from ²H NMR data (stimulated echo, T_1^{chain}, and T_1^{methyl}, and T_1^{-1}maximum, respectively) for the α-relaxation. Filled triangles are the BDS data for the β-relaxation. Open triangles and open squares are the ²H NMR data ($R(t_p)$ and T_1^{-1}, respectively) for the β-relaxation. Open squares and . Solid line and dashed lines are fits to the VFT and Arrhenius equations, respectively. Dotted line is a guide for the eye. Reproduced with permission from reference (134). Copyright 2017 American Chemical Society.

Here it has been discussed how frequency and temperature can affect the dielectric response of a system, but other factors such as pressure and the strength of the electric field can be used. These experimental variables have been covered in detail elsewhere and need not be repeated here (*135*, *136*). Suffice it to say that the independence of a dielectric spectrum on such effects (termed linearity here) is in truth never fully achieved. Nonlinear effects due to high electric fields occur at any field strength greater than zero, and hence it is expected that the experimentalist will consider these effects unavoidable and instead aim to reduce them below the precision of the measurement by controlling variables as well as is necessary. Returning to the topic of Kramers-Kronig transformations, these are quite valuable as a tool to confirm linearity. Calculating the imaginary part from the real part and then plotting against the measured imaginary part can reveal portions of the spectra that are nonlinear. This in turn can reveal, for example, when too high of an electric field has been applied or when instrument effects are negatively impacting the calculated permittivity.

On the topic of linearity, this book aims to argue to the analytical chemist that this technique can be used to study relaxations of relevance to chemical systems. In many cases such systems are undergoing chemical or physical change during the measurement process and so linearity (in terms of an unchanging dielectric response) is purposefully absent. At high fields dielectric heating may alter chemical kinetics, or nonuniform fields can cause dipolar diffusion via dielectrophoresis. Even at low fields electrode polarization or adsorption can affect the perceived chemistry. Time can become an important consideration for a given experiment. As will be discussed further in this chapter, certain dielectric measurements can take several minutes to complete and thus in the event of an ongoing chemical reaction a single spectrum may include nonlinear data. Such effects may also be true in the measurement of crystallization effects or degradation at high temperatures. It is therefore prudent to consider the order of operations in an experiment, for example how reversing the temperature or frequency sweep direction might affect the results. Although chemists have been involved in the development of BDS since the beginning and reaction kinetics have been studied for decades (*15–18*), the discussion around the use of BDS as an analytical tool is a newer one (*137, 138*). The use of dielectric relaxations to study the chemistry of a system is a challenging task, but one that is certainly possible with modern equipment and an understanding of the relaxation processes at work.

The Experiment

The classic example of a sample in between two metal electrodes measured via two leads to an instrument (Figure 6) is still commonly practiced today, but to be sure many different experimental setups exist. The exact size, shape, and type of electrodes varies widely in the literature, as does the instrumentation and other experimental factors (e.g. temperature and pressure control). A full review is therefore not possible, but instead a very basic review of commercial instrumentation is given, with the understanding that this section is not comprehensive. Its purpose is to act as a springboard for the chemist interested in developing this capability.

Measurements with Commercial Instrumentation

Commercial dielectric spectrometers have existed in some capacity for decades, albeit under various names (impedance analyzers, dielectric analyzers, etc.). For the purposes of this discussion, let it be defined that a commercial dielectric spectrometer has the ability to measure the real and imaginary part of the impedance for dielectric samples over a broad frequency range (*139*). This last caveat is important, as there are many commercially available instruments that provide values at a single frequency (typically 50/60 Hz or perhaps, as with some "DC" conductivity meters, some arbitrary frequency in the kHz range to avoid electrode polarization). Dielectric spectroscopy is valuable to the analytical chemist *because of its frequency range.* Consider early undersea telecommunication cables developed by AT&T at Bell Labs in the 1950's, where 10 microradians of Δ tanδ in the polyethylene insulation translated to \$1 million in savings (*140*). The same can be said for any kind of electrical insulation, which are mostly tested at a single frequency (e.g. 50/60 Hz for power transport) using "tan delta" testers. Understanding how dipole relaxations affect the entire signal at either side of the frequency of interest are paramount to modern technology (*141*).

Several companies produce commercial dielectric spectrometers and a complete list is not possible here. Keysight Technologies (previously Agilent and before that Hewlett-Packard) has produced bench top *LCR* meters and impedance analyzers for decades and offers several types of sample cells depending on the application. These impedance analyzers use auto-balancing bridge techniques which are in general better suited to lower impedances, higher losses, and frequencies above 10^1 Hz but are nevertheless reliable workhorses in the laboratory. Frequency response analyzers (FRAs) or digital lock-in amplifiers (with or without dielectric conversion) are available from several companies and offer lower frequencies as well as high impedance ranges. Stanford Research Systems has produced such bench top instruments for decades. Novocontrol Technologies and Solartron Analytical offer turnkey solutions specifically designed for dielectric spectroscopy, including a myriad of sample cell designs and options for a wide temperature range. In addition, precision capacitance meters exist (e.g. the Andeen-Hagerling, Inc. AH 2700A Ultra-Precision Capacitance/Loss Bridge, which covers 27 discrete frequencies from 50 Hz to 20 kHz).

Modern analyzers designed to measure impedance will offer accuracy contour plots with frequency on the *x*-axis and impedance on the *y*-axis, demonstrating a region of maximum impedance accuracy (Figure 20). However, when measuring complex permittivity it is necessary to understand the accuracy in measuring the capacitance C_p' and the loss $\tan\delta$. It can be demonstrated from the pedagogical description of an equivalent circuit (Figure 7b) that the impedance is inversely dependent on $\omega C_p'$ and thus for a given sample impedance will be highest at low frequencies and lowest at high frequencies, thus crossing the region of highest accuracy. Turnkey dielectric spectrometers may seamlessly combine impedance bridges and FRAs with dielectric converters such that the entire high-impedance low-frequency region is covered, thus allowing for a single optimum capacitance (e.g. 100 pF). Less expensive impedance analyzers will offer a narrower frequency and impedance range. It is therefore important to appreciate the accuracy and range of the chosen instrument with respect to the expected signal of the sample, and thought must be given toward the optimal sample size or whether a variety of sample sizes must be made. Using Equation 1 it can be seen that the capacitance can be increased one order of magnitude by making the sample thinner or the area larger by one order of magnitude—a task made much simpler at the beginning stages of experimental design.

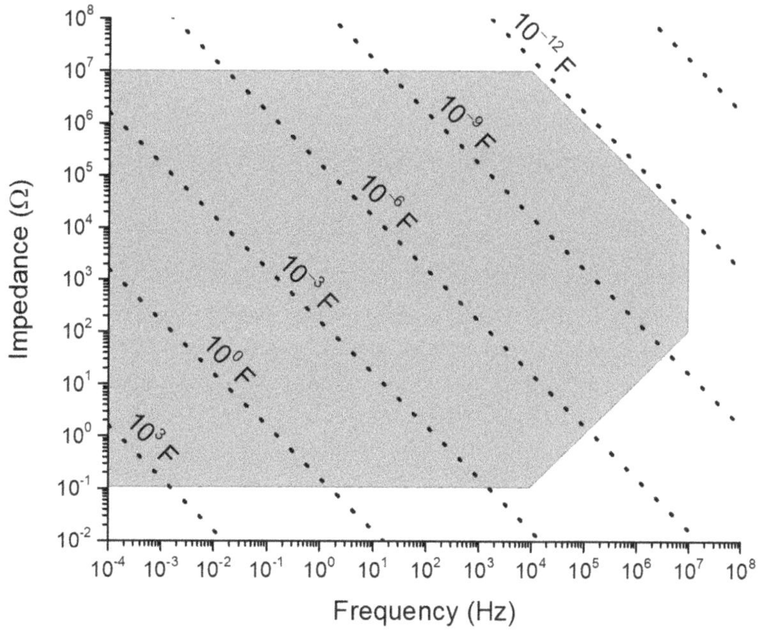

Figure 20. Example accuracy contour plot for a hypothetical impedance analyzer. Grey area represents a zone of accuracy: often multiple zones are shown to emphasize increasing accuracy towards the center of the plot.

At higher frequencies the leads used to connect the analyzer to the test cell become a limiting factor in accuracy. When measuring a low-impedance material, coaxial cables will impart an inductance that raises the lowest-possible impedance on the accuracy contour plot. When measuring a high-impedance material, the cables will act as capacitors, and will lower the highest-possible impedance. The impedance range of most instruments therefore tapers above 10 kHz, eventually creating a narrow window of acceptable accuracy. Although many analyzers can be pushed up to 10 MHz in certain situations, most published data ends at 1 MHz. Reducing the length of the cable (or eliminating it entirely) will improve resolution at higher frequencies, as will modifying test leads (separate them for high impedance measurements, shield them for low impedance). It is therefore important to understand the resolution of the instrument. A strategy for testing this is to create a mock sample out of simple electronic components (e.g. the equivalent circuits in Figures 6b and 12b). By mimicking the expected equivalent circuit of a sample with an actual mock circuit, accuracy can be appreciated and any unpredicted instrument contributions may be discovered (discussed further in this chapter).

For measurements between 1 MHz and 10 GHz often reflectometry is used. This method is covered in more detail in Chapter 2 and additional details can be found elsewhere: in brief the sample is treated as the termination of a coaxial line and the impedance of said line is measured. Sample preparations and experimental considerations are similar as for the lower frequencies but a network analyzer is used and the test cell is of a different design. Split-cylinder and split-post resonators can be used for sensitive measurements but they are limited to a single frequency per apparatus. Approaching the THz regime, resonant cavity systems can be used. One may think that at these high frequencies relaxations are absent but on the contrary: this is an increasingly valuable area of research made possible in part by instrumental advances related to the needs of information and communication technology (much the same as the GHz regime but a few decades ago). At extremely high frequencies the infrared is reached and the square root of the complex permittivity will equal the refractive index n^* as defined by Maxwell's equations:

$$\sqrt{\varepsilon^*} = n^* \tag{29}$$

It is therefore possible to create spectra that bridge the gap between the electronic and the optical regimes.

Impedance analyzer measurements are typically carried out from high to low frequencies, as the highest frequencies take the least amount of time and bad electrode contact with the sample can be quickly identified. Without fully discussing the nuances between different analyzer types, a modern frequency response analyzer converts the response signal via two reference waveforms (sine and cosine) which create a pair of waveforms. If the waveforms are integrated over time the two parts of the impedance are found. This method is advantageous because it eliminates noise which will be averaged out of the signal. However, the accuracy of the measured value is dependent on how many waveform cycles are integrated. Because of this, the time to collect a measurement is inversely related to the frequency, and the signal-to-noise is proportional to the number of cycles to the power of 0.5. With ten data points per decade of frequency, experiments from 1 MHz to 1 Hz take tens of seconds; to 100 mHz minutes; to 10 mHz tens of minutes; and to 1 mHz hours. Depending on the sample and the chemistries being studied, this may be prohibitively long. Fast Fourier-transform techniques are available which analyze multiple waveforms at once, but these measurements are still at least as long as the longest cycle. Because of this, most published BDS data ends at or above 10 mHz.

For measurements at lower frequencies the time scales of FRA tests become inconveniently long and transient techniques are used instead. These methods are covered in more detail in Chapter 2. Although different methods exist—some of which are commercially available—the basic premise is that the time-dependent permittivity is measured via an applied or released DC bias and the results are transformed into a frequency-dependent domain with the help of a computer. Transient measurements are advantageous in the study of extremely slow relaxations, such as the dynamics of a glass well below the glass transition temperature.

It is also important to consider the strength of the applied field. As discussed earlier in this chapter, the permittivity is dependent on the field with nonlinear effects at higher fields. In general a field should be chosen that is low enough to minimize nonlinear effects but high enough to minimize noise. Measurements at multiple field strengths demonstrating the absence of higher harmonics will convince the experimentalist of sufficient linearity. Studies that explore nonlinear effects may require special thin-sample or high-voltage instrumentation (*136*, *142*). In any case the applied field should be disclosed with the results.

Electrode Configurations and Sample Preparation

In some complicated electrode configurations (see examples further in this chapter and in Chapter 2) it is not trivial to determine sample thickness or area. It is often convenient instead to measure the "cell constant" $d{\cdot}A^{-1}$. This is done by measuring the capacitance of the empty cell (C_{empty}) in the presence of dry air, nitrogen, or vacuum. The permittivity of dry air is 1.0005 at room temperature, and thus ε'_{empty} is often assumed to be equal to one. The cell constant can then be calculated using Equation 1. If the cell constant is known then the permittivity of an unknown sample can be determined at any frequency. Summarizing this process:

$$\frac{C_{sample}}{C_{empty}} = \frac{\varepsilon'_{sample}\varepsilon_0 dA^{-1}}{\varepsilon'_{empty}\varepsilon_0 dA^{-1}} \approx \varepsilon'_{sample} \qquad (30)$$

As mentioned earlier in this chapter, it is usually necessary that some understanding of the system be known beforehand, otherwise it is difficult to have confidence in the data. Further in this chapter it will be discussed in detail how novel systems require complementary analytical studies. However, even with a previously studied system and careful sample preparation data misinterpretation can still occur. The following sections will discuss the basics of sample preparation, but to be sure a literature review for each system to be studied is unavoidable.

Equivalent Circuits: Revisited

It was previously discussed how a dipole relaxation can be envisioned as a simple Debye circuit, and how real-world systems can be studied and data analyzed as equivalent circuit diagrams. It was warned earlier that simply finding a "fit" to the data in the form of an equivalent circuit is meaningless, but that is not to say that mysterious circuit elements do not exist. The following considerations are useful when attempting to validate data.

The first consideration to make is that of the actual instrument. Measurements outside of the optimal range, improper shielding, and long cables can introduce circuit elements that are in series with the equivalent circuit of the sample. It is prudent to eliminate these possible artifacts via adequate grounding, shielding, connections, and cable length to ensure that the data is from the sample alone and not, metaphorically speaking, dust on the lens of the telescope. Most commercial instrumentation will allow for a calibration of the instrument, where measurements are taken with the leads shorted, open, and with one or more known *RC* elements. If one is unsure about the quality of their experimental setup, measurement of a dummy circuit can help. If such effects cannot be avoided they should at least be understood.

Equivalent circuits are also valuable in the sense that electrochemical impedance spectroscopy provides decades of research that can help explain experimental results. Ionic conduction, electrode polarization, charge injection, adsorption, or otherwise non-relaxation phenomena can be identified based on their impedance spectra (Warburg-type and Gerischer elements are popular to model diffusion, but a myriad of models and equations exist). The experiment can be redesigned to study these circuit elements for a more comprehensive study, or they can simply be recognized.

It is also possible to introduce equivalent circuits purposefully in order to reduce the effect of electrode contact on a sample. Briefly, instead of placing the sample of known thickness in between two electrodes, the sample is instead loosely placed between two electrodes of a known distance apart. This creates two capacitors in series with the sample, and some basic algebra can be used to calculate the permittivity of the sample:

$$\varepsilon' = \frac{1}{1 - \left(\frac{C_{\mathrm{samp}} - C_{\mathrm{empt}}}{C_{\mathrm{empt}}} \cdot \frac{d_0}{d}\right)} \tag{31}$$

$$\tan\delta = \tan\delta_{\mathrm{samp}} + \varepsilon'\left(\frac{d_0}{d} - 1\right)\left(\tan\delta_{\mathrm{samp}} - \tan\delta_{\mathrm{empt}}\right) \tag{32}$$

where d_0 is the total distance between the electrodes and $\tan\delta_{\mathrm{samp}}$ and $\tan\delta_{\mathrm{empt}}$ are the measured dissipation factors of the cell with and without the sample, respectively. If the distance between the electrodes or the sample thickness is not confidently known, then the sample cell can be filled with a known reference dielectric (often a silicone oil such as Dow Corning 200 Fluid). Four measurements of the cell with and without the ref. dielectric and the sample are used:

$$\varepsilon' = \varepsilon'_{\mathrm{empty}} + \frac{C_{\mathrm{samp\,\&\,ref}}\left(C_{\mathrm{samp}} - C_{\mathrm{empt}}\right) \cdot \left(\varepsilon'_{\mathrm{ref}} - \varepsilon'_{\mathrm{air}}\right)}{C_{\mathrm{samp\,\&\,ref}} \cdot \left(C_{\mathrm{samp}} - C_{\mathrm{empt}}\right) - C_{\mathrm{samp}} \cdot \left(C_{\mathrm{samp\,\&\,ref}} - C_{\mathrm{ref}}\right)} \tag{33}$$

$$\tan\delta = \tan\delta_{\mathrm{samp\,\&\,ref}} + \frac{\left(\varepsilon' \frac{C_{\mathrm{empt}}}{\varepsilon'_{\mathrm{empt}}}\right) - C_{\mathrm{samp\,\&\,ref}}}{\left(C_{\mathrm{samp\,\&\,ref}} - C_{\mathrm{ref}}\right)} \cdot \left(\tan\delta_{\mathrm{samp\,\&\,ref}} - \tan\delta_{\mathrm{ref}}\right) \tag{34}$$

These methods are known as the "air-gap method" and the "two-fluid method," respectively (air is one of the fluids) (*104, 143, 144*). Accurate thickness-independent measurements also offer the added convenience of measuring the negative temperature-dependence of permittivity for an isotropic homogeneous system. This effect can be calculated from the Clausius-Mossotti equation (*145*).

$$-\frac{d\varepsilon'}{dT} = \alpha_L\left(\varepsilon_s^2 + \varepsilon_s - 2\right) \tag{35}$$

where α_L is the linear coefficient of thermal expansion. Since the sample cell dimensions are not needed for the two-fluid method, thermal expansion of the cell and the sample will be captured by the measurement.

In summary, equivalent circuits are a useful tool when designing experiments or diagnosing m7easurement artifacts. For all experimental setups consideration should be made as to whether there are erroneous contributions to the equivalent circuit or whether the sample cell dimensions are known throughout the entire temperature range.

Sample and Electrode Dimensions

The care of sample preparation is too broad a topic to cover here, suffice it to say that very small concentrations of additives in samples can create very large changes in the impedance. The presence of water or ions are of particular concern as conductivity effects and electrode polarization become pervasive. Although extreme care must be taken when preparing samples, it is worthwhile to mention that as an analytical technique this is quite valuable. Trace analyses are possible if a system is understood well enough that changes in the dielectric properties can be associated with concentrations or chemical reactions. As a single example, the crystallization of poly(3-hexylthiophene) was observed in the dielectric response of toluene solutions down to 10 ppm (*146*).

Sample thickness and sample area can affect the measured capacitance and therefore instrument resolution, but it is also important to realize how altering the sample dimensions can benefit the experimenter by giving increased confidence in the measurement. For example, for most materials at macroscopic length scales the dipoles will not demonstrate a dependence on sample thickness or sample area. On the contrary, electrode polarization and MWS effects may. If one is suspicious that part of the signal is due to polarization effects, altering the sample thickness should shift these effects. Similarly, artifacts from the instrument itself will not be dependent on sample size. Many headaches can be avoided if the experimental plan consists of measuring materials of several different dimensions. Similarly, BDS is a sub-field of electrochemical impedance spectroscopy, and indeed at sufficiently high fields electrochemistry can occur. By adjusting such factors as the field strength and the electrode type these effects can be confirmed or refuted.

Several different types of electrodes are available commercially, from micrometer-mounted metal discs to sealed liquid cells (specific examples are given in Chapter 2). For liquids, soft materials, or to achieve thin sample thicknesses for optimized resolution, one can employ spacers such as fused silica fibers. Measurements on nanoscale thin films are possible using conductive silica with spacers, and interdigit electrodes comprising of thin (tens of microns) interwoven gold combs printed on a circuit board allow for the measurement of thin samples with one surface exposed to the environment. Recently the measurement of nanoscale films has been made commercial via conductive silica wafers with insulating spacers, allowing for the measurement of delicate cast thin films (*147–150*). In short, custom sample cells abound in the literature and new ones are constantly being invented. When doing so it is important to remember that poor electrode contact can lead to electrode polarization, which is worse with thin samples. It is therefore common to deposit thin electrodes onto a sample, thus ensuring good contact and reducing the risk of polarization effects. Gold and platinum are popular metals due to their electrochemical favorability, but by no means required. Whatever the chosen electrode configuration, care must be taken to ensure that the sample is the only thing being measured. Even if instrument effects are minimized, this is not always guaranteed.

The mathematics assume that the electric field lines exclusively pass through the sample, but this is not always true. For example, a sample between parallel plates will have field lines escaping outside of the sample edges (Figure 21).

The measured capacitance C_{sample} will therefore not be the accurate sample capacitance. It is possible to use Maxwell's equations to predict these stray fields and subtract them from the measured value (known as edge compensation or edge correction) (*151*), but most often this effect is best avoided entirely, for example by using the Kelvin guard-ring technique as described in Chapter 2. If guarded electrodes are not possible, the following strategies are summarized:

1. For flat, circular samples, if the sample and electrode sizes are explicitly known then stray fields can be calculated. The equations for various configurations are conveniently outlined in ASTM D150 "Standard Test Methods for AC Loss Characteristics and Permittivity (Dielectric Constant) of Solid Electrical Insulation" (*152*). This is also a convenient means of estimating what percentage of the signal is comprised of stray fields. For example: a 1 mm-thick sample, 40 mm in diameter, sandwiched between two 1 mm-thick electrodes will have an empty capacitance C_{empty} of 11.1 pF, but the capacitance measured will be 12.0 pF—over 8% higher than the actual value. When possible, experimental design should aim to reduce stray fields as much as possible rather than to rely on subtracting them away.
2. For some sample cells there is an enclosure that may use an insulating material to isolate the electrodes, and the stray fields may interact with this insulator. This can be true, for example, in an interdigitated electrode or an unguarded liquid cylindrical cell (Figure 22).

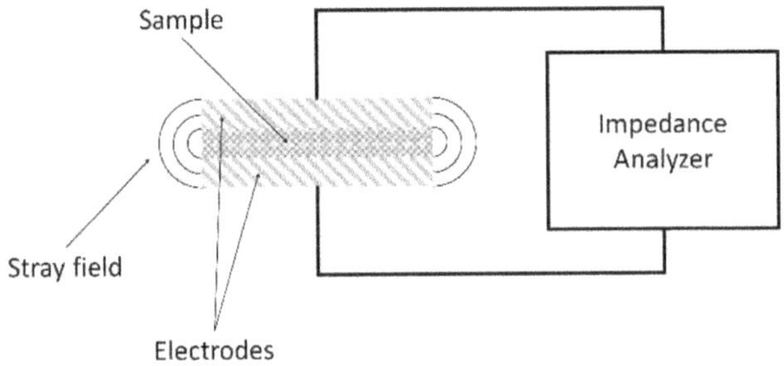

Figure 21. Cartoon of a test cell where stray fields escape outside the sample of interest.

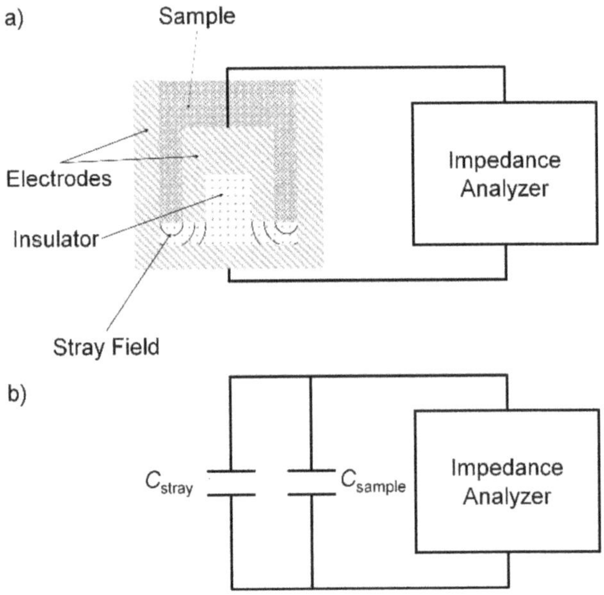

Figure 22. a) Cartoon of a liquid cylindrical test cell where stray fields can be affected by an insulator used in the test cell construction. b) Equivalent circuit of such as system.

Common materials used include fused silica, polytetrafluoroethylene, or polyethylene, and while these materials have very little loss over a large frequency range, they will nevertheless contribute to the measured capacitance. Calculating stray fields in these cases is difficult, but using equivalent circuits it is possible to empirically measure the cell constant and the capacitance of the insulation. The stray fields are treated as an additional capacitor in parallel with the sample, and a reference material with a known permittivity is measured to calculate this capacitance. This method is not ideal since the field lines will change when the sample cell is empty versus full. Nevertheless, being duly warned, the following equations are useful:

$$C_{\text{stray}} = C_{\text{empt}} + \left(\frac{C_{\text{empt}} - C_{\text{ref}}}{\varepsilon'_{\text{ref}} - \varepsilon'_{\text{empt}}}\right) \tag{36}$$

$$\varepsilon'_{\text{samp}} = \frac{C_{\text{sample}} - C_{\text{stray}}}{C_{\text{empt}} - C_{\text{stray}}} \tag{37}$$

Historical Instrumentation

This chapter has shown that dielectric spectroscopy existed for several decades before the birth of modern FRAs and commercially available sample cells. The result of this is a huge labyrinth of data spanning several fields and many journals. The following provides but a few examples of experimental protocols from the past century, such that the younger scientist can appreciate the data presented in these extremely valuable original sources.

Prior to the existence of modern turnkey analyzers many systems were custom-built, using an oscillator, a sample cell, and a bridge circuit. The frequency of the oscillator was adjusted by hand, typically using a large dial that would rotate semi-overlapping metal plates inside of the device to adjust the capacitance of the internal Wien bridge. The HP Model 200 Series Audio Oscillators were produced in some capacity for decades and could access frequencies on the orders of 1 Hz to 100 kHz (sometimes "cycles per second" or cps was used). Other oscillators offered discrete frequencies via switches, for example the General Radio 1311-A which could provide eleven frequencies between 20 Hz and 20 kHz. Sample cells were custom-built but much of their accessibility and applicability still exists today—excepting perhaps the use of mercury electrodes. Capacitance and loss were measured using a bridge circuit, for example the Cole-Gross bridge (*153*), and an oscilloscope, for example a HP Model 120 Series. Amplifiers were used throughout as necessary to increase the applied signal or the response. Although the oscillator, null detector, and measuring bridge circuits could be custom-built, commercial options were available. For example, in the 1970's General Radio offered the 1621 Precision Capacitance Measurement System which consisted of a 1316 Oscillator (continuous frequency control between 10 Hz and 100 kHz), a 1616 Precision Capacitance Bridge (12 significant figures of control), and a 1238 Detector (with amplification). With experience and care the data generated by these instruments was exemplary. Two of the best reference texts on dielectric spectroscopy of polymers were written before 1980 (*18*, *19*), and much of their information comes from studies performed using oscillators and bridge circuits.

To summarize, measurements were performed by carefully tuning the frequency of the measurement, balancing the bridge, and recording each value individually. Depending on the specifics of the study, contemporary data is often represented as spectra with frequency on the *x*-axis; an easy task thanks to modern electronics (*154*). In the literature prior to the 1970's data was sometimes represented with temperature on the *x*-axis, as a Cole-Cole plot, or with some other controlled variable (Figure 23) (*155*), and these traditional graphical methods are still found in the literature today. Sometimes only tables with peak frequencies or dipole moments are given.

Regardless, in the literature there are countless of studies on a myriad of dipolar relaxation systems probed under various conditions with data plotted in different ways. In fact, depending on the linearity of a particular system it can still be advantageous to hold the frequency constant and sweep the temperature—a task again made easy using modern instrumentation.

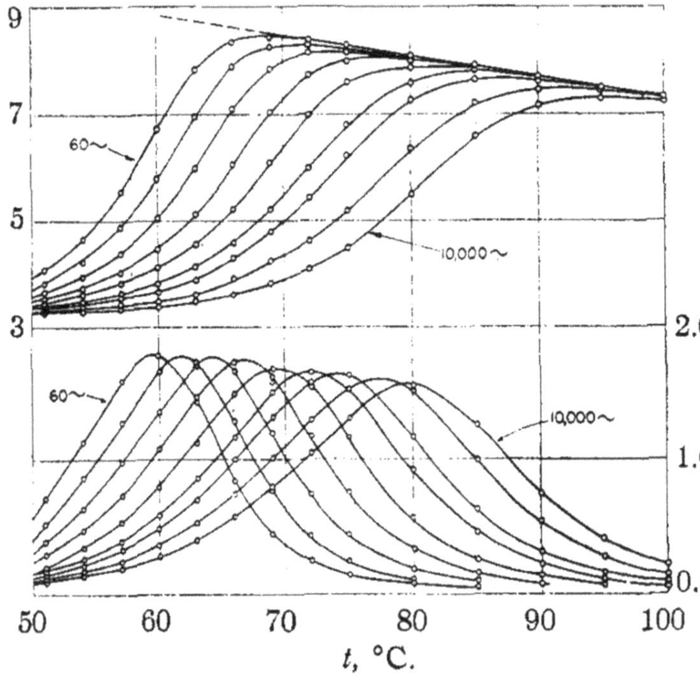

Figure 23. (Top) Real and (Bottom) imaginary permittivity spectra for poly(vinyl acetate) at frequencies from 60 Hz to 10 kHz. Note how the temperature dependence of the permittivity can be seen at elevated temperatures (Equation 35). Reproduced with permission from reference (155). Copyright 1941 American Chemical Society.

Examples of Complementary BDS Experiments

Being suitably prepared, the following section will outline several examples of experiments which utilize BDS to understand certain aspects of chemical systems. Note that in every example a second technique is used to help elucidate the signal that is measured via the dielectric spectrometer. This supports the argument that BDS is a *complementary* analytical chemistry technique, and care should be taken when analyzing data without any extra-experimental support, either from other experiments or from the literature.

Much of the existing literature has utilized complementary techniques in order to confidently identify the fundamental mechanisms of specific relaxations in complex systems. Techniques such as *x*-ray scattering and NMR can provide insights into the structural components of a system, while techniques such as rheology can provide complementary insights into the dynamics. In the ibuprofen study referenced earlier (*129*), differential scanning calorimetry was used to confirm the crystallization and glass transitions that resulted in the elimination or slowing of relaxations, respectively. As the study was performed from high to low temperatures, liquid chromatography was used to confirm that the sample did not degrade during the experiment. Electrospray ionization mass spectrometry confirmed that the ibuprofen molecule prefers to form hydrogen-bonded dimers and trimers. FTIR confirmed this hydrogen bonding both in the crystalline and the amorphous phases and determined that the carboxyl group was free to vibrate. It was therefore concluded that the α- and β- relaxations were due to the dynamics of the glass transition, the Debye relaxation at low frequencies was due to the hydrogen bonding (*156*), and the γ-relaxation was due to the molecular fluctuations of the carboxyl group. Finally, molecular dynamics simulations were used to support these hypotheses. This is but one example of the amount of work that goes into identifying the relaxations of a system. The stability of amorphous pharmaceutical systems is a particularly valuable use of BDS (*157*), since the crystallization of active pharmaceutical ingredients is a constant challenge in the industry (*158*).

Fourier-Transform Infrared Spectroscopy

FTIR has been used to complement BDS for decades, in no small part because the electromagnetic response of the two methods are related. The experimental contribution of FTIR is to determine when a particular dipolar functional group is free to rotate, and how this behavior changes under certain conditions (e.g. temperature). Similar to the ibuprofen study referenced earlier (*107*), here two examples are given in which FTIR was used to monitor the hydrogen bonding of a system. In a recent study (*159*), water confined in self-ordered nanoporous alumina was measured via BDS and FTIR during crystallization. As can be expected the crystallization of water immobilizes dipoles and reduces the permittivity (Figure 24). However, when the pore size is small enough the process switches from heterogeneous nucleation of hexagonal ice to homogeneous nucleation of cubic ice. The BDS and FTIR data were in agreement on this transition, and BDS elucidated significantly different dynamics of the confined ice which agree with earlier studies on cubic ice formed under high pressure. ^{1}H magic-angle spinning NMR experiments provided further insights into the dynamics of the ice under confinement. The behavior of water and other hydrogen-bonded systems under confinement and pressure has implications for the chemical and physical understanding of systems ranging from fundamental biology to advanced nanotechnology (*160*).

In another FTIR example it was examined why the anti-inflammatory drug etoricoxib exhibits a resistance to crystallization over similar drugs rofecoxib and celecoxib (*161*). BDS observed similar dipole relaxations for all three drugs, but FTIR and theoretical modelling suggested a tautomeric conversion of the pyridine nitrogen to a pyridinium ion by hydrogen transfer from the neighboring methyl group. This stimulated hydrogen bonding which stabilized the amorphous state over time. FTIR detected both the formation of the CH_2 group and the change in hydrogen bonding via the nitrogen atom. The relative difference in the dipole moment of the two tautomers was detected via BDS, thus supporting this hypothesis (Figure 25).

Nuclear Magnetic Resonance

NMR has been used to help identify relaxations in BDS spectra for many decades, and it is still an indispensable tool for this purpose today. Similar to FTIR, the temperature at which molecules and functional groups are free to move affects the chemical shifts of the NMR spectrum. Also similar to FTIR, the popularity of NMR has led to a broad collection of experimental techniques on systems of all compositions allowing for the study of several different phenomena. Relaxation characterization (*162*), intermolecular interactions (*163*), and diffusion are the primeval examples, such as the diffusion of 1,4-polybutadiene of varying molecular weights to identify the mechanism of polymer diffusion (*164*). In this study fast field cycling NMR and BDS were able to confirm the molecular weight at which the polymer transitions from a simple liquid to the Rouse model to the reputation model. The BDS results further confirmed that the glass transition is intimately related to these molecular weight-dependent models.

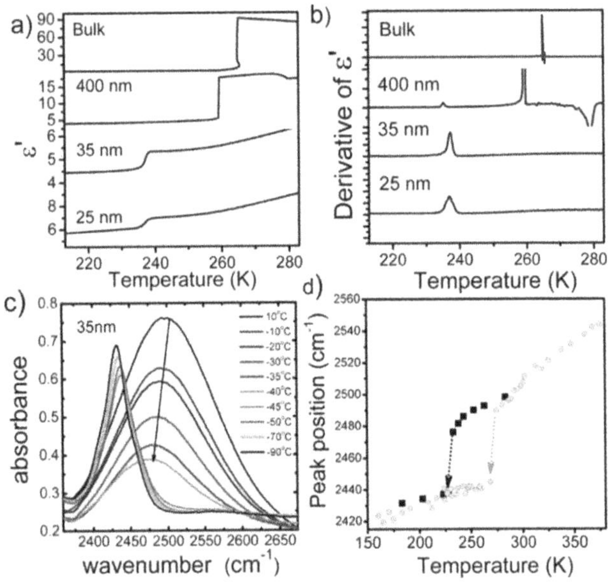

Figure 24. a) Temperature-dependent permittivity of water in nanoporopus alumina at 1 MHz as the temperature is lowered at 5 K·min^{-1}. Ice formation is registered as a sudden drop in the permittivity. b) Derivative of the permittivity highlighting the temperature of the observed phase transitions. c) Temperature-dependent FTIR of the system with 35 nm pores containing 10% D_2O in H_2O, demonstrating a change in the O-D stretching vibration as ice is formed. d) O-D stretching vibration peak position versus temperature determined via peak fitting. Reproduced with permission from reference (159). Copyright 2015 American Chemical Society.

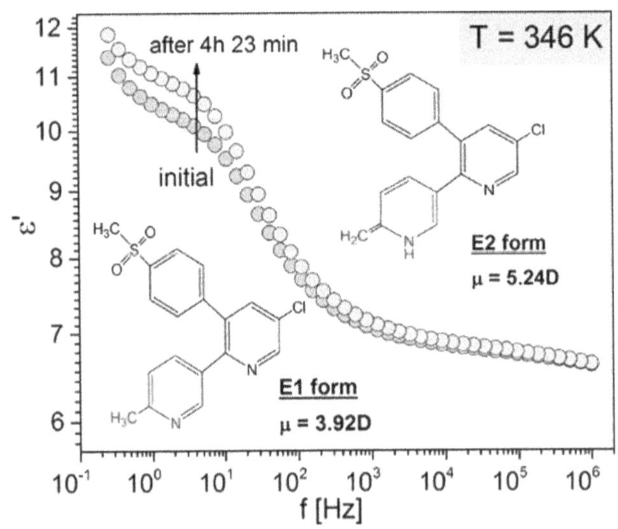

Figure 25. Dielectric spectra of etoricoxib measured before and after aging at 346 K for over four hours, demonstrating an increase in permittivity from tautomerization. Dipole moments were calculated using density functional theory. Reproduced with permission from reference (161). Copyright 2015 American Chemical Society.

The use of NMR and BDS in concert is covered in detail in Chapter 17 of Reference (*14*), but it is worth mentioning that the elemental specificity of NMR allows for detailed analyses, such as the identification of unknown relaxations in a mixture of water and dimethyl sulfoxide using deuterated water (Figure 26) (*165*).

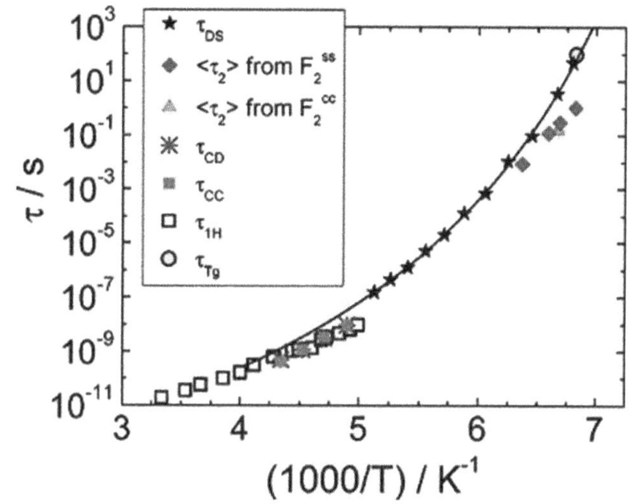

Figure 26. Activation plot of a supercooled 2:1 molar mixture of heavy water and dimethyl sulfoxide. Black stars are from BDS experiments. Red diamonds and green triangles are from ^{2}H NMR correlation functions. Blue asterisks, pink squares, and open squares are from fits to the spin-lattice relaxation times. Yellow circle is the glass transition temperature. Black line is a fit to the VFT equation. Note how the BDS and NMR data are in agreement at high temperature suggesting a coupling of the relaxation dynamics. However, the relaxation times diverge near the glass transition temperature by nearly two orders of magnitude. Since NMR only probed the water relaxation, it is posited that a dynamic process for the water splits off from the main relaxation, with a weaker dependence on temperature. A small shoulder in the relaxation peak as measured via BDS supports this hypothesis. Reproduced with permission from reference (165). Copyright 2011 American Chemical Society.

The identification of the molecular identity of dipole relaxations is a key part of BDS applications. In a recent study the relaxations of a metal-organic framework were examined (*166*). This system comprised of cubic Zn_4O nodes supported by 1,4-benzodicarboxylic acid struts (MOF-5). MOFs are of great interest today because of their ability to maintain very specific pore sizes in between the nodes, offering applications in gas filtration or storage (*167*). In systems such as MOF-5 exact molecular relaxations and the activation energy of their movement become of prime importance. By comparing BDS data with NMR data (*168*), the process labeled Region II was assigned to the rotation of the phenyl groups in the struts (Figure 27). The activation energy for the relaxations were also determined.

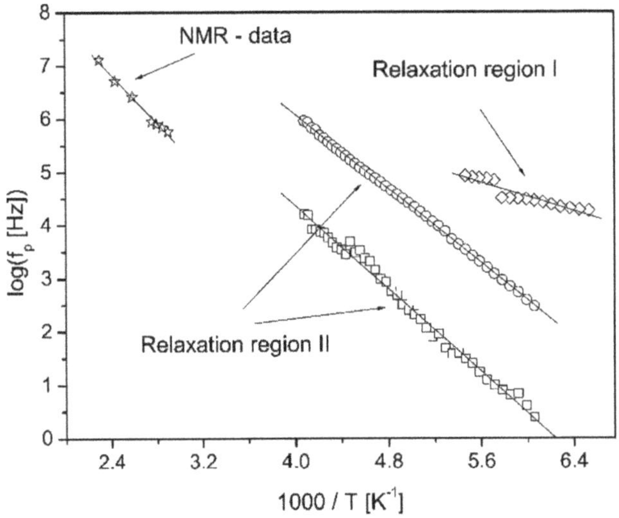

Figure 27. Activation plot of relaxations measured in MOF-5 over a wide temperature and frequency range. Two relaxation regions are seen, each with different activation energies. Earlier NMR data from reference (168) identified a relaxation as the phenyl group rotation. Extrapolation of Relaxation region II suggests that this is the same relaxation. The faster region II relaxation is hypothesized to be a different configuration of the phenyl group since the activation energy is identical. The low activation energy of the region I relaxation suggests this is from bending of the MOF cage. Reproduced with permission from reference (166). Copyright 2010 American Chemical Society.

Rheology

Rheology is so intimately related to BDS that early books discussed them in tandem (*19, 86, 140*). As BDS measures the macroscopic relaxations of a dipole in the presence of an electric field and the energy lost in this process, rheology measures the macroscopic relaxation of a material after a mechanical stress is applied. Similar to BDS, rheological measurements can be performed in the time scale or the frequency scale by incorporating an oscillatory stress (i.e. dynamic mechanical analysis or DMA), and a peak in the loss factor tanδ is observed in the presence of a relaxation. Rheological relaxations caused by steric interactions between molecules or functional groups are comparable to relaxations from molecular dipoles measured via BDS. Although the strength of the response and the absolute relaxation time may be different, the resulting activation plot is quite comparable (as with NMR). Similar to the many models used to fit BDS data, the similarities between the different measurement techniques can be more important than the differences.

Indeed, rheology "equivalent circuits" are often used pedagogically, where the resistance of the system to stress is represented by a dashpot and the energy storage (capacitance) of the system is represented by a spring. There are also commercially available rheometers which employ electrically isolated parallel plates connected to an impedance analyzer such that BDS data can be collected while a mechanical stress or constant force is applied to the sample (TA Instruments and Anton Paar).

The frequency range of a modern commercially available DMA is considerably less than an impedance analyzer, with maximum frequencies of ~100 Hz. Non-oscillating recovery experiments determine frequency-dependent data using known mathematical methods (*104*). Because of this rheology experiments often rely on master plots where data at several temperatures is combined (Figure 28) (*169, 170*). Nevertheless, DMA is an invaluable tool when performing BDS studies.

In a recent study the aprotic quaternary ammonium ionic liquid methyltrioctylammonium bis(trifluoromethylsulfonyl)imide was studied via rheology, BDS, pulsed field gradient NMR, and depolarized dynamic light scattering to determine why this ionic liquid exhibits significantly lower ionic conductivity than similar materials (Figure 29). By coupling all of these techniques together it was found that the ionic liquid had a lower than expected number density of free ions and a lower diffusion coefficient. A Debye-like relaxation at very low frequencies suggests that nanoscale hydrophobic aggregates from the alkyl groups hinder ionic motion. Studies like these are paramount to the design and understanding of ionic liquids of interest to industry (*171*).

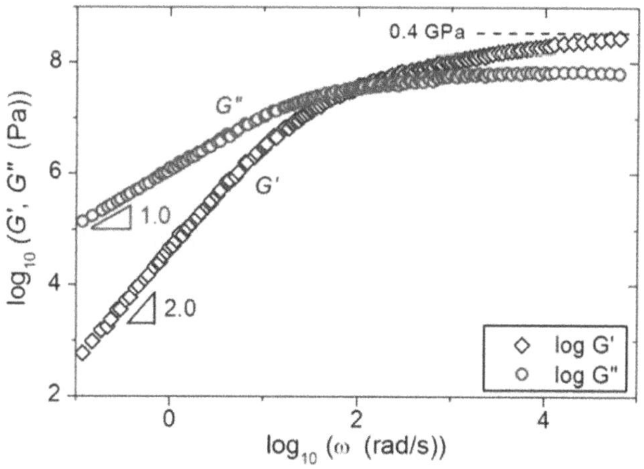

Figure 28. Complex dynamic shear modulus of ionic liquid methyltrioctylammonium bis(trifluoromethylsulfonyl)imide at 203 K. To obtain data over six decades of frequency, spectra are collected at multiple temperatures and a master plot is assembled. Note that the real and imaginary part of the modulus M' and M" are unrelated to the dielectric modulus M discussed earlier in this chapter. Dashed line is the estimated plateau shear modulus G_∞. Reproduced with permission from reference (169). Copyright 2014 American Chemical Society.*

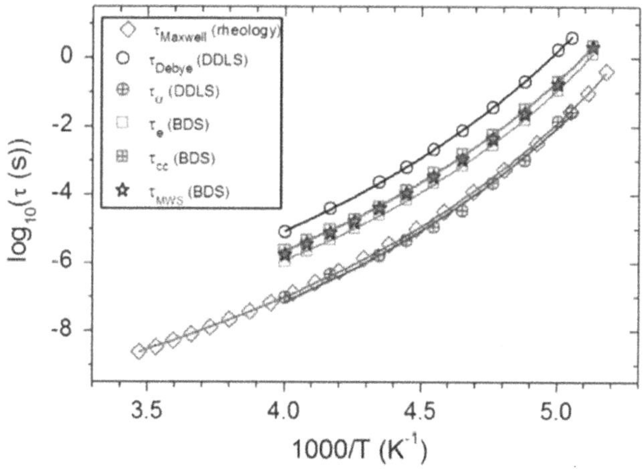

Figure 29. Activation plot of the observed relaxations in ionic liquid methyltrioctylammonium bis(trifluoromethylsulfonyl)imide as measured via BDS, rheology, and depolarized dynamic light scattering (DDLS). Solid lines are fits to the VFT equation. Reproduced with permission from reference (169). Copyright 2014 American Chemical Society.

Differential Scanning Calorimetry and Ultrafast Calorimetry

It was described earlier in this chapter how calorimetry is often used to determine phase transitions or reaction enthalpies in order to assign these features to BDS spectra. In this aspect fusion of BDS with differential scanning calorimetry (DSC) is as strong as those with IR, NMR, and DMA, and often BDS can be used to help analyze unusual DSC data (*172*). However, a new method of performing calorimetry at extremely high heating and cooling rates has become commercially available: fast-scanning calorimetry or AC chip calorimetry (*173–176*). By placing a nanogram-sized sample on a micron-scale calorimeter built onto a printed circuit board, heating rates and oscillations can be performed which are much faster than those possible with traditional DSC. As the rate of the test can have dramatic effects on such phenomena as glass transitions and crystallization, fast-scanning calorimetry can probe the effect of the rate on these transitions at the same rates that BDS can probe the dynamics. An example is shown in Figure 30 where the dynamics of linear poly(vinyl methyl ether) containing dispersed single-chain polystyrene nanoparticles was studied (*177*). The BDS spectra was mostly dominated by the amorphous polar polymer as the polystyrene nanoparticles do not exhibit a strong dipole. The calorimetry, however, captured the entire system response. BDS was also able to measure a slower relaxation which was attributed to the local motions of the poly(vinyl methyl ether) in contact with the nanoparticle surface.

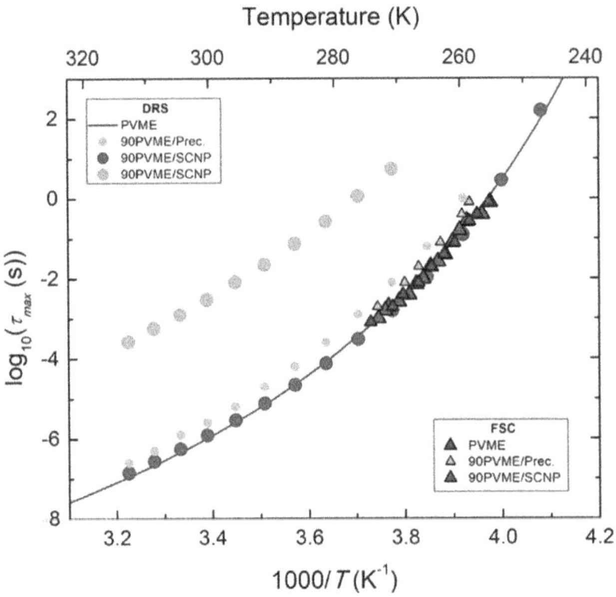

Figure 30. Activation plot of poly(vinyl methyl ether) (PVME) with dispersed single-chain polystyrene nanoparticles. Red line is the Vogel-Fulcher-Tammann fit of the relaxation of pure PVME. Yellow circles are the BDS data for PVME containing 10% linear precursor chains (i.e. not nanoparticles). Blue circles are data for PVME containing 10% nanoparticles. Green circles are the same, identifying a second, faster relaxation not seen with calorimetry. Red triangles are fast-scanning calorimetry data for pure PVME. Yellow triangles are the BDS data for PVME containing 10% linear precursor chains. Blue triangles are data for PVME containing 10% nanoparticles. Good agreement between BDS and fast-scanning calorimetry can be seen. Reproduced with permission from reference (177). Copyright 2019 American Chemical Society.

Neutron, X-Ray, and Depolarized Dynamic Light Scattering

The use of neutron scattering experiments to elucidate spatial information on specific relaxations has been used for some time (see Chapter 13 in Reference 14). With various methods it is possible to characterize and qualify the dynamics and size domain of a particular relaxation. In Figure 31 an example is shown where water absorbed in amorphous and semi-crystalline polyamide was studied. BDS experiments identified three relaxations while quasi-elastic neutron scattering identified the faster process as diffusive water motions in the polymer (*178*). Neutron scattering studies like this shed valuable light on understanding the plasticization effects of polymers used in coatings and industrial applications (*179*).

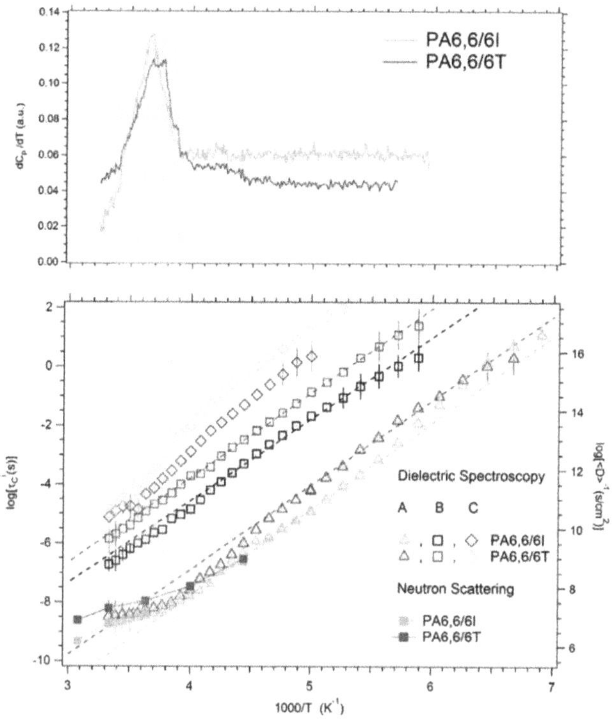

Figure 31. (Top) Derivative of the heat capacity for fully wet Nylon 66 with random isophthalic or terephthalic acid monomers incorporated (PA6,6/6I and PA6,6/6T, respectively). The calorimetric glass transition can be seen in the blue shaded area. (Bottom) Activation plot of the three relaxations measured via broadband dielectric spectroscopy and neutron scattering. Relaxation A is associated with a fast water process. Note how it deviates from linearity at the glass transition, suggesting a coupling to that relaxation. Relaxations B and C are associated with the γ and β relaxations, respectively, and are distinct due to the movement of the amide groups in the presence of water. Reproduced with permission from reference (178). Copyright 2012 American Chemical Society.

Similar to neutron scattering, wide- and small-angle x-ray scattering are often utilized in the comparison of BDS data with structural changes. In Figure 32 it is shown how the isothermal crystallization of poly(trimethylene terephthalate) affects the BDS as well as the scattering spectra (*180*). It was elucidated from H-N fitting parameters that the relaxation becomes broader but more evenly distributed during crystallization, which suggests the formation of an immobilized rigid amorphous phase between nascent lamellae. In this way crystallization dynamics are gleaned as well as the overall crystallinity—a feat not possible with x-ray scattering or microscopy alone.

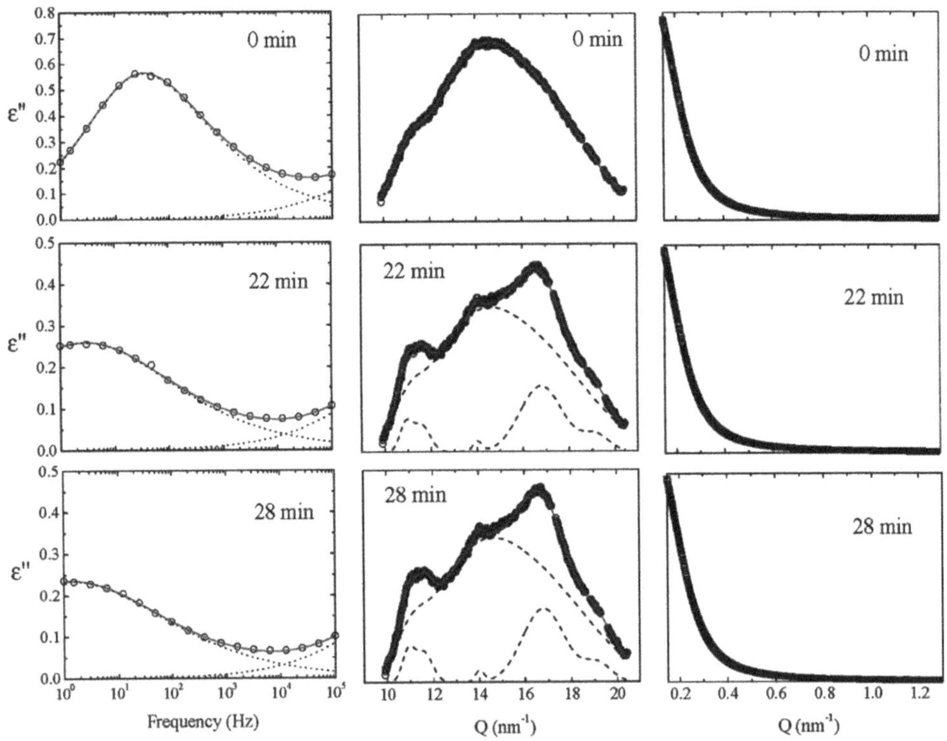

Figure 32. Isothermal crystallization of poly(trimethylene terephthalate) at 325 K. (Left) The dielectric strength is shown to decrease as the peak shifts to lower frequencies. (Center) the wide-angle x-ray scattering data demonstrates crystal formation via the formation of Bragg peaks, while (Right) the absence of structure in the small-angle x-ray scattering data suggests a lack of periodicity between the growing crystalline lamella. Reproduced with permission from reference (180). Copyright 2010 American Chemical Society.

In a similar vein, x-ray diffraction can determine crystalline structure, which can elucidate the change in relaxation processes in polymorphous materials (e.g. pinanediol, Figure 33) (*181*).

Finally, depolarized dynamic light scattering (DDLS) can provide valuable comparisons with BDS (*182*). While the latter measures the reorientation of dipoles, the former measures both reorientation and translation. Thus the two methods can be directly compared in an effort to isolate specific dielectric responses (*92, 182–184*).

Atomic Force Microscopy

BDS measures microscopic relaxations, and thus the use of microscopy has in some instances aided in the understanding of the observed relaxations (e.g. crystal nucleation or polymorphism). AFM is unique, however, in that conductive tips are commercially available, thus allowing for microscopic BDS measurements (*185–187*). This was seen as one of the most exciting new arrivals to the field of BDS twenty years ago (*188*), and indeed the technique has continued to progress (*189–191*). Although originally limited to the kHz range due to the restrictions of the cantilever, recent progress has allowed for measurements several decades higher and lower in frequency (*192–195*).

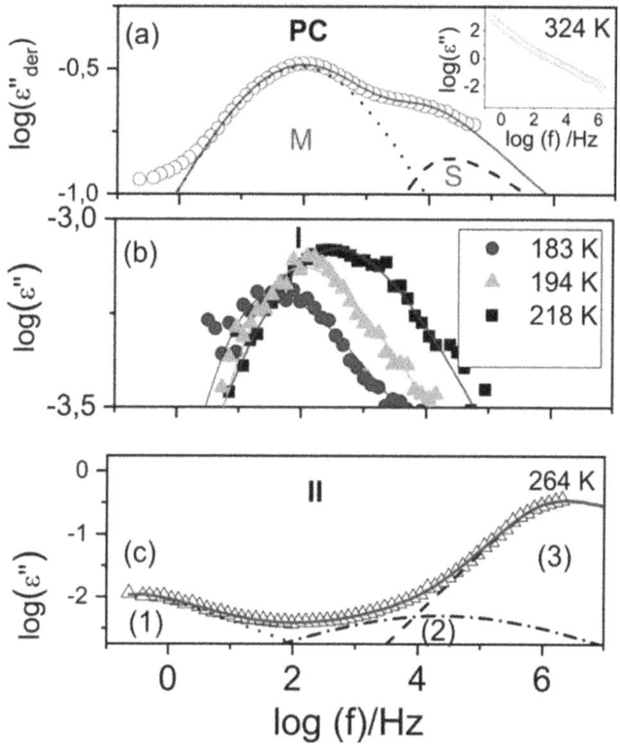

Figure 33. Dielectric spectra of the plastic crystal pinanediol at different temperatures. Sudden transitions in the dielectric spectra revealed a) the amorphous plastic crystal and b) and c) distinct polymorphs. Calorimetry, microscopy, and x-ray scattering experiments confirmed that the plastic crystal undergoes rotational relaxations, the form I exhibits relaxations due to the reorientation of hydroxyl groups, and the form II crystals exhibit relaxations due to molecular motions in the crystal lattice. Note how the dielectric loss for the plastic crystal a) is extracted from a region of high conductivity (inset). Reproduced with permission from reference (181). Copyright 2019 American Chemical Society.

In one study dielectric imaging was used to examine the cause of a broadened dielectric relaxation in poly(ethylene-vinyl acetate) random copolymers (*196*). Microscopy revealed two separate relaxation domains which were then fitted to the macroscopic dielectric spectra (Figure 34). Crystallinity was measured using calorimetry and x-ray diffraction and it was posited that the two relaxations are due to the noncrystalline vinyl-acetate containing phases which are either a) constrained by polyethylene nanocrystals or b) large amorphous regions of the random copolymer.

The Future

It is worthwhile to conclude this chapter with a note on the future of broadband dielectric spectroscopy. At its current pace this technique promises to become a ubiquitous tool in the modern analytical chemistry laboratory; one that is capable of providing insight into a wide variety of projects and materials. That being said, the technique has evolved considerably since the 1990's and it would be judicious for the current practitioners of BDS to think about what this technique will look like in thirty years—and what should be done to promote it to the levels of IR and NMR spectroscopies (Figure 1). There are several aspects that remain to be developed in order to truly propel BDS to the celebrity of the other techniques to which it often complements.

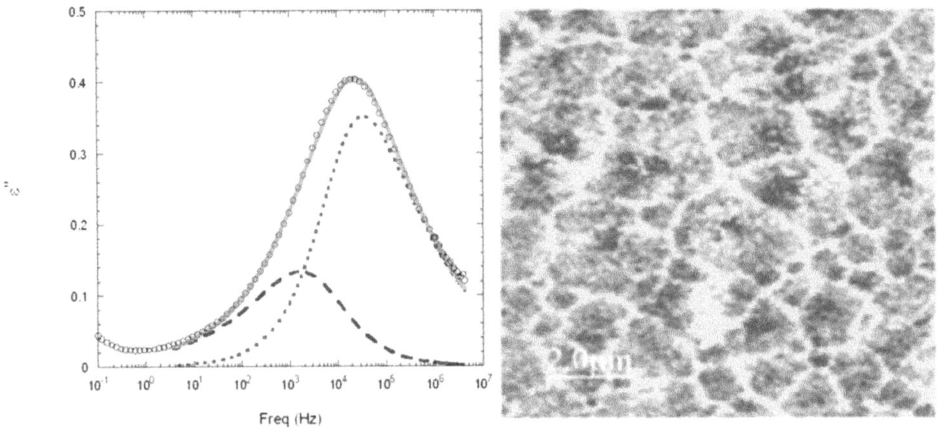

Figure 34. (Left) Dielectric spectrum of EVA40 at 270 K. Circles are from experimental data. Dashed blue line is the contribution from constrained system. Dotted red line is the contribution from the nonconstrained system. Green line is the fit to the data using both contributions. (Right) Dielectric image from single-pass electrostatic force microscopy of EVA40, demonstrating the two domains. Reproduced with permission from reference (196). Copyright 2013 American Chemical Society.

First and foremost, BDS is rarely taught in undergraduate classes dedicated to analytical chemistry. Some of the more popular text books on instrumentation barely mention it (or electrochemical impedance spectroscopy) at all. The incorporation of a benchtop impedance analyzer into a laboratory curriculum is inexpensive and would also provide training in electrochemistry and electrical engineering. On the graduate level the forecast is promising: sessions at large technical meetings dedicated to BDS are on the rise, and conferences dedicated to frequency-dependent electrical properties are also growing (*81*, *197–203*, *204*). However, it is the experience of this author that the average chemist is still relatively unfamiliar with BDS, and many laboratories that regularly measure dielectric constant, dissipation factor, or conductivity do so at a single frequency with little concern for what lies to either side.

Second, modern commercially available dielectric spectrometers lack the autosampler functionality that accompanies most other common analytical techniques. The role of the modern industrial analytical chemist is to set foot in lab as little as possible—automation is a key factor for industrial adoption. To be sure, BDS experiments can last hours or days and careful sample preparation is key to successful measurements. Why worry about a few minutes setting up the next experiment? The same can be said of NMR, DMA and AFM, all of which have commercially available autosampler functions. Several commercial BDS sample holders have micrometers built in with options for digital connections to the software but these sample holders have temperature limitations and fully automatic thickness measurements are rare, which makes compensation for sample thickness at variable temperatures difficult. This is surprising given the attempted marriage between some DMA instruments and impedance analyzers (mentioned earlier in this chapter), and in fact the DEA 2970 (TA Instruments, Inc.) was a dedicated dielectric spectrometer in the 1990's with a variety of disposable electrode types which could control both electrode spacing and applied force during a measurement. Today no such commercial instrumentation exists.

Third, the analysis of large amounts of frequency-dependent data would benefit from a computer learning or chemometrics approach. Chemometrics is the application of statistics to translate large amounts of data into useful chemical information. As discussed earlier in this chapter, a firm understanding of the system under study and a prediction for the equivalent circuit model is required in order to tease any value out of a BDS experiment. Although machine learning has been used on many occasions to predict product quality using dielectric spectroscopy (usually at a high frequency range, for food quality control purposes) (*205–215*), to date no attempt has been made to develop a predictive tool for studying dielectric spectra of unknown materials. The decades of experiments and online databases available are a rich data source that is likely under-leveraged today (i.e. features in the data could contain patterns that machine learning algorithms could find). This effort might prove quite valuable for data analysis software and could allow BDS to join other spectroscopies, chromatographies, and spectrometries as a commonly used in-line process control method for various industries.

In summary, the field of broadband dielectric spectroscopy continues to grow at a rapid pace and the technique has proven itself a valuable capability in the modern analytical chemistry toolbox. It often provides unique solutions to difficult problems in conjunction with complementary techniques, and the studied analytical chemist can produce valuable data using modern instrumentation and careful experimental considerations. The following chapters, contributed by top names in the field, will provide concrete examples of this.

Additional Sources of Information

For the reader interested in bolstering their knowledge with more in-depth reviews on specific experimental setups, the author recommends the following readily available texts (Table 1) (*14, 18, 19, 35, 53, 81, 86, 116, 128, 135, 136, 150, 216–222*). There are also several modern sources available on specific aspects of phenomena commonly measured via BDS, such as dipole relaxations and electrical properties of specific materials.

Table 1. Additional Sources of Information

Title	*Description*
Broadband Dielectric Spectroscoy (F. Kremer & A. Schönhals, Eds.) Springer: 2002	This is the most up-to-date review of BDS today. Much of what has been written in this chapter can be found here in more comprehensive detail.
Impedance Spectroscopy (V. Lvovich) Wiley: 2012	As BDS is a sub-field of impedance spectroscopy, there is much to be learned from texts on the latter (*51, 54, 55, 223*). This book in particular provides several modern examples of applied impedance spectroscopy.
Advances in Dielectrics (F. Kremer, Series Ed.) Springer	This series includes monographs dedicated to BDS, with such topics as crystallization, nonlinear dielectric spectroscopy, ionic liquids, geometrical confinement, glassy systems, and the scaling of relaxation processes (*116, 128, 135, 136, 150*).
Dielectrics Newsletter: Scientific Newsletters for Dielectric and Impedance Spectroscopy, Novocontrol Technologies	Published every few years, this newsletter has provided a consistent combination of state-of-the-art instrument development and fundamental experimental practices since the first issue in 1994.

Table 1. (Continued). Additional Sources of Information

Title	*Description*
Application Notes, Solartron Analytical	Predating the Dielectrics Newsletter, this collection of several dozen updated documents targets specific experiments or techniques, including many on electrochemical impedance spectroscopy. In a similar vein, HP Application Notes date back to the 1960's and are becoming easier to find on the internet.
Anelastic and Dielectric Effects in Polymeric Solids (N. McCrum, B. Read, G. Williams) Wiley: 1967	These books collectively make up the four most comprehensive polymer-centric texts dedicated to BDS. Many common polymers were studied and documented in the first two examples, all before modern instrumentation.
Dielectric Spectroscopy of Polymers (P. Hedvig) Halsted Press: 1977	
Dielectric Spectroscopy of Polymeric Materials (J. Runt & J. Fitzgerald, Eds.) ACS: 1997	
Dielectric and Mechanical Relaxation in Materials (S. Havriliak & S. Havriliak) Hanser: 1997	
Non-Equilibrium Phenomena in Confined Soft Matter (S. Napolitano, Ed.) Springer: 2015	As its resolution increases with decreasing sample thickness, BDS has been used extensively in the study of confined systems. This monograph is a comprehensive review of the state-of-the-art.
Recent Advances in Broadband Dielectric Spectroscopy (Y. Kalmykov, Ed.) Springer: 2011	Several modern examples of high-frequency dielectric spectroscopy, including fundamental studies on water and ion diffusion, can be found here.
Dielectric Relaxation in Biological Systems (V. Raicu & Y. Feldman)	A treatise on dielectric spectroscopy in aqueous and biological systems, with a detailed historical introduction and a comprehensive discussion of MWS polarization.

Acknowledgments

The author would like to thank Professor John Berberian for many helpful discussions and for assisting with the historical accuracy of the content. Many thanks also to Professor Friedrich Kremer and Drs. Tom Fielitz, James Wade, Bill Winniford, and Meaghan Woodward for critically reviewing the manuscript.

References

1. Debye, P. Einige Resultate einer kinetischen Theorie der Isolatoren. *Physik Z.* **1912**, *13* (3), 97–100.
2. Debye, P. *Polare Molekeln*; S. Hirzel: Leipzig, 1929.
3. Debye, P. J. W. *Polar Molecules*; Chemical Catalog Company: New York, 1929.
4. Ball, P.Opinion: The Crucible. *Chemistry World*, Feb. 26, 2010.
5. Tarnacka, M.; Madejczyk, O.; Dulski, M.; Wikarek, M.; Pawlus, S.; Adrjanowicz, K.; Kaminski, K.; Paluch, M. Kinetics and dynamics of the curing system. High pressure studies. *Macromolecules* **2014**, 47 (13), 4288–4297.

6. Hamming, R. *The Art of Doing Science and Engineering: Learning to Learn*; Stripe Press: San Francisco, 2020.
7. Böttcher, C. J. F.; van Belle, O. C.; Bordewijk, P.; Rip, A. *Theory of Electric Polarization*, 2nd ed.; Elsevier: New York, 1973; Vol. 1.
8. Böttcher, C. J. F.; Bordewijk, P. *Theory of Electric Polarization*, 2nd ed.; Elsevier: New York, 1978; Vol. 2.
9. Von Hippel, A. R. *Dielectrics and Waves*; John Wiley & Sons: New York, 1954.
10. Smyth, C. P. *Dielectric Behaviour and Structure*; McGraw-Hill: New York, 1955.
11. Fröhlich, H. *Theory of Dielectrics*; Clarendon Press: Oxford, 1949.
12. Hill, N. E.; Vaughan, W. E.; Price, A. H.; Davies, M. *Dielectric Properties and Molecular Behaviour*; Van Nostrand Reinhold: New York, 1969.
13. Scaife, B. K. P. *Principles of Dielectrics*; Oxford University Press: Oxford, 1988.
14. Kremer, F.; Schönhals, A. *Broadband Dielectric Spectroscopy*; Springer: Berlin, 2002.
15. Fitzky, H. G. Applications of high-frequency spectroscopy in the chemical industry with examples of electron-spin, nuclear quadrupole resonance, and dielectric spectroscopy. *GIT Fachzeitschrift fuer das Laboratorium* **1976**, *20* (3), 205–216.
16. Fitzky, H. G. Applications of high-frequency spectroscopy in the chemical industry with examples of electron-spin, nuclear quadrupole resonance, and dielectric spectroscopy. Part II. *GIT Fachzeitschrift fuer das Laboratorium* **1976**, *20* (4), 341–345.
17. Ehrlich, P.; De Lollis, N. J. Dielectric relaxation in a styrene-acrylonitrile copolymer during and after its polymerization. *J. Res. Natl. Bur. Stand. (U.S.)* **1953**, *51* (3), 145–154.
18. Hedvig, P. *Dielectric Spectroscopy of Polymers*; John Wiley & Sons: Hoboken, NJ, 1977.
19. McCrum, N. G.; Read, B. E.; Williams, G., *Anelastic and Dielectric Effects in Polymeric Solids*. John Wiley & Sons: Hoboken, NJ, 1967.
20. Tanaka, J. History of the Dielectrics and Electrical Insulation Society. *IEEE Trans. Electr. Insul.* **1990**, *25* (1), 3–16.
21. Tanaka, J. History of the Dielectrics and Electrical Insulation Society (update). *IEEE Trans. Dielectr. Electr. Insul.* **2002**, *9* (5), 666–667.
22. Williams, G. The dielectrics discussion group (UK) and the dielectrics society platforms for dielectrics research. *IEEE Trans. Dielectr. Electr. Insul.* **2011**, *18* (2), 570–578.
23. This statement does not include a 1963 Symposium organized by R. Cole in honor of P. Debye winning the Priestley Medal.
24. Approximately twice as many publications can be found in the *Journal of Chemical Physics* than the *Journal of Physical Chemistry*.
25. *Dielectrics Newsletter*, March 1994; pp 1–8.
26. Schaumburg, G. New integrated dielectric analyzer extends accuracy and impedance range for material measurements. *Dielectrics Newsletter*, May 1999; pp 4–6.
27. Lunkenheimer, P.; Schneider, U.; Brand, R.; Loid, A. Glassy dynamics. *Contemp. Phys.* **2000**, *41* (1), 15–36.
28. Vallerien, S. U.; Kremer, F.; Boeffel, C. Broadband dielectric spectroscopy on side group liquid crystal polymers. *Liq. Cryst.* **1989**, *4* (1), 79–86.

29. Kremer, F.; Vallerien, S. U.; Zentel, R.; Kapitza, H. Broad-band dielectric spectroscopy on a set of combined main-chain side-group liquid-crystalline polymers. *Macromolecules* **1989**, *22* (10), 4040–4045.
30. Vallerien, S. U.; Kremer, F.; Kapitza, H.; Zentel, R.; Frank, W. Field dependent soft and goldstone mode in a ferroelectric liquid crystal as studied by dielectric spectroscopy. *Phys. Lett. A* **1989**, *138* (4), 219–222.
31. Vallerien, S. U.; Kremer, F.; Hüser, B.; Spiess, H. W. Broadband dielectric spectroscopy on a discotic liquid crystalline polymer. *Colloid Polym. Sci.* **1989**, *267* (7), 583–586.
32. Kremer, F.; Boese, D.; Meier, G.; Fischer, E. W. Broadband dielectric spectroscopy on the molecular dynamics in polymer model-systems: polyphenylmethylsiloxane and linear and star-branched poly(cis-1,4-isoprene). In *Relaxation in Polymers*; Pietralla, M., Pechhold, W., Eds.; Steinkopff: Darmstadt, 1989; Vol. 80, pp 129−139.
33. Gaiduk, V. I. *Dielectric Relaxation and Dynamics of Polar Molecules*; World Scientific Publishing: Singapore, 1999.
34. Richert, R. Supercooled liquids and glasses by dielectric relaxation spectroscopy. *Adv. Chem. Phys.* **2015**, *156*, 101–195.
35. *Dielectric Relaxation in Biological Systems: Physical Principles, Methods, and Applications*; Oxford University Press: Oxford, 2015; p 464.
36. Riande, E.; Díaz-Calleja, R. *Electrical Properties of Polymers*; Marcel Dekker: New York, 2004.
37. Relative permittivity is a unitless value because it is in fact the absolute permittivity divided by ε0, thus making it "relative to vacuum." This may seem tedious, but it is important to pay attention to terminology. Reported permittivity values on the order of 10−11 F·m−1 are in fact absolute permittivities, as for practical purposes no measured relative permittivity will have a value less than one.
38. Aston, F. *Nobel Lecture, Chemistry 1922-1941*; Elsevier: New York, 1966.
39. Smyth, C. P. *Dielectric Constant and Molecular Structure*; Chemical Catalog Company: New York, 1931.
40. Daniel, V. V. *Dielectric Relaxation*; Academic Press: New York, 1967.
41. *Dielectric and Related Molecular Processes*; Royal Society of Chemistry: Cambridge, 1972; Vol. 1.
42. *Dielectric and Related Molecular Processes*; Royal Society of Chemistry: Cambridge, 1975; Vol. 2.
43. *Dielectric and Related Molecular Processes*; Royal Society of Chemistry: Cambridge, 1977; Vol. 3.
44. Kohlrausch, R. Theorie des elektrischen Rückstandes in der Leidener Flasche. *Ann. Phys.* **1854**, *167* (2), 179–214.
45. Williams, G.; Watts, D. C. Non-symmetrical dielectric relaxation behaviour arising from a simple empirical decay function. *Trans. Faraday Soc.* **1970**, *66*, 80–85.
46. The relaxation follows a stretched exponential function called the Kohlrausch-Williams-Watts (KWW) function. It is named after R. Kohlrausch, the German Physicist who derived the function in 1854, and G. Williams and D. Watts, two scientists at the University of Wales who reintroduced it in 1970.
47. Some scientists are brought up on admittance as opposed to impedance. An understanding of the math allows one to convert between the two seamlessly. Impedance is convenient because of the scores of books dedicated to impedance spectroscopy and the equivalent circuits seen in common experiments. Admittance is convenient when calculating permittivity of Debye relaxations (i.e. Equations 2 through 8 are much simpler).

48. j and i are used interchangeably in the literature to represent the imaginary unit $\sqrt{(-1)}$.
49. The argument δ is sometimes represented by a different Greek letter or is the adjacent angle and can be expressed in degrees or radians but for dielectric spectroscopy δ in degrees is common.
50. "Dia-electric" is from the Greek for "through-field," in that it allows a field to pass through but not particles. "Di-pole" is Greek for "two poles."
51. *Impedance Spectroscopy: Theory, Experiment and Applications*, 3rd ed.; John Wiley & Sons: Hoboken, NJ, 2018.
52. Sengupta, D. L.; Sarkar, T. K. Maxwell, Hertz, the Maxwellians, and the early history of electromagnetic waves. *IEEE Antennas and Propagation Magazine* **2003**, *45* (2), 13–19.
53. Lvovich, V. F. *Impedance Spectroscopy: Applications to Electrochemical and Dielectric Phenomena*; John Wiley & Sons: Hoboken, NJ, 2012.
54. Orazem, M. E.; Tribollet, B. *Electrochemical Impedance Spectroscopy*, 2nd ed.; John Wiley & Sons: Hoboken, NJ, 2017.
55. Lasia, A. *Electrochemical Impedance Spectroscopy and its Applications*; Springer: Berlin, 2002.
56. Recall from the impedance equations above that the permittivity at frequencies approaching zero is called the static permittivity (εs), where $\Delta\varepsilon = \varepsilon s - \varepsilon\infty$.
57. It is common to use frequency as the x-axis and not angular frequency, but to be sure the mathematics of the classical mechanics introduced previously employ the latter for all relationships (relaxation times, complex permittivity, etc.). This discrepancy is mixed in the literature, and so care must be taken to identify when a value is or is not multiplied by 2π.
58. Alvarez, F.; Alegra, A.; Colmenero, J. Relationship between the time-domain Kohlrausch-Williams-Watts and frequency-domain Havriliak-Negami relaxation functions. *Phys. Rev. B* **1991**, *44* (14), 7306.
59. Alvarez, F.; Alegría, A.; Colmenero, J. Interconnection between frequency-domain Havriliak-Negami and time-domain Kohlrausch-Williams-Watts relaxation functions. *Phys. Rev. B* **1993**, *47* (1), 125.
60. Berberian, J. G.; Cole, R. H. Approach to glassy behavior of dielectric relaxation in 3-bromopentane from 298 to 107 K. *J. Chem. Phys.* **1986**, *84* (12), 6921–6927.
61. Havriliak, S.; Negami, S. A complex plane representation of dielectric and mechanical relaxation processes in some polymers. *Polymer* **1967**, *8*, 161–210.
62. When b = 1 Equation 15 becomes the Cole-Cole equation, originally proposed by K. Cole (Columbia University) and R. Cole (Harvard University) in 1941. When a = 1 Equation 15 becomes the Cole-Davidson equation, proposed by R. Cole and D. Davidson (Brown University) in 1950, and when a = b = 1 Equation 15 becomes the Debye equation. In the event of a Debye relaxation, a plot of $\varepsilon'' / (\varepsilon' - \varepsilon\infty)$ versus ω will yield a slope of τ.
63. Cole, K. S.; Cole, R. H. Dispersion and Absorption in Dielectrics I. Alternating Current Characteristics. *J. Chem. Phys.* **1941**, *9* (4), 341–351.
64. Davidson, D. W.; Cole, R. H. Dielectric Relaxation in Glycerine. *J. Chem. Phys.* **1950**, *18* (10), 1417–1417.
65. The Kohlrausch-Williams-Watts (KWW) function is also common in the literature, both today but especially prior to the Havriliak-Negami function was proposed in the late 1960's. A discussion on their relative advantages is not given here, suffice it to say that both produce similar relaxation times but the H-N function has been referenced approximately twice as frequently since 1990.

66. Grosse, C. A program for the fitting of Debye, Cole–Cole, Cole–Davidson, and Havriliak–Negami dispersions to dielectric data. *J. Colloid Interface Sci.* **2014**, *419*, 102–106.
67. Volkov, A.; Koposov, G.; Perfil'ev, R.; Tyagunin, A. Analysis of experimental results by the Havriliak–Negami model in dielectric spectroscopy. *Opt. Spectrosc.* **2018**, *124* (2), 202–205.
68. Kobayashi, K.; Suzuki, T. S. Development of Impedance Analysis Software Implementing a Support Function to Find Good Initial Guess Using an Interactive Graphical User Interface. *Electrochemistry* **2020**, *88* (1), 39–44.
69. The experimental technique used to demonstrate dynamic heterogeneity is known as nonresonant dielectric hole burning. This is a valuable sub-field of dielectric spectroscopy which utilizes high fields and is beyond the scope of this chapter. For more details, see the references at the end of this sentence in addition to Chapter 14 of Reference 14 and Chapter 4 of this book.
70. Chamberlin, R. V.; Böhmer, R.; Richert, R. Nonresonant spectral hole burning in liquids and solids. In *Nonlinear Dielectric Spectroscopy*; Richert, R., Ed.; Springer: Berlin, 2018; pp 127−185.
71. Böhmer, R.; Diezemann, G., Principles and applications of pulsed dielectric spectroscopy and nonresonant dielectric hole burning. In *Broadband Dielectric Spectroscopy*; Kremer, F., Ed.; Springer: Berlin, 2003; pp 523−569.
72. El Goresy, T.; Kircher, O.; Böhmer, R. Nonresonant hole burning spectroscopy of the relaxor ferroelectric PLZT. *Solid State Commun.* **2002**, *121* (9−10), 485–488.
73. Kircher, O.; Diezemann, G.; Böhmer, R. Nonresonant dielectric hole-burning spectroscopy on a titanium-modified lead magnesium niobate ceramic. *Phys. Rev. B* **2001**, *64* (5), 054103.
74. Schiener, B.; Chamberlin, R.; Diezemann, G.; Böhmer, R. Nonresonant dielectric hole burning spectroscopy of supercooled liquids. *J. Chem. Phys.* **1997**, *107* (19), 7746–7761.
75. Schiener, B.; Böhmer, R.; Loidl, A.; Chamberlin, R. Nonresonant spectral hole burning in the slow dielectric response of supercooled liquids. *Science* **1996**, *274* (5288), 752–754.
76. Jeffrey, K. R.; Richert, R.; Duvvuri, K. Dielectric hole burning: Signature of dielectric and thermal relaxation time heterogeneity. *J. Chem. Phys.* **2003**, *119* (12), 6150–6156.
77. Duvvuri, K.; Richert, R. Dielectric hole burning in the high frequency wing of supercooled glycerol. *J. Chem. Phys.* **2003**, *118* (3), 1356–1363.
78. Richert, R. Dielectric hole burning in an electrical circuit analog of a dynamically heterogeneous system. *Phys. A* **2003**, *322*, 143–154.
79. Chamberlin, R.; Schiener, B.; Böhmer, R. Slow dielectric relaxation of supercooled liquids investigated by nonresonant spectral hole burning. In *Materials Research Society Symposium Proceedings*; Materials Research Society: 1997; pp 117−126.
80. Not to belabor this point, but all data analysis should rely on as few fitting parameters as possible, and no fitting parameter has physical significance until the theory predicts the experiment.
81. *Dielectric Spectroscopy of Polymeric Materials: Fundamentals and Applications*; American Chemical Society: Washington, DC, 1997.
82. In fact for many studies in analytical chemistry it is unreasonable to pursue the one true relaxation time, as τ is dependent on the model assumed to fit the data and hence a Debye relaxation τD will not be the same as a Havriliak-Negami relaxation τHN. The same is true for other aspects of the relaxation determined using different models as different numbers of fitting

parameters are employed. Literature sources will not always use subscripts to identify which model was used, but R. Cole originally demonstrated that the effect is small.

83. Wübbenhorst, M.; Van Turnhout, J. Analysis of complex dielectric spectra. I. One-dimensional derivative techniques and three-dimensional modelling. *J. Non-Cryst. Solids* **2002**, *305* (1–3), 40–49.
84. van Turnhout, J.; Wübbenhorst, M. Analysis of complex dielectric spectra. II: Evaluation of the activation energy landscape by differential sampling. *J. Non-Cryst. Solids* **2002**, *305* (1–3), 50–58.
85. Wübbenhorst, M. *Short Course: Analysis and modelling of dielectric relaxation data using DCALC*; KU Leuven: 2008.
86. Havriliak, S.; Havriliak, S. J. *Dielectric and Mechanical Relaxation in Materials: Analysis, Interpretation and Application to Polymers*; Hanser: Cincinnati, 1997.
87. Johari, G. P.; Goldstein, M. Viscous liquids and the glass transition. II. Secondary relaxations in glasses of rigid molecules. *J. Chem. Phys.* **1970**, *53* (6), 2372–2388.
88. Goldstein, M. Viscous liquids and the glass transition: a potential energy barrier picture. *J. Chem. Phys.* **1969**, *51* (9), 3728–3739.
89. Schneider, U.; Brand, R.; Lunkenheimer, P.; Loidl, A. Excess wing in the dielectric loss of glass formers: A Johari-Goldstein β relaxation? *Phys. Rev. Lett.* **2000**, *84* (24), 5560.
90. Singh, L. P.; Richert, R. Watching hydrogen-bonded structures in an alcohol convert from rings to chains. *Phys. Rev. Lett.* **2012**, *109* (16), 167802.
91. Fragiadakis, D.; Roland, C.; Casalini, R. Insights on the origin of the Debye process in monoalcohols from dielectric spectroscopy under extreme pressure conditions. *J. Chem. Phys.* **2010**, *132* (14), 144505.
92. Gabriel, J.; Pabst, F.; Blochowicz, T. Debye process and β-relaxation in 1-propanol probed by dielectric spectroscopy and depolarized dynamic light scattering. *J. Phys. Chem. B* **2017**, *121* (37), 8847–8853.
93. Gabriel, J.; Pabst, F.; Helbling, A.; Böhmer, T.; Blochowicz, T. Nature of the Debye-process in monohydroxy alcohols: 5-methyl-2-hexanol investigated by depolarized light scattering and dielectric spectroscopy. *Phys. Rev. Lett.* **2018**, *121* (3), 035501.
94. Huang, Y.; Saiz, E.; Ezquerra, T. A.; Guzmán, J.; Riande, E. Broad-Band Dielectric Spectrocopy of Polymer Chains Containing Structurally Complex Side Groups. *Macromolecules* **2002**, *35* (5), 1785–1790.
95. Sillars, R. The properties of a dielectric containing semiconducting particles of various shapes. *J. Inst. Electr. Eng.* **1937**, *80* (484), 378–394.
96. Wagner, K. Electricity of the dielectric behaviour on the basis of the Maxwell theory. *Arch. Elektrotechnik* **1914**, *2*, 371–387.
97. Maxwell, J. C. *A Treatise on Electricity and Magnetism*; Clarendon Press: Oxford, 1873; Vols. I & II.
98. Richert, R. Dielectric spectroscopy and dynamics in confinement. *Eur. Phys. J.: Spec. Top.* **2010**, *189* (1), 37–46.
99. Lunkenheimer, P.; Bobnar, V.; Pronin, A. V.; Ritus, A.; Volkov, A.; Loidl, A. Origin of apparent colossal dielectric constants. *Phys. Rev. B* **2002**, *66* (5), 052105.

100. McKenzie, R.; Zurawsky, W.; Mijovic, J. A molecular interpretation of Maxwell–Wagner–Sillars processes. *J. Non-Cryst. Solids* **2014**, *406*, 11–21.
101. Van Beek, L. The Maxwell-Wagner-Sillars effect, describing apparent dielectric loss in inhomogeneous media. *Physica* **1960**, *26* (1), 66–68.
102. These words of caution are understated. Even air bubbles in a viscous dielectric liquid can cause MWS polarization, resulting in a false relaxation at lower frequencies.
103. Warburg, E. Polarization capacity of platinum. *Ann. Phys.* **1901**, *6*, 125–135.
104. Gillespie, R.; Cole, R. The dielectric constant of sulphuric acid. *Trans. Faraday Soc.* **1956**, 52, 1325–1331.
105. Schwan, H.; Onaral, B. Linear and nonlinear properties of platinum electrode polarisation III: equivalence of frequency- and time-domain behaviour. *Med. Biol. Eng. Comput.* **1985**, *23* (1), 28–32.
106. Schwan, H. Alternating current electrode polarization. *Biophysik* **1966**, 3 (2), 181–201.
107. Schwan, H. Electrode polarisation and its influence on determination of electrical properties of solutions and biological materials. *Z. Naturforsch.* **1951**, *6b*, 121–129.
108. Samet, M.; Levchenko, V.; Boiteux, G.; Seytre, G.; Kallel, A.; Serghei, A. Electrode polarization vs. Maxwell-Wagner-Sillars interfacial polarization in dielectric spectra of materials: Characteristic frequencies and scaling laws. *J. Chem. Phys.* **2015**, *142* (19), 194703.
109. Hamon, B. An approximate method for deducing dielectric loss factor from direct-current measurements. *Proc. IEE - Part IV: Institution Monographs* **1952**, *99* (3), 151–155.
110. Dyre, J. C. The random free-energy barrier model for ac conduction in disordered solids. *J. Appl. Phys.* **1988**, *64* (5), 2456–2468.
111. Kramers, H. A. La diffusion de la lumiere par les atomes. *Atti Cong. Intern. Fisica (Transactions of Volta Centenary Congress)*, Como, 1927; pp 545—557.
112. de L. Kronig, R. On the Theory of Dispersion of X-Rays. *J. Opt. Soc. Am.* **1926**, *12* (6), 547–557.
113. Atorngitjawat, P.; Runt, J. Dynamics of Sulfonated Polystyrene Ionomers Using Broadband Dielectric Spectroscopy. *Macromolecules* **2007**, *40* (4), 991–996.
114. Steeman, P.; Van Turnhout, J. A numerical Kramers-Kronig transform for the calculation of dielectric relaxation losses free from Ohmic conduction losses. *Colloid Polym. Sci.* **1997**, *275* (2), 106–115.
115. Schröter, K.; Unger, R.; Reissig, S.; Garwe, F.; Kahle, S.; Beiner, M.; Donth, E. Dielectric Spectroscopy in the αβ Splitting Region of Glass Transition in Poly(ethyl methacrylate) and Poly(n-butyl methacrylate): Different Evaluation Methods and Experimental Conditions. *Macromolecules* **1998**, *31* (25), 8966–8972.
116. *Dielectric Properties of Ionic Liquids*; Springer: Berlin, 2016.
117. Sangoro, J.; Serghei, A.; Naumov, S.; Galvosas, P.; Kärger, J.; Wespe, C.; Bordusa, F.; Kremer, F. Charge transport and mass transport in imidazolium-based ionic liquids. *Phys. Rev. E* **2008**, *77* (5), 051202.
118. Serghei, A.; Tress, M.; Sangoro, J.; Kremer, F. Electrode polarization and charge transport at solid interfaces. *Phys. Rev. B* **2009**, *80* (18), 184301.

119. Wang, Y.; Sun, C. N.; Fan, F.; Sangoro, J. R.; Berman, M. B.; Greenbaum, S. G.; Zawodzinski, T. A.; Sokolov, A. P. Examination of methods to determine free-ion diffusivity and number density from analysis of electrode polarization. *Phys. Rev. E* **2013**, *87* (4), 042308.
120. Sangoro, J. R.; Iacob, C.; Agapov, A.; Wang, Y.; Berdzinski, S.; Rexhausen, H.; Strehmel, V.; Friedrich, C.; Sokolov, A.; Kremer, F. Decoupling of ionic conductivity from structural dynamics in polymerized ionic liquids. *Soft Matter* **2014**, *10* (20), 3536–3540.
121. Wang, Y.; Fan, F.; Agapov, A. L.; Saito, T.; Yang, J.; Yu, X.; Hong, K.; Mays, J.; Sokolov, A. P. Examination of the fundamental relation between ionic transport and segmental relaxation in polymer electrolytes. *Polymer* **2014**, *55* (16), 4067–4076.
122. Maksym, P.; Tarnacka, M.; Dzienia, A.; Matuszek, K.; Chrobok, A.; Kaminski, K.; Paluch, M. Enhanced Polymerization Rate and Conductivity of Ionic Liquid-Based Epoxy Resin. *Macromolecules* **2017**, *50* (8), 3262–3272.
123. Piecha-Bisiorek, A.; Jakubas, R.; Medycki, W.; Florek-Wojciechowska, M.; Wojciechowski, M.; Kruk, D. Dynamics of Ferroelectric Bis(imidazolium) Pentachloroantimonate(III) by Means of Nuclear Magnetic Resonance 1H Relaxometry and Dielectric Spectroscopy. *J. Phys. Chem. A* **2014**, *118* (20), 3564–3571.
124. Vogel, H. The law of the relation between the viscosity of liquids and the temperature. *Phys. Z.* **1921**, *22*, 645–6.
125. Fulcher, G. S. Analysis of recent measurements of the viscosity of glasses. *J. Am. Ceram. Soc.* **1925**, *8*, 339–355.
126. Tammann, G.; Hesse, W. The dependence of viscosity upon the temperature of supercooled liquids. *Z. Anorg. Allg. Chem* **1926**, *156*, 245–257.
127. Richert, R. Scaling vs. Vogel–Fulcher-type structural relaxation in deeply supercooled materials. *Phys. A* **2000**, *287* (1), 26–36.
128. *The Scaling of Relaxation Processes*; Springer: Berlin, 2018.
129. Brás, A. R.; Noronha, J. P.; Antunes, A. M. M.; Cardoso, M. M.; Schönhals, A.; Affouard, F.; Dionísio, M.; Correia, N. T. Molecular Motions in Amorphous Ibuprofen As Studied by Broadband Dielectric Spectroscopy. *J. Phys. Chem. B* **2008**, *112* (35), 11087–11099.
130. Ngai, K. An extended coupling model description of the evolution of dynamics with time in supercooled liquids and ionic conductors. *J. Phys.: Condens. Matter* **2003**, *15* (11), S1107.
131. Ngai, K.; Kamińska, E.; Sekuła, M.; Paluch, M. Primary and secondary relaxations in bis-5-hydroxypentylphthalate revisited. *J. Chem. Phys.* **2005**, *123* (20), 204507.
132. Johari, G. P.; Goldstein, M. Viscous liquids and the glass transition. III. Secondary relaxations in aliphatic alcohols and other nonrigid molecules. *J. Chem. Phys.* **1971**, *55* (9), 424–4252.
133. Ngai, K.; Casalini, R.; Capaccioli, S.; Paluch, M.; Roland, C. Do theories of the glass transition, in which the structural relaxation time does not define the dispersion of the structural relaxation, need revision? *J. Phys. Chem. B* **2005**, *109* (37), 17356–17360.
134. Körber, T.; Mohamed, F.; Hofmann, M.; Lichtinger, A.; Willner, L.; Rössler, E. A. The Nature of Secondary Relaxations: The Case of Poly(ethylene-alt-propylene) Studied by Dielectric and Deuteron NMR Spectroscopy. *Macromolecules* **2017**, *50* (4), 1554–1568.
135. Floudas, G.; Paluch, M.; Grzybowski, A.; Ngai, K. *Molecular Dynamics of Glass-Forming Systems: Effects of Pressure*; Springer: Berlin, 2010.
136. *Nonlinear Dielectric Spectroscopy*; Springer: Berlin, 2018.

137. Fitz, B. D.; Mijovic, J. In Situ Monitoring of Chemical Reactions in and Molecular Dynamics of Model Compounds by Dielectric Relaxation Spectroscopy. *J. Phys. Chem. B* **2000**, *104* (51), 12215–12223.
138. Mijovic, J.; Fitz, B. Dielectric Spectroscopy of Reactive Polymers. *Application Note Dielectrics 2* **1998**, 1–24.
139. As discussed earlier, scientists interested in electrode polarization (EIS) use impedance while those interested in molecular motion (BDS) use admittance. The instrument does not care: it simply measures the response of a device under test to a reference signal. Bridges and analyzers may be designed to have a region of accuracy better suited for impedance or admittance measurements, but they can report any equivalent circuit element (Cp, Cs, Gp, Rs, Lp, Ls, tanδ, etc.). "Impedance spectroscopy" or "admittance spectroscopy" is a matter of pushing a button.
140. Matsuoka, S. *Relaxation phenomena in polymers*; Hanser: Munich, 1992.
141. Relative permittivity is often called "dielectric constant" or "Dk" in the literature or chemical databases. The reader should now be aware that there is nothing constant about this value. In addition to frequency and temperature, many other experimental variables can affect the permittivity. Nevertheless, one should be aware that when a dielectric constant is reported without an explicit frequency, it is expected to be at frequency below all relaxations but above any MWS or polarization effects (i.e. Dk = εs). This point is important for industrial applications such as radar tank level sensors, which operate at GHz frequencies where relaxations may exist.
142. Brun, C.; L'Hôte, D.; Ladieu, F.; Crauste-Thibierge, C.; Biroli, G.; Tarzia, M., Nonlinear Susceptibility Experiments in a Supercooled Liquid: Evidence of Growing Spatial Correlations Close to Tg. In *Recent Advances in Broadband Dielectric Spectroscopy*; Kalmykov, Y. P., Ed.; Springer: Berlin, 2013; pp 89–102.
143. Scott, A. H. Techniques for using the air-gap method for the precise determination of the dielectric constant and loss angle of solid disk specimens; *Annual Report 1963 Conference on Electrical Insulation*; IEEE: 1963; pp 74–76.
144. Harris, W. P.; Scott, A. H. Precise measurement of dielectric constant by the two-fluid technique. *Annual Report 1962 Conference on Electrical Insulation*; IEEE: 1962; pp 51–53.
145. Cockbain, A.; Harrop, P. The temperature coefficient of capacitance. *J. Phys. D: Appl. Phys.* **1968**, *1* (9), 1109.
146. Monnaie, I.; Napolitano, S.; Nies, E.; Wübbenhorst, M. Exploiting interfacial polarization to detect phase transitions in dilute solutions: Crystallization and melting of P3HT in toluene down to ppm contents. *Polymer* **2011**, 52 (2), 239–242.
147. Serghei, A.; Tress, M.; Kremer, F. The glass transition of thin polymer films in relation to the interfacial dynamics. *J. Chem. Phys.* **2009**, *131* (15), 154904.
148. Tress, M.; Erber, M.; Mapesa, E. U.; Huth, H.; Muller, J.; Serghei, A.; Schick, C.; Eichhorn, K.-J.; Voit, B.; Kremer, F. Glassy dynamics and glass transition in nanometric thin layers of polystyrene. *Macromolecules* **2010**, *43* (23), 9937–9944.
149. Tress, M.; Mapesa, E. U.; Kossack, W.; Kipnusu, W. K.; Reiche, M.; Kremer, F. Glassy Dynamics in Condensed Isolated Polymer Chains. *Science* **2013**, *341* (6152), 1371.
150. *Dynamics in Geometrical Confinement*; Springer: Berlin, 2014.

151. Woolley, D. Edge Correction in Calculation of Dielectric Constant. *J. Test. Eval.* **2011**, *39* (2), 140–149.
152. *Standard Test Methods for AC Loss Characteristics and Permittivity (Dielectric Constant) of Solid Electrical Insulation*; ASTM D150-18; ASTM International: West Conshohocken, PA, 2018.
153. Cole, R. H.; Gross, P. M. A Wide Range Capacitance-Conductance Bridge. *Rev. Sci. Instrum.* **1949**, *20* (4), 252–260.
154. "Even though microprocessors were used in desktop computers in the 1980's, they were not fast enough for the kinds of calculations needed for network analysis. One of the first uses of the microprocessor in laboratory instrumentation was in phase-lock detectors. However, because of the speed issue, the initial bandwidth was small." Private communication with J. Berberian, October 2020
155. Mead, D. J.; Fuoss, R. M. Electrical Properties of Solids. XI. Polyvinyl Acetate and Polyvinyl Chloroacetate. *J. Am. Chem. Soc.* **1941**, *63* (10), 2832–2840.
156. Huth, H.; Wang, L.-M.; Schick, C.; Richert, R. Comparing calorimetric and dielectric polarization modes in viscous 2-ethyl-1-hexanol. *J. Chem. Phys.* **2007**, *126* (10), 104503.
157. Sibik, J.; Löbmann, K.; Rades, T.; Zeitler, J. A. Predicting crystallization of amorphous drugs with terahertz spectroscopy. *Mol. Pharmaceutics* **2015**, *12* (8), 3062–3068.
158. Baghel, S.; Cathcart, H.; O'Reilly, N. J. Polymeric amorphous solid dispersions: a review of amorphization, crystallization, stabilization, solid-state characterization, and aqueous solubilization of biopharmaceutical classification system class II drugs. *J. Pharm. Sci.* **2016**, *105* (9), 2527–2544.
159. Suzuki, Y.; Steinhart, M.; Graf, R.; Butt, H.-J.; Floudas, G. Dynamics of Ice/Water Confined in Nanoporous Alumina. *J. Phys. Chem. B* **2015**, *119* (46), 14814–14820.
160. Byrappa, K.; Adschiri, T. Hydrothermal technology for nanotechnology. *Prog. Cryst. Growth Charact. Mater.* **2007**, *53* (2), 117–166.
161. Rams-Baron, M.; Wojnarowska, Z.; Grzybowska, K.; Dulski, M.; Knapik, J.; Jurkiewicz, K.; Smolka, W.; Sawicki, W.; Ratuszna, A.; Paluch, M. Toward a Better Understanding of the Physical Stability of Amorphous Anti-Inflammatory Agents: The Roles of Molecular Mobility and Molecular Interaction Patterns. *Mol. Pharmaceutics* **2015**, *12* (10), 3628–3638.
162. Lusceac, S. A.; Gainaru, C.; Vogel, M.; Koplin, C.; Medick, P.; Rössler, E. A. Secondary Relaxation Processes in Polybutadiene Studied by 2H Nuclear Magnetic Resonance and High-Precision Dielectric Spectroscopy. *Macromolecules* **2005**, *38* (13), 5625–5633.
163. Rios de Anda, A.; Sotta, P.; Modjinou, T.; Langlois, V.; Versace, D.-L.; Renard, E. Multiscale Structural Characterization of Biobased Diallyl–Eugenol Polymer Networks. *Macromolecules* **2020**, *53* (6), 2187–2197.
164. Kariyo, S.; Brodin, A.; Gainaru, C.; Herrmann, A.; Hintermeyer, J.; Schick, H.; Novikov, V. N.; Rössler, E. A. From Simple Liquid to Polymer Melt. Glassy and Polymer Dynamics Studied by Fast Field Cycling NMR Relaxometry: Rouse Regime. *Macromolecules* **2008**, *41* (14), 5322–5332.
165. Lusceac, S. A.; Gainaru, C.; Ratzke, D. A.; Graf, M. F.; Vogel, M. Secondary Water Relaxation in a Water/Dimethyl Sulfoxide Mixture Revealed by Deuteron Nuclear Magnetic Resonance and Dielectric Spectroscopy. *J. Phys. Chem. B* **2011**, *115* (40), 11588–11596.

166. Frunza, S.; Schönhals, A.; Frunza, L.; Ganea, P.; Kosslick, H.; Harloff, J.; Schulz, A. Molecular Relaxation Processes in a MOF-5 Structure Revealed by Broadband Dielectric Spectroscopy: Signature of Phenylene Ring Fluctuations. *J. Phys. Chem. B* **2010**, *114* (40), 12840–12846.
167. *Metal-Organic Frameworks: Applications from Catalysis to Gas Storage*; Wiley-VCH: Weinheim, 2011.
168. Gould, S. L.; Tranchemontagne, D.; Yaghi, O. M.; Garcia-Garibay, M. A. Amphidynamic character of crystalline MOF-5: Rotational dynamics of terephthalate phenylenes in a free-volume, sterically unhindered environment. *J. Am. Chem. Soc.* **2008**, *130* (11), 3246–3247.
169. Griffin, P. J.; Holt, A. P.; Wang, Y.; Novikov, V. N.; Sangoro, J. R.; Kremer, F.; Sokolov, A. P. Interplay Between Hydrophobic Aggregation and Charge Transport in the Ionic Liquid Methyltrioctylammonium Bis(trifluoromethylsulfonyl)imide. *J. Phys. Chem. B* **2014**, *118* (3), 783–790.
170. Ferry, J. D. *Viscoelastic Properties of Polymers*; John Wiley & Sons: Hoboken, NJ, 1980.
171. Plechkova, N. V.; Seddon, K. R. Applications of ionic liquids in the chemical industry. *Chem. Soc. Rev.* **2008**, *37* (1), 123–150.
172. Koh, J. H.; Zhu, Q.; Asano, Y.; Maher, M. J.; Ha, H.; Kim, S.-S.; Cater, H. L.; Mapesa, E. U.; Sangoro, J. R.; Ellison, C. J.; Lynd, N. A.; Willson, C. G. Unusual Thermal Properties of Certain Poly(3,5-disubstituted styrene)s. *Macromolecules* **2020**, *53* (13), 5504–5511.
173. *Fast Scanning Calorimetry*; Springer: Berlin, 2016.
174. Zhou, D.; Huth, H.; Gao, Y.; Xue, G.; Schick, C. Calorimetric glass transition of poly(2,6-dimethyl-1,5-phenylene oxide) thin films. *Macromolecules* **2008**, *41* (20), 7662–7666.
175. Huth, H.; Minakov, A.; Serghei, A.; Kremer, F.; Schick, C. Differential AC-chip calorimeter for glass transition measurements in ultra thin polymeric films. *Eur. Phys. J.: Spec. Top.* **2007**, *141* (1), 153–160.
176. Huth, H.; Minakov, A.; Schick, C. Differential AC-chip calorimeter for glass transition measurements in ultrathin films. *J. Polym. Sci., Part B: Polym. Phys.* **2006**, *44* (20), 2996–3005.
177. Robles-Hernández, B.; Monnier, X.; Pomposo, J. A.; Gonzalez-Burgos, M.; Cangialosi, D.; Alegría, A. Glassy Dynamics of an All-Polymer Nanocomposite Based on Polystyrene Single-Chain Nanoparticles. *Macromolecules* **2019**, *52* (18), 6868–6877.
178. Laurati, M.; Sotta, P.; Long, D. R.; Fillot, L. A.; Arbe, A.; Alegrı`a, A.; Embs, J. P.; Unruh, T.; Schneider, G. J.; Colmenero, J. Dynamics of Water Absorbed in Polyamides. *Macromolecules* **2012**, *45* (3), 1676–1687.
179. Goracci, G.; Arbe, A.; Alegría, A.; García Sakai, V.; Rudić, S.; Schneider, G. J.; Lohstroh, W.; Juranyi, F.; Colmenero, J. Influence of solvent on poly (2-(dimethylamino) ethyl methacrylate) dynamics in polymer-concentrated mixtures: A combined neutron scattering, dielectric spectroscopy, and calorimetric study. *Macromolecules* **2015**, *48* (18), 6724–6735.
180. Sanz, A.; Nogales, A.; Ezquerra, T. A.; Soccio, M.; Munari, A.; Lotti, N. Cold Crystallization of Poly(trimethylene terephthalate) As Revealed by Simultaneous WAXS, SAXS, and Dielectric Spectroscopy. *Macromolecules* **2010**, *43* (2), 671–679.
181. Silva, J. F. C.; Rosado, M. T. S.; Jasiurkowska-Delaporte, M.; Silva, M. R.; Piedade, M. F. M.; Dryzek, E.; Eusébio, M. E. S. Ordered and Plastic Crystals in the Complex Polymorphism of Pinanediol. *Crys. Growth Des.* **2019**, *19* (11), 6127–6135.

182. Brodin, A.; Rössler, E. A. Depolarized light scattering study of glycerol. *Eur. Phys. J. B* **2005**, *44* (1), 3–14.
183. Pabst, F.; Helbling, A.; Gabriel, J.; Weigl, P.; Blochowicz, T. Dipole-dipole correlations and the Debye process in the dielectric response of nonassociating glass forming liquids. *Phys. Rev. E* **2020**, *102* (1), 010606.
184. Petzold, N.; Rössler, E. Light scattering study on the glass former o-terphenyl. *J. Chem. Phys.* **2010**, *133* (12), 124512.
185. *Conductive Atomic Force Microscopy: Applications in Nanomaterials*; Wiley-VCH: Weinheim, 2017.
186. Russell, E. V.; Israeloff, N.; Walther, L.; Gomariz, H. A. Nanometer scale dielectric fluctuations at the glass transition. *Phys. Rev. Lett.* **1998**, *81* (7), 1461.
187. Russell, E. V.; Israeloff, N. Direct observation of molecular cooperativity near the glass transition. *Nature* **2000**, *408* (6813), 695–698.
188. Kremer, F. Dielectric spectroscopy - yesterday, today and tomorrow. *J. Non-Cryst. Solids* **2002**, *305* (1−3), 1–9.
189. Miccio, L. A.; Kummali, M. M.; Schwartz, G. A.; Alegría, Á.; Colmenero, J. AFM based dielectric spectroscopy: Extended frequency range through excitation of cantilever higher eigenmodes. *Ultramicroscopy* **2014**, *146*, 55–61.
190. Miccio, L. A.; Schwartz, G. A. Study of relaxation and transport processes by means of AFM based dielectric spectroscopy. *AIP Conference Proceedings*; American Institute of Physics: 2014; pp 150−153.
191. Kummali, M. M.; Miccio, L. A.; Schwartz, G. A.; Alegría, A.; Colmenero, J.; Otegui, J.; Petzold, A.; Westermann, S. Local mechanical and dielectric behavior of the interacting polymer layer in silica nano-particles filled SBR by means of AFM-based methods. *Polymer* **2013**, *54* (18), 4980–4986.
192. Cadena, M. J.; Sung, S. H.; Boudouris, B. W.; Reifenberger, R.; Raman, A. Nanoscale mapping of dielectric properties of nanomaterials from kilohertz to megahertz using ultrasmall cantilevers. *ACS Nano* **2016**, *10* (4), 4062–4071.
193. Labardi, M.; Bertolla, A.; Sollogoub, C.; Casalini, R.; Capaccioli, S. Lateral resolution of electrostatic force microscopy for mapping of dielectric interfaces in ambient conditions. *Nanotechnology* **2020**, *31* (33), 335710.
194. Labardi, M.; Lucchesi, M.; Prevosto, D.; Capaccioli, S. Broadband local dielectric spectroscopy. *Appl. Phys. Lett.* **2016**, *108* (18), 182906.
195. Labardi, M.; Prevosto, D.; Nguyen, K. H.; Capaccioli, S.; Lucchesi, M.; Rolla, P. Local dielectric spectroscopy of nanocomposite materials interfaces. *J. Vac. Sci. Technol., B* **2010**, *28* (3), C4D11–C4D17.
196. Kummali, M. M.; Alegría, A.; Miccio, L. A.; Colmenero, J. Study of the Dynamic Heterogeneity in Poly(ethylene-ran-vinyl acetate) Copolymer by Using Broadband Dielectric Spectroscopy and Electrostatic Force Microscopy. *Macromolecules* **2013**, *46* (18), 7502–7512.
197. Williams, G. Polymer dynamics and broadband dielectric spectroscopy. In *New Polylmeric Materials*; Korugic-Karasz, L. S., MacKnight, W. J., Martuscelli, E., Eds.; ACS Publications: Washington, DC, 2005; pp 268−281.

198. Di Noto, V.; Negro, E.; Lavina, S. Broadband dielectric spectroscopy and conductivity mechanism of Nafion 117 and Nafion/[ZrO2] hybrid inorganic-organic membranes. In *Fuel Cell Chemistry and Operation*; Herring, A. M., Zawodzinski, T. A., Hamrock, S. J., Eds.; ACS Publications: Washington, DC, 2010; pp 97−111.
199. Hassan, M. K.; Mauritz, K. A. Broadband Dielectric Spectroscopic Studies of Nafion®/Silicate Membranes. In *Fuel Cell Chemistry and Operation*; Herring, A. M., Zawodzinski, T. A., Hamrock, S. J., Eds.; ACS Publications: Washington, 2010; pp 113−124.
200. Hefter, G.; Buchner, R.; Hunger, J.; Stoppa, A. Chemical speciation in ionic liquids and their mixtures with polar solvents using dielectric spectroscopy. In *Ionic Liquids: From Knowledge to Application*; Plechkova, N. V., Rogers, R. D., Seddon, K. R., Eds.; ACS Publications: Washington, DC, 2010; pp 61−74.
201. Dewhurst, J. E.; Drayton-Elder, A. S.; Gao, X.; Santosusso, T. M.; Tien, C.-f.; Wickmann, T. L. Property development during film formation of two component waterborne polyurethane using dielectric spectroscopy. In *Film Formation in Coatings*; ACS Symposium Series 790; Provder, T., Urban, M. W., Eds.; ACS Publications: Washington, DC, 2001; pp 124−140.
202. Pratt, G.; Smith, M. Dielectric Spectroscopy of Bisphenol-A Polycarbonate and Some of Its Blends. In *Structure and Properties of Glassy Polymers*; Tant, M. R., Jill, A. J., Eds.; ACS Publications: Washington, DC, 1998; pp 144−157.
203. Lunkenheimer, P.; Pimenov, A.; Dressel, M.; Gorshunov, B.; Schneider, U.; Schiener, B.; Loidl, A. High-frequency dielectric spectroscopy of glass-forming liquids. In *Supercooled Liquids*; Fourkas, J. T., Kivelson, D., Mohanty, U., Nelson, K. A., Eds.; ACS Publications: Washington, DC, 1997; pp 168−180.
204. In 1994 an ACS Symposium was co-sponsored by the Divisions of Polymeric MS&E and Polymer Chemistry. This symposium led to the book "Dielectric Spectroscopy of Polymeric Materials" in 1997. While there have been individual presentations at National Meetings which led to chapters in ACS Symposium Series books, it has been more than two decades since the last dedicated Symposium at a National Meeting. Promisingly, in 2019 there was a Session at the American Physical Society March Meeting in Boston dedicated solely to BDS of polymers and soft matter.
205. Guo, W.; Shang, L.; Zhu, X.; Nelson, S. O. Nondestructive detection of soluble solids content of apples from dielectric spectra with ANN and chemometric methods. *Food Bioprocess Technol.* **2015**, *8* (5), 1126–1138.
206. Liu, J.; Li, Z.; Hu, F.; Chen, T.; Du, Y.; Xin, H. Identification of GMOs by terahertz spectroscopy and ALAP–SVM. *Opt. Quantum Electron.* **2015**, *47* (3), 685–695.
207. Soltani, M.; Omid, M. Detection of poultry egg freshness by dielectric spectroscopy and machine learning techniques. *LWT-Food Sci. Technol.* **2015**, *62* (2), 1034–1042.
208. Mireei, S. A.; Bagheri, R.; Sadeghi, M.; Shahraki, A. Developing an electronic portable device based on dielectric power spectroscopy for non-destructive prediction of date moisture content. *Sens. Actuators, A* **2016**, *247*, 289–297.
209. Liu, W.; Liu, C.; Hu, X.; Yang, J.; Zheng, L. Application of terahertz spectroscopy imaging for discrimination of transgenic rice seeds with chemometrics. *Food Chem.* **2016**, *210*, 415–421.
210. Rashvand, M.; Omid, M.; Mobli, H.; Firouz, M. S. Adulteration detection in olive oil using dielectric technique and data mining. *Sensing Bio-Sensing Res.* **2016**, *11*, 33–36.

211. Nicholson, D. J.; Kell, D. B.; Davey, C. L. Deconvolution of the dielectric spectra of microbial cell suspensions using multivariate calibration and artificial neural networks. *Bioelectrochem. Bioenerg.* **1996,** *39* (2), 185–193.
212. Sanaeifar, A.; Jafari, A.; Golmakani, M.-T. Fusion of dielectric spectroscopy and computer vision for quality characterization of olive oil during storage. *Comput. Electron. Agric.* **2018,** *145*, 142–152.
213. Khaled, A. Y.; Abd Aziz, S.; Bejo, S. K.; Nawi, N. M.; Seman, I. A. Spectral features selection and classification of oil palm leaves infected by Basal stem rot (BSR) disease using dielectric spectroscopy. *Comput. Electron. Agric.* **2018,** *144*, 297–309.
214. Naderi-Boldaji, M.; Mishra, P.; Ahmadpour-Samani, M.; Ghasemi-Varnamkhasti, M.; Ghanbarian, D.; Izadi, Z. Potential of two dielectric spectroscopy techniques and chemometric analyses for detection of adulteration in grape syrup. *Measurement* **2018,** *127*, 518–524.
215. Tiitta, M.; Tiitta, V.; Heikkinen, J.; Lappalainen, R.; Tomppo, L. Classification of Wood Chips Using Electrical Impedance Spectroscopy and Machine Learning. *Sensors* **2020,** *20* (4), 1076.
216. Schaumburg, G. On the Accuracy of Dielectric Measurements. *Dielectrics Newsletter*, Nov. 1997; pp 5–10.
217. *Non-Equilibrium Phenomena in Confined Soft Matter*; Springer: Berlin, 2015.
218. *Recent Advances in Broadband Dielectric Spectroscopy*; Springer: Berlin, 2012.
219. Habasaki, J.; Leon, C.; Ngai, K. *Dynamics of Glassy, Crystalline and Liquid Ionic Conductors: Experiments, Theories, Simulations*; Springer: Berlin, 2016.
220. Ngai, K. L. *Relaxation and Diffusion in Complex Systems*; Springer: Berlin, 2011.
221. *Disorder Effects on Relaxational Processes*; Springer: Berlin, 2012.
222. *Crystallization as Studied by Broadband Dielectric Spectroscopy*; Springer: Berlin, 2020.
223. Martinsen, Ø. G.; Grimnes, S. *Bioimpedance and Bioelectricity Basics*, 3rd ed.; Academic Press: New York, 2015.

Chapter 2

A Brief History and Introduction to Admittance/Impedance Measurements

John G. Berberian*

Department of Chemistry, Saint Joseph's University, Philadelphia, Pennsylvania 19131, United States

***Email: johngberberian@gmail.com**

A brief history of and introduction to impedance/admittance spectroscopy measurements is presented in terms of the timely development of the instrumentation. Most of the instruments presented are bridges though some direct measurement equipment is mentioned. Stories personally known to this author regarding some of these developments are presented with the researchers involved, which gives animation to progress in instrumentation. An example is given in which dielectric and calorimetric data provide complementary information regarding a molecular event.

Introduction

I want to thank the ACS organizers for inviting me to give the introductory presentation. I will present a brief history of and a simple introduction to dielectric or permittivity [$\varepsilon^*(\omega) = \varepsilon'(\omega) - i\varepsilon''(\omega)$] measurement techniques from DC transient measurements to broadband bridges and to transmission line measurements. The purpose of this paper is to present the reader with a view of the information that dielectric measurements can reveal. It is written for the scientist not in the field of dielectrics but who is contemplating using dielectric relaxation measurements as an analytic tool. Some applications will be presented showing the usefulness of permittivity measurements in elucidating molecular motion and in some commercial applications.

In this paper, I try to cover the salient features of dielectric spectroscopy, often referred to as Broadband Dielectric Spectroscopy (BDS). Like other fields, such as calorimetry, kinetics, rheology, etc., there is a theory backdrop to the experimental methods used in these specific fields. The purpose of this paper is to give readers a sense of how BDS might be helpful in viewing their research from a different angle. A good example of two research areas focusing on the same property is the crystallization of a supercooled polar liquid below its cold crystallization region using dielectric permittivity and modulated differential scanning calorimetry (discussed in more detail in the last section of this chapter).

The first section of this paper is on the transient measurement, which is somewhat fundamental to dielectric measurements, and the principle involved in the transient measurements. Next is a discussion of dielectric cells, their configurations and electrical characteristics. The mapping of transient measurements into the frequency domain and a discussion of the direct measurement of admittance/impedance using bridge techniques are presented in the AC bridge measurement section. The next section on measuring dielectric response is in the time domain spectroscopy section with emphasis on the time domain reflection technique. Lastly are the analysis, display of data, and some commercial uses of Broadband Dielectric Spectroscopy.

The Transient Measurement

As a start, let's look at what a dielectric measurement actually measures. Consider a system of bound charges arranged asymmetrically about the center of charge so as to form a dipole. Let's say this system of bound charges is a molecule (or part of a molecule) and place this dipolar molecule between two parallel metal plates (see Figure 1). If we apply a potential difference between the two parallel plates, a uniform electric field will be established between the plates, and the molecule will reorient under the influence of the applied electric field. If there is a net charge on the molecule, the molecule will also migrate toward the plate with opposite charge (free charges will give rise to DC conductance, and, unless the metal plates can perform as reversible electrodes, there will be electrode polarization).

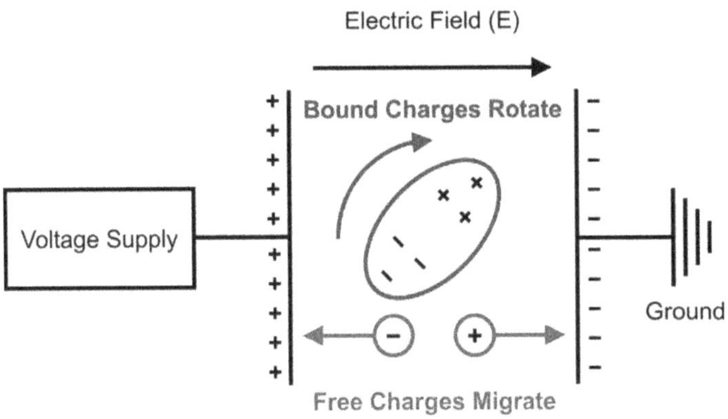

Figure 1. Bound charge in the form of a molecular dipole. The free charge gives rise to ionic conduction.

Consider only a neutral polar molecule in between two parallel metal plates (a parallel plate capacitor). When a voltage is applied to the plates, the polar molecule will try to align with the applied electric field at a specific rate determined by its molecular geometry, magnitude of the dipole moment, the environment surrounding the molecule, and other such elements of interaction that affect the movement of the molecule. As the molecule aligns with the electric field, the charges on the plates change with time, dq/dt, which is a manifestation of the elements involved in the alignment process.

Similarly, when the field is turned off by shorting the plates (see Figure 2), the dipole will return to random orientation with a consummate change in flow of charge. Again, this time evolution of charge is characteristic of the molecular motion in its environment. Figure 3 shows the charge and applied voltage profile for the circuit in Figure 2. The almost instantaneous change in charge (vertical solid line) at zero time (when the voltage (dashed line) is turned off) is due to the atomic and electronic polarization. For times greater than zero in Figure 3, the change in charge due to the motion of polar moiety of the molecule generates a finite conductance that reduces to zero at long times. If this time dependence is exponential in time (as it is in Figure 3), the dependence is referred to as Debye behavior (named after Peter Debye) (*1*) and has a characteristic decay time called the relaxation time, τ. Non-interacting dipoles are a basic premise of the Debye behavior that results in an exponential change in charge with applied voltage on or off.

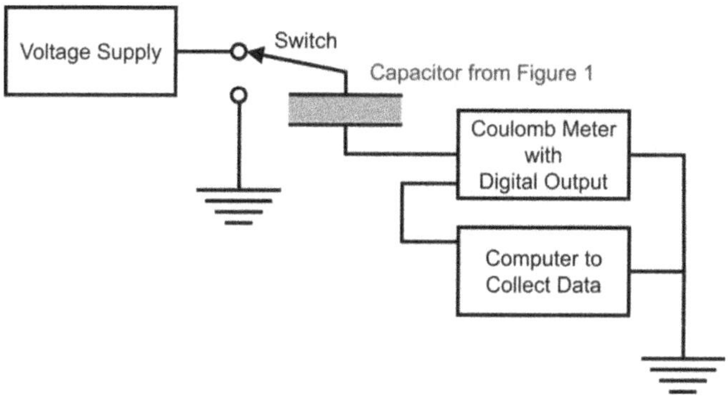

Figure 2. Charge/discharge circuit for a parallel plate capacitor filled with a dielectric.

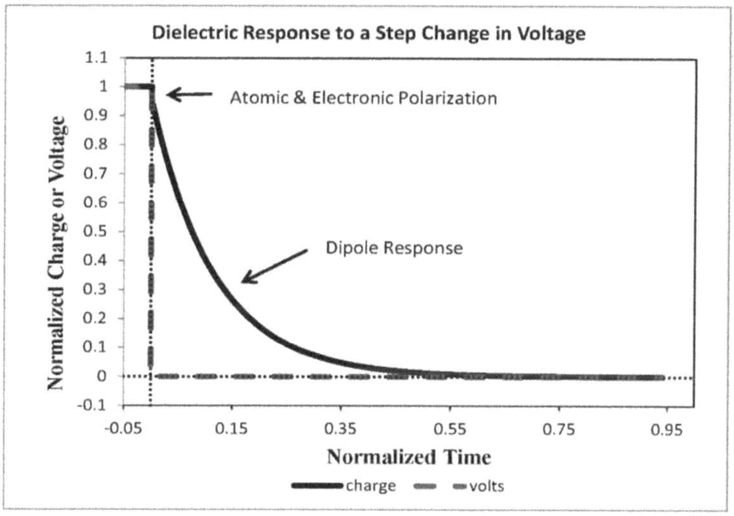

Figure 3. Charge and applied voltage versus time for the discharging circuit in Figure 2.

This simple Debye relaxation of charge alignment is much like first order chemical kinetics where the decay of charge alignment to randomness in time is mathematically analogous to the loss of reactant in time in a reaction that starts with all reactant with no perceptible reverse reaction. The reciprocal of the relaxation time is analogous to the kinetic rate constant. Similarly, just as the kinetic rate constant (or half-life) contains information about the mechanism of the reaction, the dielectric relaxation time, τ, contains information about the relaxation process of the motion of the dipole in the applied field determined by a voltage, V, to zero field at long times (times greater than, say, 5τ). More the case than not, many dielectrics that exhibit dielectric relaxation with time have relaxations that are not simple exponential decays. These dielectric materials still generate relaxation times which are characteristic of the molecular motion involved in the relaxation process along with other parameters that model the deviation from exponential behavior.

Although the field/no field experiment is a fundamental measurement in real time (called transient measurements), analysis of the characteristic parameters (such as τ and, as will be shown later, α, β, etc.) in the time domain is somewhat difficult, especially for non-exponential behavior. However, if we transform this data from the time domain [q(t)] to the frequency domain [q(ω)], the analysis of these characteristic parameters is just a matter of algebraic manipulation. The Laplace transform, $\mathcal{L}[\ ]$, is the mathematical tool that performs this transformation as given in Equation 1,

$$q(\omega) = \int_0^{\infty} q(t)e^{-\iota\omega t}dt = \mathcal{L}[q(t)] \qquad (1)$$

Basically, Equation 1 maps a time domain function, q(t), into the frequency domain, q(ω), at chosen frequencies, $f = \omega/2\pi$. Before looking at the frequency domain representation of this data, it is appropriate to discusse the dielectric cell, in the case above, the parallel plates.

The Dielectric Cell

The usual geometry of capacitance cells is either parallel plates or concentric cylindrical electrodes. The use of Gauss's theorem shows that the equation for electrical capacitance for a parallel plate capacitance cell is given by the following equation,

$$C_x = \varepsilon_o \frac{A}{d} \varepsilon_x \qquad (2)$$

where A is the area of the plates (meters^2), d is the distance between the plates (meters), ε_o is the permittivity of free space ($8.85419 * 10^{-12}$ Farad/meter), and ε_x is the relative permittivity of the material between the plates. If the capacitance of a cell filled with a substance of known permittivity, ε_r (a reference, usually air), the measured capacitance, C_r, is called the cell constant, and when the cell is filled with a material of unknown permittivity, ε_x, the measured capacitance is called C_x. If we divide this measured capacitance (C_x) by reference capacitance (C_r), we get ε_x in terms of the reference capacitance as the cell geometry cancels, i.e.,

$$\frac{C_x}{C_r} = \frac{\varepsilon_o \frac{A}{d} \varepsilon_x}{\varepsilon_o \frac{A}{d} \varepsilon_r} = \frac{\varepsilon_x}{\varepsilon_r} \quad so \quad \varepsilon_x = \varepsilon_r \frac{C_x}{C_r} \qquad (3)$$

where ε_r for air is 1.0006 (for a vacuum, $\varepsilon_r \equiv 1$). This makes measuring the permittivity of an unknown a simple matter of measuring the capacitance of the empty cell and then the cell filled with the substance of interest, thus eliminating the need for the cell geometry if calculating ε_x using Equation 2. The caveat here is that the electric field geometry between the cell plates must be the same for the empty cell and the full cell. To ensure that materials of various permittivities are subject to the same uniform field between the electrodes, one of the electrodes is guarded. An example of a guard for a circular parallel electrode cell is a concentric ring around one electrode with a small gap between the center disk electrode and the concentric guard ring; the outside diameter of the guard ring electrode should be larger than the diameter of the center disk electrode, and the guard gap is filled with an insulator such as Kel F or Teflon. Figure 4 is a diagram of a guarded parallel plate capacitor. There are specific dimensions for the electrode, plate separation and guard gap width for insuring a uniform field in the sample between the electrodes; these specifics are described in an American Society for Testing and Materials document (2).

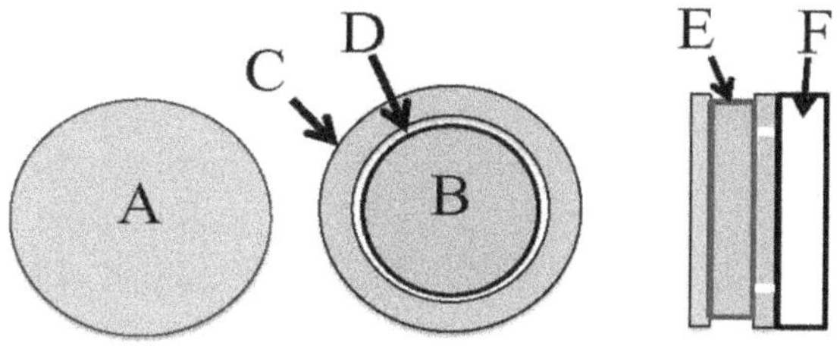

Figure 4. Diagram of parallel plate capacitor with a guard ring (C), guard gap (D), guarded electrode (B) and non-guarded electrode (A). Plate A and plate (B, C, and D) are "open" on the left for clarity. The half section side view, with a sample (E), is on the right. The insulator support (F) for B and C is also the guard gap filler.

The equation for a concentric cylindrical cell is given by

$$C_r = \frac{2\pi\varepsilon_o L\varepsilon_r}{ln\left(\frac{b}{a}\right)} \qquad (4)$$

where a is the radius of the inner cylinder, b is the radius of the outer cylinder and L is the length of the guarded cylinder. Figure 5 is a half section of a cylindrical capacitor. See Jacobs and Greer for a detailed description with measurements of a guarded cylindrical electrode (3).

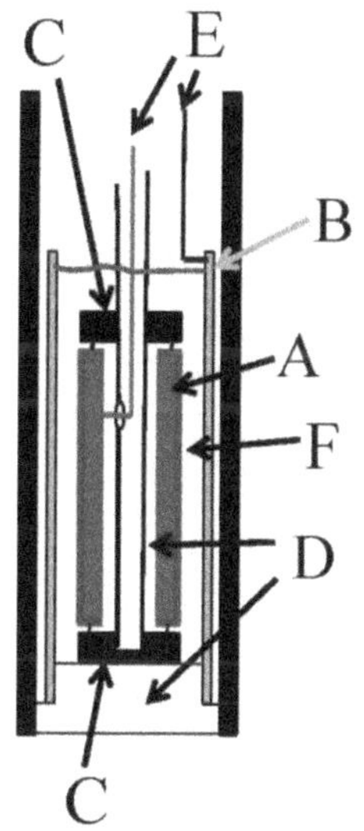

Figure 5. The half section of a cylindrical capacitor: (A) guarded electrode, (B) unguarded electrode, (C) guard caps, (D) insulator, (E) electrode leads, (F) sample.

The equation for a hemispherical cell is given by

$$C_r = \frac{2\pi\varepsilon_o\varepsilon_r}{\frac{1}{a} - \frac{1}{b}} \qquad (5)$$

where a is the radius of the inner sphere and b is the radius of the outer sphere. Again, there is no need for the geometric details if the cell capacitance is measured when the cell is empty and then full; the sample permittivity is given by, using Equation 2, the ratio of the two measurements and ε_r.

Figure 6 is (a) a picture of a hemispherical dielectric cell with a guarded electrode, (b) a half section view of the cell and (c) a plot of the sample capacitances as measured in a hemispherical cell versus their known permittivity (4). To ensure electric field uniformity for a range of permittivities, the guard gap should be small compared to the electrode gap. A test that the electric field remains the same in the guarded region of the cell for the measurement of samples of varying permittivity is the linearity of a plot of the measured capacitance of a number of samples of known permittivity versus their known permittivity, as in Figure 6c. The slope of this plot is the cell constant and should be independent of frequency over a wide range of frequencies; for this cell, the slope, (C_r) is 3.1815 pF up to 100 kHz.

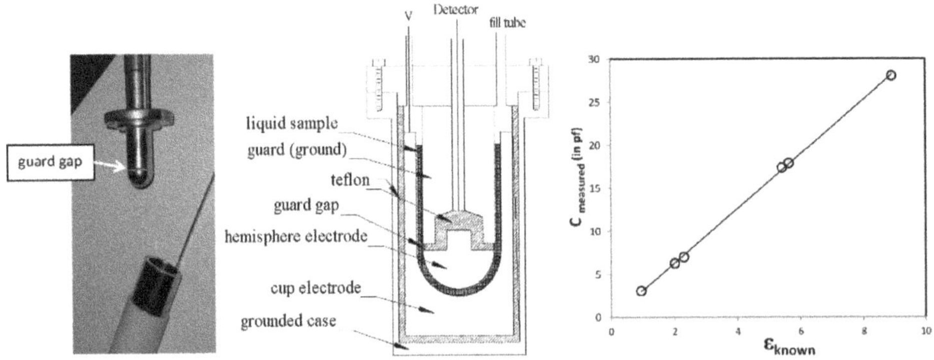

(a) (b) (c)

Figure 6. (a) picture of the hemispherical dielectric cell taken apart, (b) half section of cell (hemisphere electrode is 1.91 cm in diameter) and (c) plot of the measured capacitance of dielectrics versus their known permittivity in the hemispherical cell.

AC Bridge Measurements

As stated previously, data analysis in the time domain, especially for non-exponential data, is somewhat difficult; however, if transformed into the frequency domain, the mathematics is a matter of algebra. Given that $C = q/V_o$, we go back to the Laplace transform (Equation 1) to numerically calculate a complex permittivity (ε^*) at user selected frequencies [$f=\omega/(2\pi)$] from data obtained in the time domain.

Operationally, Equation 1 becomes

$$\varepsilon^*(\omega) = \frac{\varepsilon_r \Delta t}{C_r V_o}\left[\sum_{k=0}^{all\ data} \frac{dq(t)}{dt}\cos(\omega k \Delta t) - i \sum_{k=0}^{all\ data} \frac{dq(t)}{dt} sin(\omega k \Delta t)\right] \quad (6)$$

If the charge, q, is acquired at uniform time intervals, Δt, Equation 6 can be rewritten as

$$\varepsilon^*(\omega) = \frac{\varepsilon_r}{C_r V_o}\left[\sum_{k=1}^{all\ data-1} \frac{(q_{k+1} - q_{k-1})}{2}\cos(\omega k \Delta t) - i \sum_{k=1}^{all\ data-1} \frac{(q_{k+1} - q_{k-1})}{2} sin(\omega k \Delta t)\right] \quad (7)$$

where V_o is the applied voltage and C_r is the empty cell capacitance ($\varepsilon_r = \varepsilon_{air} \approx 1.000$). Here the contribution to ε* from the ε component of the transient decay of charge, $f[q_k(t)]$, is summation (integration) of q_k for all data from the start of the decay, q_o, using the average of q_k. Alternatively, one can use other integration summations such as Simpson's rule.

Figure 7 is the time domain data shown in Figure 3 transformed into the frequency domain using Equation 7. Here the open circles represent the real part of the complex permittivity, ε′(ω), and the filled squares represent the imaginary part of the complex permittivity, ε″(ω), the dielectric loss. The complex permittivity can be measured directly in the frequency domain using AC bridges. In general, a bridge measures the unknown values of electrical elements (capacitance, resistance and inductance) by comparing them with known values (called standards) of the elements. The key feature of a bridge is the ability to achieve high accuracy and resolution in the measurement of admittance/impedance. The drawback of bridge measurements is that each measurement is done at a single frequency, and there is a practical bandwidth of frequencies to which a bridge can respond (in general, bridges span 3 to 3½ decades of frequencies in the range from 50 Hz to 1 MHz). An excellent treatise on the history of impedance instruments and measurements (5) by Henry Hall is available on a website supported by IETLabs (6). Another source of information on AC bridges is the classic book by Hague, "Alternating Current Bridge Methods" (7).

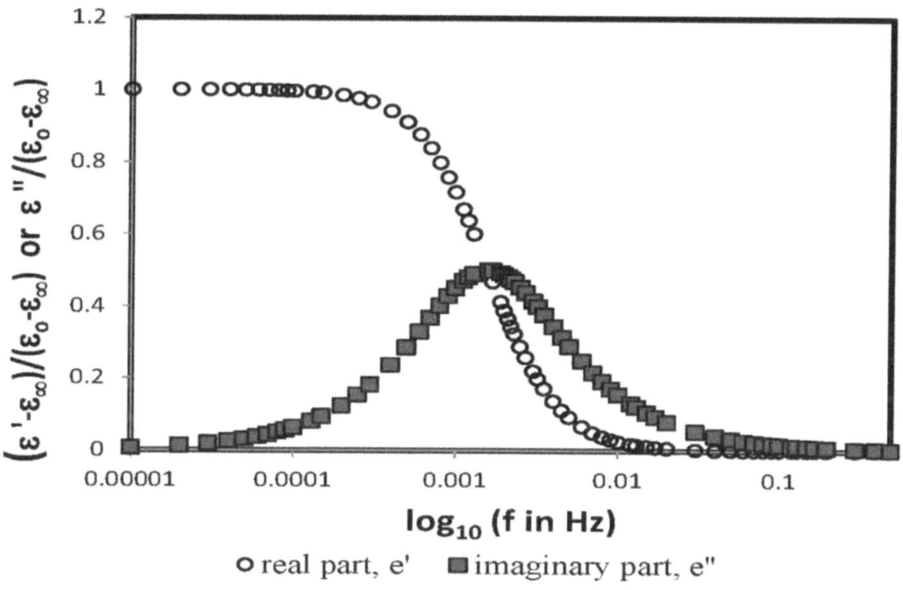

Figure 7. Normalized plot of the data in Figure 3 transformed into the frequency domain.

Before embarking on bridge measurements, it is interesting to note the history of the use of bridges and transient measurements, which is based on the specific limitations of each method. Transient measurements require data in digital form and numerical transformation for accurate frequency domain results. In the early 1970s, computing was mostly done on mechanical calculators or data entry using punched cards on mainframe computers. Thus, transient data were either analyzed in the time domain or put into digital form from measured coordinates obtained from a printed plot. With the advent of the desktop computer, improved measuring equipment (coulombmeters with analog to digital converters and data buses) and user friendly software, data acquisition and analysis became commonplace. Initially, the slow signal response of the equipment limited transformation for short time data. Thus, the transient measurement had its place in the dielectric world by providing low-frequency permittivity data from transformed transient data that complemented the high frequency bridge measurements.

As mentioned above, the hallmark feature of the bridge is its high accuracy, which is based on the quality standards used in measuring impedance/admittance. This high accuracy is a function of the measurement resolution on the standard elements (capacitance, resistance. and/or inductance) and the sensitivity of the detector. High accuracy and resolution are necessary features for permittivity measurements involving gases, as the relative permittivities of gases are a little more than unity and differ from one another in the range of parts per thousand. Another important feature of a bridge is its usable frequency range. Most bridges have a frequency range from about 100 Hz up to 100 kHz. It should be noted that a bridge requires a source (such as an oscillator of suitable frequency range) and a null detector (usually a wide band high-gain amplifier).

One problem with bridge measurements is stray impedances, such as impedance of coaxial lead cables that connect the cell to the bridge. In the early days, this was solved by a substitution measurement protocol. A better solution came about with the use of a center tapped transformer as two arms of the bridge and with the two arms of the transformer mutually inductively coupled, which essentially eliminates the lead impedance (*8*); General Radio used this technology in their Type1615-A Capacitance bridge (an interesting story about the transformer in the endnotes) (*9*). A bridge entry for low frequency measurements is the low frequency bridge (range from 0.001 Hz up to 10 KHz) designed and built by Harris and based on phase lag oscillators and standard variable capacitors (*10*). With the advent of the microprocessor and "fast" software, the measurement of impedance or admittance measurements is now done by signal processing.

These microprocessor based impedance devices are usually called LRC meters, or a variant thereof; the measurement technique is basically a four terminal measurement (Kelvin sensing method). A known current is passed through the unknown, and voltage probe leads measure the voltage directly across the unknown. The amplitude and phase of the current are known, and the voltage amplitude and relative phase are measured at a frequency chosen by the user. From this information, capacitance, resistance, and/or inductance can be calculated in a number of configurations (series, parallel, loss tangent, etc.). Additionally, the frequency range of these instruments is 20 Hz to 2 MHz, and measurements can usually be made at over 200 selected frequencies in this range, automatically in about 3 minutes. This feature, along with the high resolution and speed of the analog to digital converter, makes the LRC meter a formidable competitor to the bridge's accuracy.

The specific output of a bridge measurement (series resistance and reactance, parallel conductance and capacitance, etc.) depends on the configuration of the bridge. Generally, admittance bridges measure parallel conductance and capacitance ($Y_X = G_X + i\omega C_X$), so that

$$\varepsilon^* = \frac{Y_x}{i\omega C_o} = \varepsilon' - i\varepsilon'' \qquad (8)$$

where C_o is the empty cell capacitance ($\varepsilon_{air} \approx 1$). This gives the working equations for the real and imaginary parts of the permittivity as

$$\varepsilon' = \frac{C_x}{C_o} \text{ and } \varepsilon'' = \frac{G_x}{\omega C_o} \qquad (9)$$

Time Domain Spectroscopy Measurements

In late 1960s to the early 1970s there were significant advances in electronic instrumentation, most notably, the development of sampling oscilloscopes and sampling heads, along with fast rising pulses generated by tunnel diodes. Transmission line theory would have Time Domain Spectroscopy (TDS) divided into two areas: transmission and reflection. Experimentally, the reflection method is preferred as it the more convenient of the two methods and is referred to as Time Domain Reflectometry (TDR).

Figure 8 is a circuit for TDR measurements (*11*). The sample cell (Figure 9) is basically a connector with an extended center conductor; this center conductor and the outer coaxial shell form the sample cell with cylindrical geometry. Basically, an incident voltage [$V(t)_0$ from the tunnel diode] is propagating through the sampling head down the transmission line which is terminated with the dielectric of interest. $R(t)_X$ is the voltage reflected from the termination and back through the sampling head; $R(t)_X$ contains the dielectric information about the sample. Both of these signals, $V(t)_0$ and $R(t)_X$, are recorded in time for analysis, in this case by the sampling oscilloscope/storage oscilloscope. Figure 10 is a voltage profile with time as seen by the sampling head.

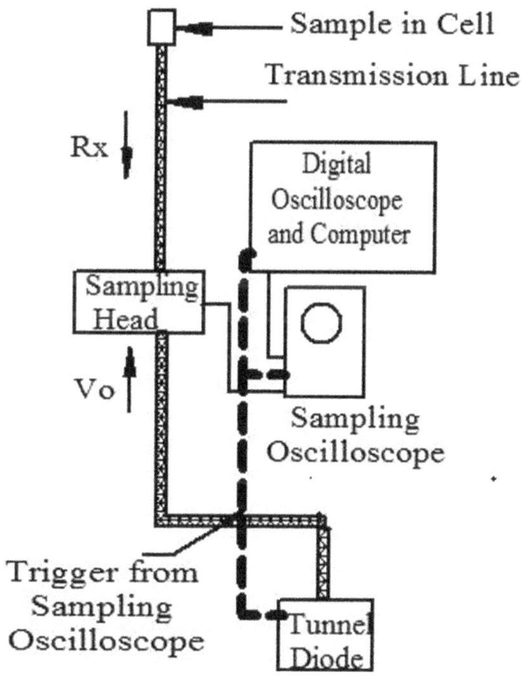

Figure 8. Time-domain reflectometry measuring circuit. Reproduced with permission from Reference (11). Copyright 2002 Elsevier.

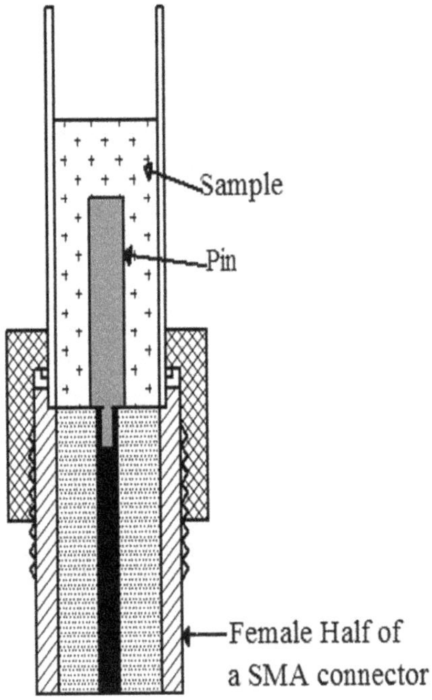

Figure 9. Time-domain reflectometry sample cell. Reproduced with permission from Reference (11). Copyright 2002 Elsevier.

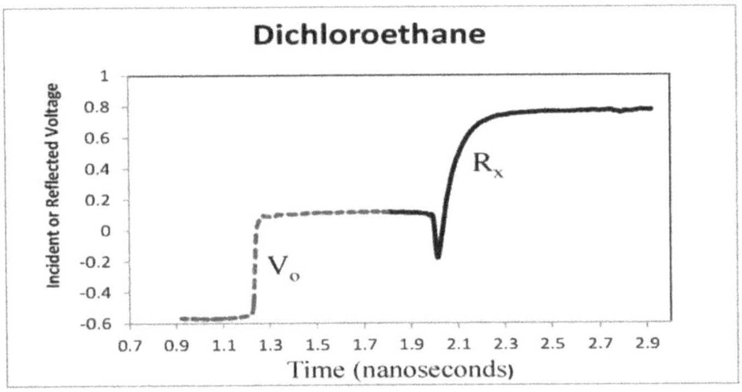

Figure 10. Incident and reflected voltage from cell fill with unknown ε_x. Reproduced with permission from Reference (11). Copyright 2002 Elsevier.

The unknown admittance of the sample in the sample cell is related to the incident and reflected voltage by the ratio of current to voltage

$$y(\omega)_x = \frac{g_c(v(\omega)_o - r(\omega)_x)}{v(\omega)_o + r(\omega)_x} = \frac{g_c\mathcal{L}[V(t)_o - R(t)_x]}{\mathcal{L}[V(t)_o + R(t)_x]} \quad (10)$$

where g_c is the characteristic admittance of the coaxial line (a real constant), and $v(\omega)_o$ and $r(\omega)_x$ are the Laplace transforms of $V(t)_o$ and $R(t)_x$, respectively. Some articles by Cole have a complete description of the various TDR methods available and some cell designs for liquids and solids in addition to details for calculations of permittivity up to 10 GHz (*12, 13*).

Within the last ten years or so, the development of high impedance analyzers in the form of network analyzers have become available for permittivity measurements up to and beyond 100 GHz, depending on the measuring conditions. Currently, there is a turnkey system commercially available that has temperature regulation (−160 °C to 400 °C) over a wide range of frequencies (3 μHz up to 3 GHz) with data acquisition equipment and analysis software; the system includes sample cells for the various frequency ranges (*14, 15*).

Data Analysis

As mentioned in the transient section of this paper, motion of polar molecules (or polar segments) gives rise to a change in permittivity with the electric field on/off. In the transformed time domain of the exponential (Figure 3) to the frequency domain (Figure 7), this Debye behavior shows a peak in the imaginary part (ε'') of the transform and a commensurate decrease in the real part (ε'). The exponential origin implies independent moment of the polar moieties. Cole and Cole introduced plots of the imaginary part (ε'') against the real part (ε') of the permittivity (*16*); the Debye data in Figure 7 are normalized and plotted in this fashion in Figure 11 (open squares). The loci of points in Figure 11 describing a semicircle (open squares) is the Debye, and the loci of points are represented by the Debye equation in the frequency domain as

$$\varepsilon^*(\omega) = \varepsilon_\infty + \frac{\varepsilon_o - \varepsilon_\infty}{1 + i\omega\tau} \qquad (11)$$

where ε_∞ is the contribution to $\varepsilon^*(\omega)$ from the electronic and atomic response to the electric field, ε_o is the contribution to $\varepsilon^*(\omega)$ from the entire molecular system to the electric field, ω is the radian frequency and τ is the relaxation time. The quantity $\varepsilon_o - \varepsilon_\infty$ represents the contribution to $\varepsilon^*(\omega)$ only from the permanent polarity of the molecular system. Also plotted in Figure 11, are low temperature dielectric data on 3-bromopentane, a simple alkyl halide that supercools. It is clear that the 3-bromopentane data do not behave like a Debye dielectric. The data are a good fit to the Cole and Davison relaxation expression developed to fit dielectric data on glycerine (*17*); the expression is

$$\varepsilon^*(\omega) = \varepsilon_\infty + \frac{\varepsilon_o - \varepsilon_\infty}{(1 + i\omega\tau)^\beta} \qquad (12)$$

where all symbols have the same meaning as given above and β is an asymmetry parameter which has been shown to be a function of temperature. For 3-bromopentane in Figure 11, $\beta = 0.59$ and $\tau = 592$ seconds at a temperature of 106 K; for comparison when $\beta = 1$ and $\tau = 12$ picoseconds the skewed arc becomes a Debye semicircle that does not fit the data as well. A number of the simple alkyl halides show this Cole-Davidson skewed arc behavior, especially at low temperatures.

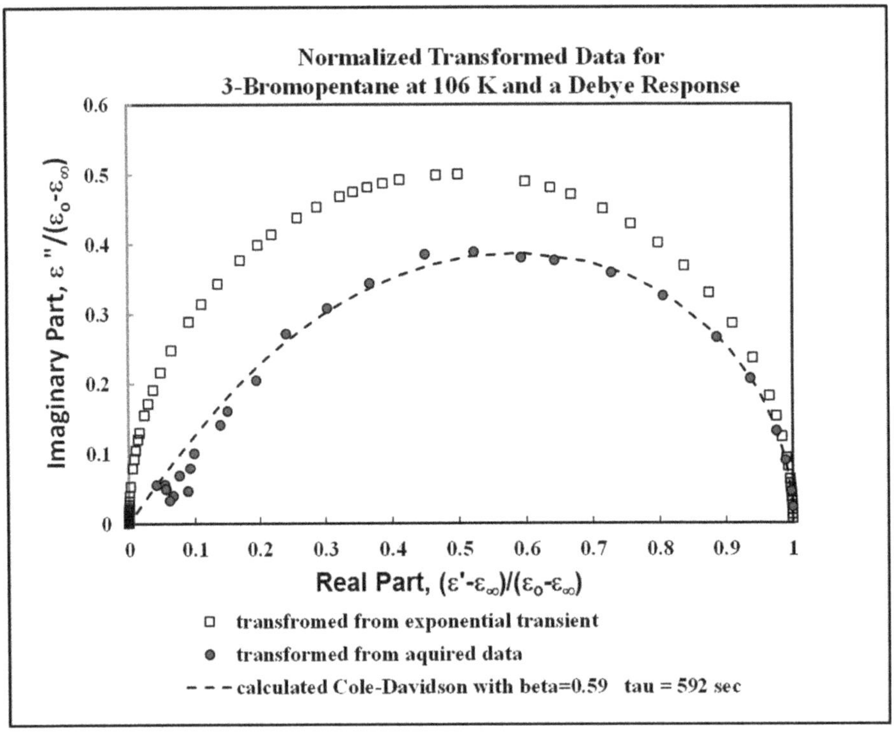

Figure 11. Normalized Cole-Cole plot for a Debye response and normalized data for 3-bromopentane.

There are a number of expressions, in addition to the Cole-Davidson skewed arc, that have been developed to fit dielectric data. A few of the more common ones are below.

The Cole-Cole suppressed arc (Equation 13)

$$\varepsilon^*(\omega) = \varepsilon_\infty + \frac{\varepsilon_0 - \varepsilon_\infty}{1 + (i\omega\tau)^{1-\alpha}} \qquad (13)$$

is a semicircle in the complex plane like the Debye semicircle, but with the center of the semicircle suppressed below the ε' axis, with the semicircle intersecting the ε' at less than a 90 degree angle, with ε' axis at both ends of the semicircle. For the limit of $\alpha=0$, the suppressed arc reduces to a Debye response.

The Havriliak-Negami function (*18*) is

$$\varepsilon^*(\omega) = \varepsilon_\infty + \frac{\varepsilon_0 - \varepsilon_\infty}{(1 + (i\omega\tau)^{\alpha})^{\beta}} \qquad (14)$$

and the Kohlrausch–Williams–Watts function (*19*) (also called the stretched exponential) is

$$f_\gamma(t) = e^{-t^\gamma} \qquad (15)$$

There are no known solutions for the Laplace transform of the stretched exponential, but numerical tables have been calculated for analysis in the frequency domains of interest. One should keep in mind that these expressions, if they fit the data, just provide an analytical expression for convenience of presentation. The link of any fit (or data it represents) to theory needs to be done from detailed analysis on the molecular level, A good example of such a link is Glarum's one dimensional defect diffusion model for the relaxation of isoamyl bromide (*20*), which has a skewed arc behavior, common to simple alkyl halides. The Glarum model produces results that closely resemble and, under certain conditions, predict the skewed arc function.

Other Applications of Dielectric Permittivity

Clearly, dielectric permittivity is used to delineate motion in molecular systems in order to understand the molecular dynamics. But dielectric permittivity can be and has been used in some pragmatic and important applications. For example, Hager *et al.* have used TDR to monitor the curing of cement (*21*). The food industry uses dielectric permittivity to monitor the age of foods like fruit and meat (*22*). Venkatesh and Raghavan present impedance measuring techniques analogous to this presentation with a focus on agricultural foods and biological materials (*23*).

Dielectric measurements (which "see" motion of polar moieties) have been complemented by heat capacity measurements (which "see" vibration and rotational motion, translational motion being silent in liquids and solids) by Takahara *et al.* (*24*) who used heat capacity data and dielectric relaxation data to calculate the number of molecules in the cooperative domain in the Adams-Gibbs theory (*25*).

Both dielectric and heat capacity data were used in a study of crystallization in supercooled 3-bromopentane in this author's laboratory. Using a modulated differential scanning calorimeter at temperatures below the cold crystallization region (≈132 K) but above T_g (109.0 ± 0.2K), isothermal heat capacity measurements were periodically made until the sample completely crystallized. As the sample crystallized, the heat capacity decreased and caused a reduction in the amount of sample not crystallized, as represented by $(C_P(t) - C_P(\text{crystal}))/(C_P(t=0) - C_P(\text{crystal}))$. Isothermal dielectric relaxation measurements (made with an LRC meter with auto frequency scan) were also made in this temperature range. For the permittivity measurements, the isothermal loss maximum decreased with time, as crystallized dipoles do not contribute to the relaxation dynamics. Therefore, the amount of sample not crystallized represented by the dielectric data is $\varepsilon''_{\max}(t)/\varepsilon''_{\max}(t=0)$. Figure 12 is a plot of the amount of sample not crystallized by both methods over a period of several days. These data suggest that the crystallization mechanism affects heat capacity and molecular relaxation in the same manner, with the possible exception of long times.

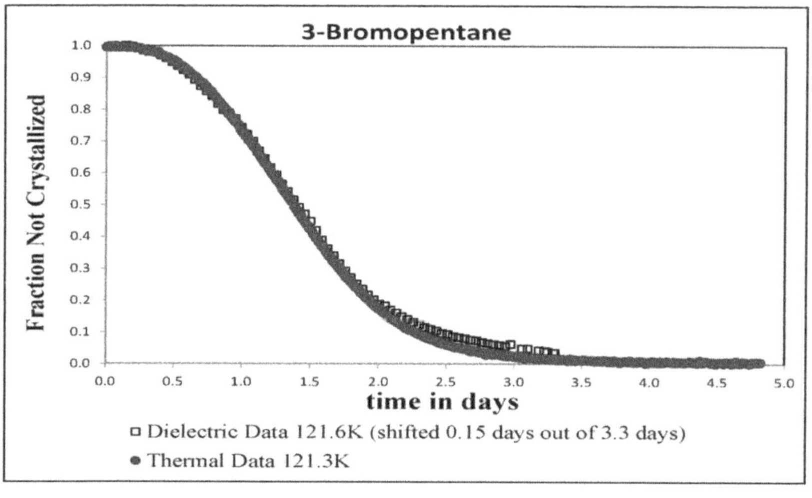

Figure 12. Complementary data for dielectric relaxation and heat capacity.

Summary

The field of dielectrics provides a very powerul tool that examines the electrical properties of materials by monitoring the motion of charge under the stress of a changing electric field. The genesis of this charge in motion is grounded in the dynamic molcular structure of the material. Time scales of dielectric response range from picoseconds to long times depending on conditions (supercooled liquids have relaxation times in the hundreds of seconds at low temperatures). This range of times is covered by three methods for measuring permittivity: transient, bridge and transmission line measurements. Each method has its advantages and disadvantages to be considered when choosing: transient (low transformed frequencies and large number of transformed frequencies in short measuring time); bridge (measurement at each frequency, adequate frequency range and excellent accuracy); transmission line (high frequencies and measurement protocol is computer intensive).

Additional references on theory and experiment are provided (*11*, *26–33*).

References

1. Debye, P. J. W. *Polar Molecules*; Chemical Catalog Company: New York, 1929.
2. *Standard Test Methods for AC Loss Characteristics and Permittivity (Dielectric Constant) of Solid Electrical Insulation*; ASTM D150-18; ASTM International: West Conshohocken, PA, 2018.
3. Jacobs, D.; Greer, S. Capacitance cell for liquids. *Rev. Sci. Instrum.* **1980**, *51* (7), 994–995.
4. Hayden, J. S.; Berberian, J. G. Hemispherical dielectric permittivity cell. *Rev. Sci. Instrum.* **1978**, *49* (7), 1014–1015.
5. Hall, H. P. *A History of Impedance Measurements*. This history is available as a download (pdf) from the website https://www.ietlabs.com/pdf/GenRad_History/A_History_of_Z_Measurement.pdf (accessed Nov. 2, 2020).
6. IETLabs is a source of dielectric instruments and cells. They have taken over the line of General Radio products.
7. Hague, B.; Foord, T. R. *Alternating Current Bridge Methods*, 6th ed.; Pitman: London, 1971.
8. Cole, R. H.; Gross, P. M. A Wide Range Capacitance-Conductance Bridge. *Rev. Sci. Instrum.* **1949**, *20* (4), 252–260.
9. In 1962, General Radio announced their newest bridge, the Type 1615-A Capacitance bridge. In the summer of 1963, I, the author, was a first year graduate student with Professor Robert H. Cole as my thesis advisor at Brown University. Bob was intrigued by the transformer in the 1615-A bridge as it was a decade tapped transformer, the arms of the center-tapped transformer were mutually inducted and, though its frequency response was lower than the transformer constructed by Cole and his student, Paul Gross (*8*), he wanted to build a new bridge around this transformer. Bob arranged a "field trip" for himself, another student and me to General Radio in West Concord, Mass. through a friend he knew in the upper management at General Radio. After a tour of the plant, Bob placed an order for just the transformer. He was somewhat surprised where he was charged for the transformer, I think $120. The next day Bob received a call from his upper management friend who found out about the charge for the transformer; he apologized and said that an entire 1615 bridge was being shipped to him "gratis". When Bob received the bridge, he promptly took it apart to get the transformer he wanted. He also used the lever switches, part of the front plate and the capacitance standards from the gifted bridge in his new bridge.

10. Harris, W. P. In *A new ultralow-frequency bridge for dielectric measurements*; Annual Report 1966 Conference on Electrical Insulation & Dielectric Phenomena; IEEE: 1966; pp 72–74.
11. Berberian, J. G.; King, E. An overview of time domain spectroscopy. *J. Non-Cryst. Solids* **2002**, *305* (1–3), 10–18.
12. Cole, R.; Berberian, J.; Mashimo, S.; Chryssikos, G.; Burns, A.; Tombari, E. Time domain reflection methods for dielectric measurements to 10 GHz. *J. Appl. Phys.* **1989**, *66* (2), 793–802.
13. Cole, R. H. Time domain reflectometry. *Annu. Rev. Phys. Chem.* **1977**, *28* (1), 283–300.
14. *Dielectrics Newsletter*, Mrach 1994; pp 1–8.
15. Schaumburg, G. New integrated dielectric analyzer extends accuracy and impedance range for material measurements. *Dielectrics Newsletter*, May 1999; pp 4–6.
16. Cole, K. S.; Cole, R. H. Dispersion and Absorption in Dielectrics I. Alternating Current Characteristics. *J. Chem. Phys.* **1941**, *9* (4), 341–351.
17. Davidson, D. W.; Cole, R. H. Dielectric Relaxation in Glycerine. *J. Chem. Phys.* **1950**, *18* (10), 1417–1417.
18. Havriliak, S.; Negami, S. A complex plane representation of dielectric and mechanical relaxation processes in some polymers. *Polymer* **1967**, *8*, 161–210.
19. Williams, G.; Watts, D. C. Non-symmetrical dielectric relaxation behaviour arising from a simple empirical decay function. *Trans. Faraday Soc.* **1970**, *66*, 80–85.
20. Glarum, S. H. Dielectric relaxation of isoamyl bromide. *J. Chem. Phys.* **1960**, *33* (3), 639–643.
21. Hager, N., III; Domszy, R. Monitoring of cement hydration by broadband time-domain-reflectometry dielectric spectroscopy. *J. Appl. Phys.* **2004**, *96* (9), 5117–5128.
22. Nelson, S.; Trabelsi, S. Dielectric spectroscopy measurements on fruit, meat, and grain. *Trans. Am. Soc. Agric. Biol. Eng.* **2008**, *51* (5), 1829–1834.
23. Venkatesh, M.; Raghavan, G. An overview of dielectric properties measuring techniques. *Can. Biosyst. Eng.* **2005**, *47* (7), 15–30.
24. Takahara, S.; Yamamuro, O.; Matsuo, T. Calorimetric study of 3-bromopentane: Correlation between structural relaxation time and configurational entropy. *J. Phys. Chem.* **1995**, *99* (23), 9589–9592.
25. Adam, G.; Gibbs, J. H. On the temperature dependence of cooperative relaxation properties in glass-forming liquids. *J. Chem. Phys.* **1965**, *43* (1), 139–146.
26. Macdonald, J. R. Impedance spectroscopy. *Ann. Biomed. Eng.* **1992**, *20* (3), 289–305.
27. *Impedance Spectroscopy: theory, experiment, and applications*, 3rd ed.; John Wiley & Sons: Hoboken, NJ, 2018.
28. *Broadband Dielectric Spectroscopy*; Springer: Berlin, 2002.
29. McCrum, N. G.; Read, B. E.; Williams, G. *Anelastic and Dielectric Effects in Polymeric Solids*; John Wiley & Sons: Hoboken, NJ, 1967.
30. Riande, E.; Díaz-Calleja, R. *Electrical Properties of Polymers*; Marcel Dekker: New York, 2004.
31. Feldman, Y.; Ermolina, I.; Hayashi, Y. Time domain dielectric spectroscopy study of biological systems. *IEEE Trans. Dielectr. Electr. Insul.* **2003**, *10* (5), 728–753.
32. Feldman, Y.; Andrianov, A.; Polygalov, E.; Ermolina, I.; Romanychev, G.; Zuev, Y.; Milgotin, B. Time domain dielectric spectroscopy: An advanced measuring system. *Rev. Sci. Instrum.* **1996**, *67* (9), 3208–3216.

33. Berberian, J. G.; Cole, R. H. Approach to glassy behavior of dielectric relaxation in 3-bromopentane from 298 to 107 K. *J. Chem. Phys.* **1986**, *84* (12), 6921–6927.

Chapter 3

Broadband Dielectric Characterization from 10 mHz to 100 GHz of a 3D Printable Material

Steven Perini,[1] Maryam Sarkarat,[*,1] Danny Zhu,[3] Brian Foley,[2] and Michael Lanagan[1,4]

[1]Materials Research Institute, Penn State, University Park, Pennsylvania 16802, United States

[2]Department of Mechanical Engineering, Penn State, University Park, Pennsylvania 16802, United States

[3]Department of Electrical Engineering and Computer Science, United States Military Academy, West Point, New York 10996, United States

[4]Department of Engineering Science and Mechanics, Penn State, University Park, Pennsylvania 16802, United States

[*]Email: mus60@psu.edu

This work presents an array of experimental techniques that have been used to characterize dielectric properties (permittivity and dissipation factor) of 3D printed acrylate-based polymer over a wide frequency range. At frequencies below 10 MHz, a parallel plate capacitor method provides accurate permittivity and loss results. At frequencies above 10 MHz, there are two general types of measurement techniques, resonant and transmission/reflection. Resonant measurements are at discrete frequencies and have high accuracy for low loss dielectrics with dissipation factors below 1%. Transmission/reflection methods have the advantage of being broadband and cover a large frequency range (1 MHz to 110 GHz); however, the accuracy limits measurement to high loss samples. A reflection method for the 1 MHz to 2 GHz frequency range was specifically developed for polymers and polymer composites. In this study, parallel plate, resonant and transmission line methods have been used to characterize the dielectric properties of 3D printed acrylic based polymer from 10 mHz to 100 GHz. A relaxation peak, which is associated from the motion of polar groups of side chains, was observed as a loss peak at the frequency range of 10^4-10^{10} Hz and shows the need for characterization methods over a broad frequency range.

Introduction

The drive toward faster data transmission and 5G comunications has deepened interest in microwave characterization of dielectric materials for transmission line substrates, radomes and antennas (*1–3*). Polymers and polymer/oxide particle composites, operating in the 10-100 GHz frequency range, are important for electronic packaging and antenna structures for 5G telecommunications and automotive guidance systems (*4*). Additive manufacturing has been proposed to rapidly explore 5G system designs and a large range of printable polymers are available (*5*). RF and mm-wave characterization of polymers, including additively manufactured polymers, is critical to incorporating new materials into advanced telecommunication systems.

Dielectric properties of polymers can be influenced by various types of polarizations including space charge, dipolar, ionic and electronic polarizations which appear as a loss peak at a specific frequency. At low frequencies (f <100 Hz), space charges, dc conductivity and an α relaxation that corresponds to the main chain movement (related to T_g) controls the dielectric response; however, at higher frequencies side group rotation dominates the dielectric properties (*6*). These responses can be quite broad and hence can easily overlap in the microwave range. Few studies have been conducted to investigate dielectric properties of polymers above 1 MHz because of the challenging measurement conditions at microwave frequency. Jackson et al. showed a very broad distribution of relaxation times for oxidized polyethylene that started from 10^4 Hz and extended to microwave frequencies (*7*). In polystyrene, a high frequency relaxation dispersion was observed at 10^{12} Hz due to the asymmetry at the phenyl side group (*8*).

Conventional impedance spectroscopy, involving parallel plate capacitor structures, are only suitable for frequencies below 10 MHz, so additional high frequency characterization methods need to be established. A summary of the dielectric measurement regimes across the electromagnetic spectrum is shown in Figure 1. Low frequency dielectric measurements use impedance analyzers while high frequency dielectric measurements rely on vector network analyzers that measure the magnitude and phase of transmitted and reflected microwaves. The microwave dielectric characterization techniques integrate electronic and photonic concepts to measure dielectric permittivity and loss from 10 MHz to 100 GHz.

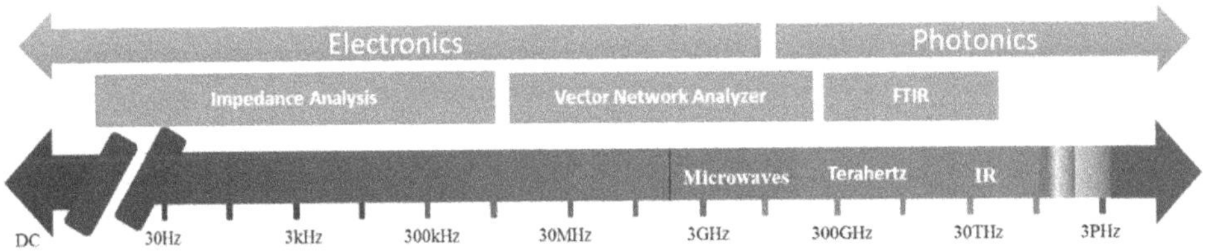

Figure 1. Frequency ranges for the primary measurement techniques for electromagnetic characterization of materials. The proposed vector network analyzer techniques will fill a critical gap between impedance analysis and optical characterization methods in the overall spectrum of test methods.

A dielectric is a material that can be polarized by applying an electric field. Complex permittivity (ε_r*) is the parameter which describes the materials behavior when it is subjected to an electric field. Complex Permittivity is defined by:

$$\varepsilon_r^* = \varepsilon_r' - j\varepsilon_r''$$

where:

ε_r' is the real part of the complex dielectric constant

ε_r'' is the imaginary part of the complex dielectric constant

j is the imaginary constant $\sqrt{-1}$

The dielectric constant, often referred to as relative permittivity and designated as ε_r, is the ability of a material to polarize under the influence of an electric field measured by the ratio of the capacitance of a capacitor with the material as dielectric to its capacitance with vacuum as dielectric.

$$\varepsilon_r = \varepsilon'_r \, \varepsilon_0$$

where ε_0 is the permittivity of free space (8.854e-12 F/m)

The dissipation factor, DF, or loss tangent, tanδ, is the measure of the electrical energy absorbed and lost (power dissipation) when electrical current is applied to the dielectric material, and is defined by the following equation:

$$\mathrm{DF} = \tan\delta = \frac{\varepsilon''_r}{\varepsilon'_r}$$

At high frequencies the inverse of the dissipation factor, referred to as the Quality Factor, or Q, is often used. Q is given by:

$$Q = \frac{1}{DF}$$

and is defined as the energy loss divided by stored energy over a charge - discharge cycle.

In the age of additive manufacturing, a large number of dielectric materials can be 3D printed in a wide range of shapes and sizes. Such flexibility in fabrication has many uses in various electrical applications, including circuits, optics, antennas and metamaterials, which often rely on broad knowledge of the dielectric's dispersive electrical properties. In polymeric components, dielectric response can be influenced by various types of polarizations including interfacial, dipolar, ionic and electronic polarizations under the influence of an applied electric field. As the frequency of applied electric field is increased, the dielectric response, namely the dielectric permittivity and dielectric loss, change drastically due to the relaxation and resonance. Hence, precise techniques are required for characterizing materials in each portion of the electromagnetic spectrum. In this study, the broadband measurement of an example 3D printed material across the broad frequency spectrum are examined in the following characterization techniques section.

Experimental

In this work, a variety of sample sizes and geometries were 3D printed using a Stratasys Objet260 Connex3 printer, utilizing the PolyJet printing technique. The method deposits liquid photopolymer directly onto a build plate and UV-cures the resin in 16 μm layers, yielding a solid and rigid sample with a smooth, opaque and non-porous surface. While more complex structures often require dissolving support structures with a sodium hydroxide solution, characterizing the material can be done using samples with simple geometries that require minimal post-processing. In these cases, residue from support material are simply pressure washed off, and the samples are left to dry. Although a wide variety of liquid photopolymers are available and can be blended to achieve a range of material properties, the samples shown in Figure 2 were fabricated purely out of a acrylate based polymer (Stratasys's VeroWhitePlus) material. Sample dimensions were selected to minimize measurement errors during broadband dielectric characterization of the material using a variety of measurement techniques.

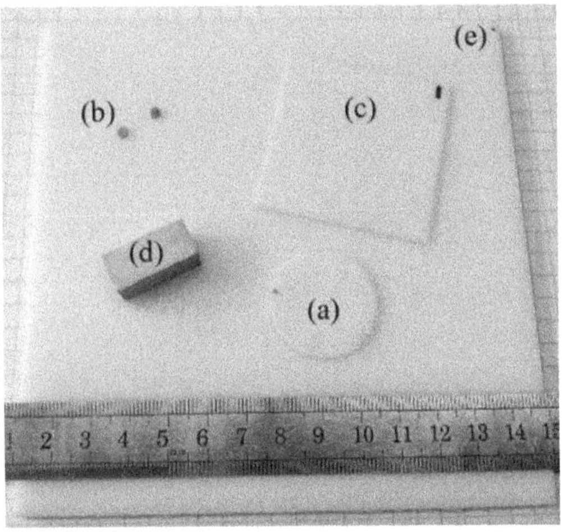

Figure 2. 3D printed acrylate based polymer (VeroWhitePlus) sample dimensions for each technique are a) 30 mm x 1mm parallel plate, b) 3 mm x 1 mm reflection, c) 22.86 mm x 10.16 mm x 12.77 x-band transmission, d) 50 mm x 50 mm x 1 mm split cavity and e) 140 mm x 140 mm x 5 mm open resonator and Swiss to 12. Photo courtesy of Steven Perini.

Characterization Techniques

Multiple characterization techniques were utilized to cover the thirteen-decade frequency span from 0.01 Hz to 100 GHz. These included the parallel plate capacitance method, the 3 mm reflection technique, an X-Band transmission technique, a Split Cavity resonance technique, an Open Resonator Fabry Perot measurement, and a Swiss to 12 MCK transmission technique. A variety of sample forms and dimensions were required to perform these measurements, and the process of 3D printing provided an easy and cost-effective way to create the various samples. The individual measurement techniques that were performed are described in more detail in the following paragraphs.

The parallel plate capacitor structure was used for measurements below 50 MHz. The sample made for these measurements was a thin disk, 30 mm in diameter and 1 mm thick. This sample was electroded on both sides by sputtering 100 nm of gold on each side of the sample, effectively forming a parallel plate capacitor. The sample was placed in an in-house designed and fabricated fixture that connects the sample to the meter/analyzer front panel connectors and the capacitance and loss are measured. The low frequency measurements from 0.01 Hz to 100 Hz were measured using an Ametek Modulab XM Materials Test System. For the frequency range from 100 Hz to 1 MHz, an Agilent E4980A LCR meter was utilized. Dielectric properties for frequencies from 1 MHz to 50 MHz were measured using an Agilent 4192 Impedance Analyzer. Prior to the sample measurement an "open and short" calibration was performed on the Agilent E4980A LCR meter and for the Agilent 4192 Impedance Analyzer an "open, short and 50-ohm load" calibration was performed. Figure 3 shows the sample in the fixture.

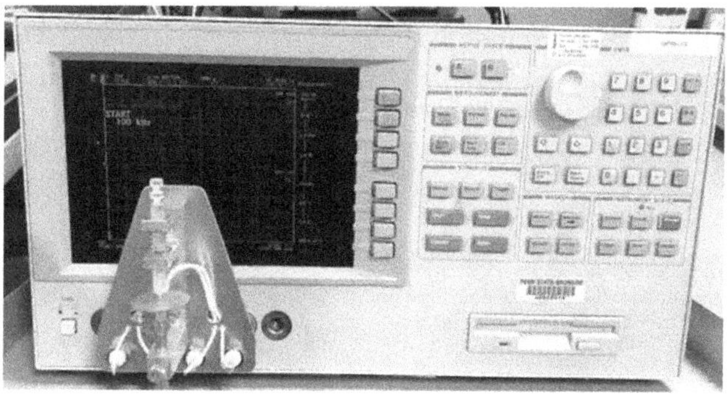

Figure 3. Agilent 4194A Impedance Analyzer with Parallel Plate Sample fixture. Sample placed between spring loaded opposing pins. Photo courtesy of Steven Perini.

The material's dissipation factor is given directly from the measurement. The dielectric constant is calculated from the measured capacitance using the equation

$$C = \frac{\varepsilon_o \varepsilon_r A}{t}$$

where:

C is the capacitance,

ε_0 is the permittivity of free space = 8.854×10^{-12} F/m

ε_r is the relative permittivity

A is the sample area

t is the sample thickness

For high frequencies above 50 MHz a Vector Network Analyzer was utilized. A Vector Network Analyzer, commonly referred to as a VNA, is an instrument that measures the reflection and transmission of an incident signal upon the sample or device under test. Three different VNA's were used in this study: a Keysight P9374A, an Agilent 37369D, and a Keysight 5724B. The VNA provides an accurate measurement of the ratio of the reflected signal to the incident signal, and the transmitted signal to the incident signal. These ratios are complex numbers consisting of the magnitude and phase of the signal and are referred to "S" parameters. The convention used to designate S parameters is "Sxy", where "x" is the port where the signal is measured and "y" is the port where the signal originated. For example, S21 is the ratio of the signal measured at port 2 to the signal originated at port 1, which is the transmitted signal. Likewise, S11 is the ratio of the reflected signal measured at port 1 to the incident signal originating from port 1. For a two-port network analyzer S11, S21, S12 and S22 can all be measured. The following techniques were used to characterize the material using the VNA.

The 3 mm Reflection Technique, which derives its name from the sample size, 3 mm in diameter, measures the impedance of the sample from 40 MHz to 2 GHz. Sample are generally around 1 mm in thickness. This technique was previously developed to characterize ceramic samples (9). To perform this measurement, the sample is placed on the end of an APC-7 connector which is connected to the Vector Network Analyzer (VNA), and then a shorting plug is placed on top. Figure 4 shows this fixture.

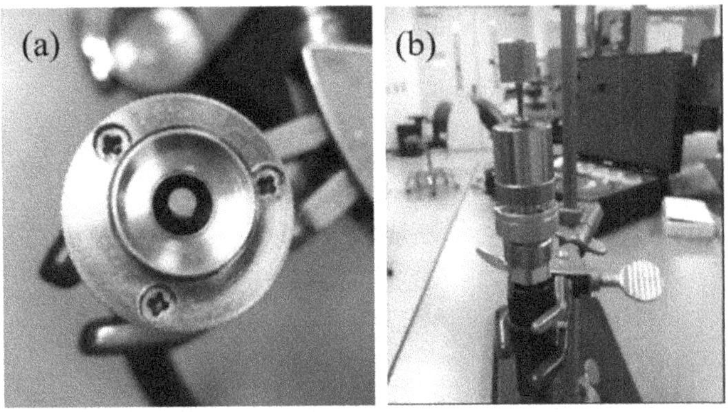

Figure 4. (a) 3 mm sample placed on center of APC-7 connector (b) shorting plunger attached. Photo courtesy of Steven Perini.

Prior to the measurement, a one-port "open, short and 50 ohm load" calibration is performed on the VNA at the end of the APC-7 connector. The complex S11 scattering paremeter is recorded across the frequency span and the material permittivity and loss data is extracted from this S11 sweep using a Bessell function field model for non-uniform electromagnetic fields. The measurement frequency range for this technique is generally from 300KHz to 2GHz, although sample permittivity can limit the measurement range. The impedance of samples with low permittivity becomes too high for the VNA at lower frequencies, since the samples impedance is given by

$$Z = \frac{1}{2\pi f C}$$

Where f is frequency and C is the sample's capacitance. For samples with high permittivity, the quarter-wave electrical length in samples begins to match the diameter of the sample causing a resonance which limits the ability to measure the sample at the higher frequencies. The electrical wavelength (one sinusoidal wave) in a dielectric, λ_d, is inversely proportional to the square root of its permittivity, ε_r as given by the equation

$$\lambda_d = \frac{\lambda_o}{\sqrt{\varepsilon_r}}$$

where λ_o is the wavelength in vacuum.

For the VeroWhitePlus sample, data below 900 MHz was unusable due to the high impedance of the sample.

An X-Band Rectangular Waveguide operates between 8.2 GHz and 12.4 GHz. For this broadband reflection/transmission measurement, the sample must be made to fit precisely in the waveguide (22.86 mm x10.16 mm) with no gaps. The samples that were printed were 12.67 mm thick. The VNA was calibrated prior to the measurement by performing a full-two port calibration across the frequency span. An in-house LabVIEW software program developed from HP application Note 8510-3 captures the reflected S11 and transmitted S21 real and imaginary data from the VNA and converts it into the material's complex dielectric properties (*10*). Figure 5 shows the samples (insert) and the waveguide which is connected to the VNA by coaxial cables.

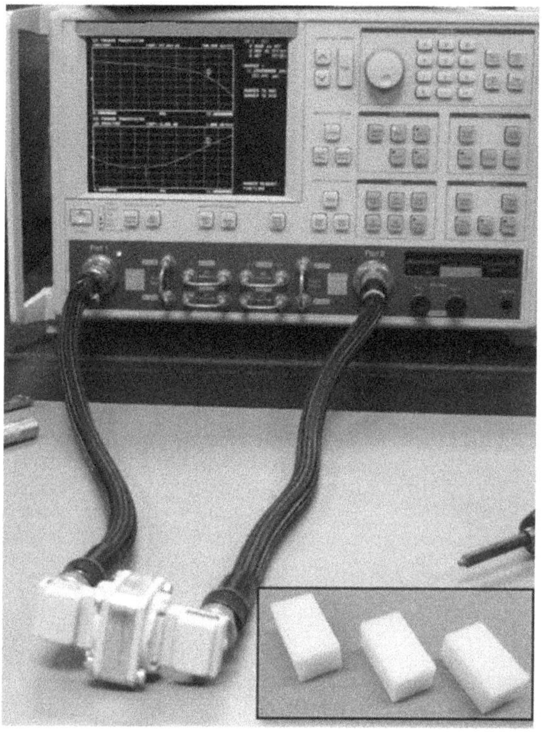

Figure 5. X-band transmission technique with VeroWhitePlus samples. Photo courtesy of Steven Perini.

The Split Cavity resonator technique was developed by Gorden Kent (*11*) and enhanced by Michael Janezic (*12*) from NIST laboratories. In this measurement, the sample must be flat, uniformly thick (approximately half a millimeter to a few millimeters), and at least 1 inch diameter (or square), and is placed between a cylindrical cavity that is separated into two halves. The sample used in this study was 50 mm x 50 mm x 1 mm. The Split Cavity resonator is connected to a VNA and a coupling loop in each half of the cavity excites a TE_{O11} resonance. The empty cavity resonance for the fixture used is this study is 20.2 GHz, and this resonance decreases when the sample is placed between the cavity halves. This resonance is displayed on the S21 transmitted power trace. Knowing the sample thickness and measuring the frequency and the -3db bandwidth of this resonance peak, the permittivity and loss of the material is calculated using NIST SplitC software. This is a single frequency measurement that is determined by the cavity size and the sample's thickness and permittivity. A higher permittivity and/or a thicker sample will cause a greater shift in the frequency. Compared to reflection/transmission techniques, resonant methods have a higher accuracy and sensitivity. The Split Cavity measurement technique with the sample inserted is shown in Figure 6.

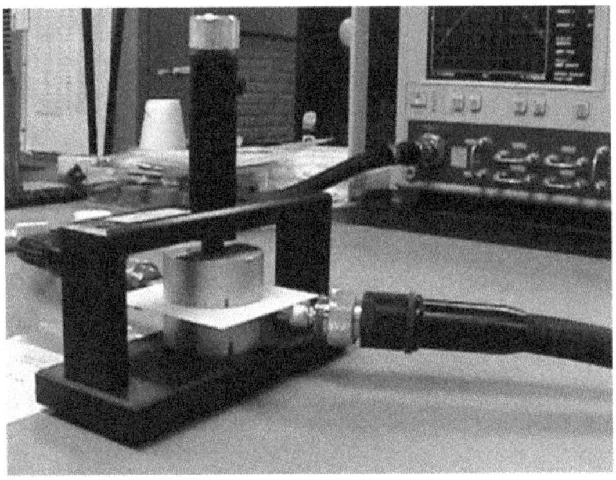

Figure 6. Split Cavity Resonator with sample inserted between halves. Photo courtesy of Steven Perini.

The Damaskos, Inc., Model 600T Open Resonator is a commercial Fabry Perot type resonant system that measures the complex dielectric constant of thin low loss dielectric materials over the approximate frequency range of 12 GHz to 70 GHz. Samples for this technique must be flat, electrically thin (approximately less than a quarter of an electrical wavelength), and roughly 11cm square or in diameter. This large area is generally not an issue for polymer materials but can be a problem when fabricating other materials like ceramics. The 3D printed Verowhite sample measured with this technique was 140 mm square and 5 mm thick. The resonator consists of two spherical reflectors and measurements are made by placing the sample in the high electric field located at the midpoint between the reflectors. Similar to the Split Cavity, the resonant frequency with the sample inserted is shifted from the empty cavity resonance. Although this is a resonant technique, multiple resonant modes can be measured across the frequency span producing a quasi-broadband measurement. CAVITY software, provided by Damaskos, Inc., is used to obtain the resonant frequency and Q values of the resonant response to the given modes. The software processes these data and extracts the permittivity that satisfies the canonical solution. Figure 7 shows the Open Resonator set-up with the sample placed between the reflectors.

Figure 7. Damsakos 600T Open Resonator connected to an Anritsu 37369D VNA. The sample is centered between the two parabolic resonant reflectors. Photo courtesy of Steven Perini.

A SWISS to 12 MCK-WR-10 is a commercially available fixture employing a reflection/transmission technique that was used to measure a material's dielectric property over a broadband frequency range from 67 GHZ to 110 GHz. The MCK is comprised of two conical waveguide horn transitions and two sections of corrugated waveguide. The MCK is connected to a VNA which must have a time domain option, and the S-parameters are measured. The sample is placed in the adjustable gap between the two corrugated waveguides. The recommended minimum sample size is 50 mm x 50 mm, with thickness in the few millimeter range (multiple electrical wavelengths preferable), however larger samples can easily be measured. As such, the same large sample that was measured in the Open Resonator method was also measured in this fixture. The SWISS to 12 WebMCK software automatically extracts the material properties from the reflected and transmitted S parameters. Figure 8 shows the Swiss to 12 MKS-WR-10 fixture with the sample inserted.

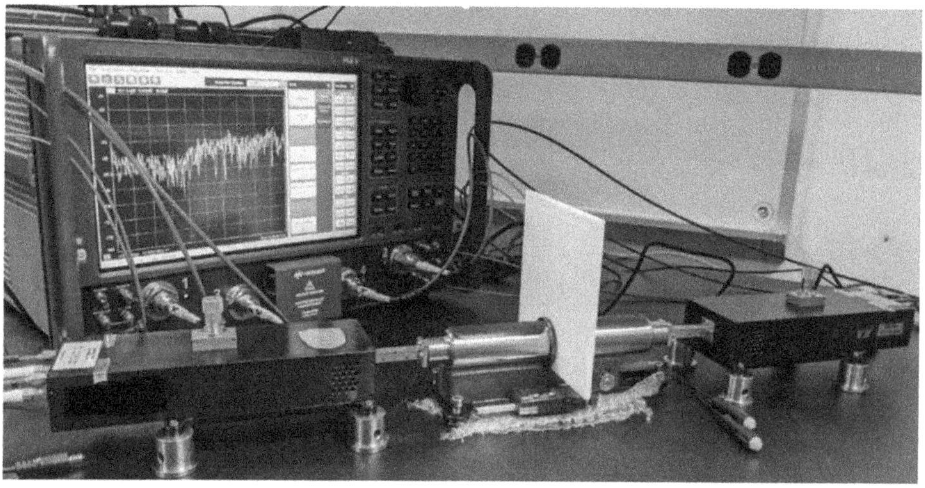

Figure 8. SWISS to 12 MCK-WR-10 measurement apparatus used to perform high frequency (67 GHz – 110 GHz) characterization of a VeroWhitePlus sample. Photo courtesy of Steven Perini.

Table 1 summarizes all the experimental techniques that are used in the present work.

Table 1. Summary of Measurement Techniques

Fixture/Measurement Method	*Instrument Used*	*Frequency Range*
Parallel Plate	Amatek Modulab XM	10 mHz to 1 MHz
Parallel Plate	LCR Meter	100 Hz to 1 MHz
Parallel Plate	LCR Meter	1 MHz to 50 MHz
3mm Reflection	Vector Network Analyzer	40 MHz to 2 GHz
Waveguide Reflect/Transmission	Vector Network Analyzer	8 GHz to 12 GHz
Split Cavity Resonance	Vector Network Analyzer	10 GHz to 20 GHz
Open Resonator Resonance	Vector Network Analyzer	12 GHz to 40 GHz
Swiss to 12 Reflect/Transmission	Vector Network Analyzer	70 GHz to 110 GHz

Results

The broadband dielectric spectra of 3D printed Verowhite at room temperature is shown in Figure 9. The high permittivity and loss below 100 Hz is attributed to ionic conductivity where there is the potential for water absorption in the commercial polymer composite. The dielectric relaxation above 1 MHz is attributed to a beta transition where there is a broad distribution of side group relaxations (*13*). As can be seen from the data, accurate loss measurements have a large error range with the 3 mm Reflection technique; however, it is evident from the broadband dielectric spectra shown in Figure 9 that the beta transition of the acrylate-based polymer spans the entire MHz range and extends into the GHz range. A previous study for a closely related polymer, poly(methyl methacrylate) showed two relaxations. An alpha relaxation with a characteristic frequency of 1 Hz and a beta relaxation with a characteristic frequency of 1 MHz. The photocurable acrylate-based polymer used in this study has a significantly higher characteristic frequency range than for PMMA (*8*). In addition the breadth of the high frequency relaxation (10^4 to 10^{10} Hz) is broader than for phase pure polymers (polystyrene, polyethylene oxide) that have beta relaxations over 2-3 orders of magnitude. The exceptionally broad relaxation is likely due to additives (crosslinker, antioxidants, etc.) to the 3D printable polymer system.

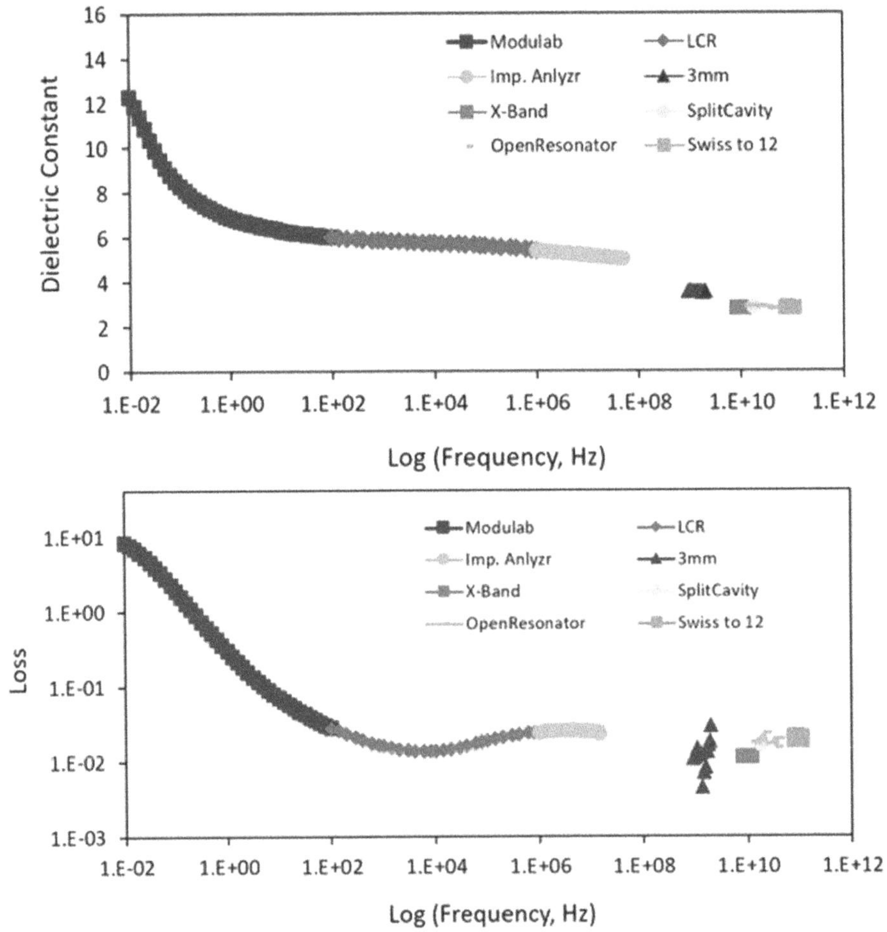

Figure 9. Broadband dielectric constant and loss of VeroWhitePlus based on various characterization methods.

Figure 10 presents dielectric constant and loss of 3D printed Verowhite polymer based on various characterization methods above 8 GHz. The dielectric constant remains relatively constant at 2.8 across this frequency range with only slight variation (2%) from the Fabry Perot resonator technique. The dielectric permittivity and loss at mm-frequency are in the same range as other 3D printable polymers such as acrylonitrile butadiene styrene (ABS), polyoxymethylene (POM) and polylactic acid (PLA) (*14*).

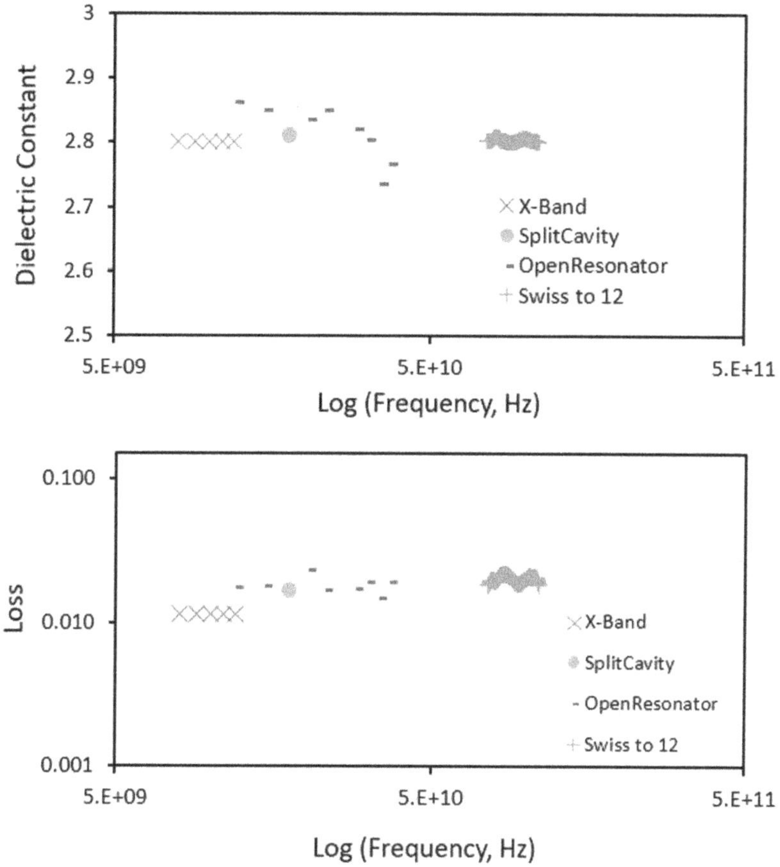

Figure 10. GHz range Broadband dielectric constant and dielectric loss of VeroWhitePlus based on various characterization methods.

Conclusion

In this paper, the broadband dielectric properties of a commercial acrylate based 3D photoprintable material were characterized over a wide range of frequency from 10 mHz to 100 GHz. Multiple techniques for frequencies above 1 MHz were utilized for this study. There were two primary features in the dielectric spectra as the frequency spanned thirteen orders of magnitude. The high loss in the mHz region is attributed to space charge formation and DC conduction. As the frequency increased to MHz region a characteristic dipolar beta relaxation was observed. This relaxation however diminished beyond 100 MHz and hence we see precipitous decrease in both dielectric constant and loss at higher frequencies. The parallel plate capacitor, resonant, and transmission measurement techniques were utilized to cover the frequency range were reviewed. We envision that the proposed characterization techniques at high frequencies will provide extensive information in developing new polymers and polymer composites suitable for high frequency 5G applications.

Acknowledgments

This material is based upon work supported by the National Science Foundation, as part of the Center for Dielectrics and Piezoelectrics under Grant Nos. IIP-1361571 and 1361503. The authors would like to thank the National Science Foundation for support of the Center for Dielectrics and Piezoelectrics as well as the member companies for funding this research.

References

1. Arunima, M. S.; Bhuvana, N. S.; Unnikrishna, M. An Overview of Millimeter-wave Antennas. *IEEE Inter. Conf. Commun. Electron. Syst. (ICCES)* **2020**, 395–400; DOI: 10.1109/ICCES48766.2020.91379852020.
2. Sghaier, T.; El Gibari, M.; Guiffard, B. New low loss soft thermoplastic blends for microwave applications. *J. Phys. D: Appl. Phys* **2020**, 53 (13), 135003; DOI: 10.1088/1361-6463/ab5756.
3. Imran, D.; Farooqi, M. M.; Khattak, M. I.; Ullah, Z.; Khan, M. I.; Khattak, M. A.; Dar, H. Millimitere Wave Microstrip Patch Antena for 5G Mobile communication. *IEEE Inter. Conf. Eng. Emerg. Technol.* **2018**, 65, 4864–4868; DOI: 10.1109/ICEET1.2018.8338623.
4. Omodara, G.; Myllymäki, S.; Kallioinen, J.; Juuti, J. Characterization of PMMA/$BaTiO_3$ composite layers through printed capacitor structures for microwave frequency applications. *IEEE Trans. Microw. Theory Tech.* **2018**, *66* (4), 1736–1743; DOI: 10.1109/TMTT.2017.2781694.
5. Parsons, P.; Larimore, Z.; Muhammad, F.; Mirotznik, M. Fabrication of low dielectric constant composite filaments for use in fused filament fabrication 3D printing. *Addit. Manuf.* **2019**, 39, 100888; DOI: org/10.1016/j.addma.2019.100888.
6. Kassap S. O. *Principles of Electrical Engineering Materials and Devices*, 3rd ed.; McGraw Hill: 1994.
7. Jackson, W.; Forsyth, J. S. A. The development of polythene as a high-frequency dielectric. *J. Inst. Electr. Eng., Part 3* **1945**, 92, 23–33; DOI: 10.1049/ji-3-2.1945.0007.
8. Bur, A. J. Dielectric properties of polymers at microwave frequencies: a review. *Polymer* **1985**, *26* (7), 963–977; DOI: 10.1016/0032-3861(85)90216-2.
9. Sharpe, C. B.; Brockus, C. G. Method for Measuring the Dielectric Constant of Ferroelectric Ceramics at S-Band Frequencies. *J. Am. Ceram. Soc.* **1960**, 43 (6), 302–305; DOI: 10.1111/j.1151-2916.1960.tb13658.x.
10. Hewlett Packard. *Measuring the dielectric constant of solids with the HP8510 network analyzer*; Product Note 8510-3; 1985; pp 1−11.
11. Kent, G. Nondestructive Permittivity Measurement of Substrates. *IEEE Trans. Instrum. Measure.* **1996**, 45 (1), 102–106; DOI: 10.1109/19.481319.
12. Janezic, M. D. Full-Wave Analysis of a split-cylinder Resonator for Nondestructive Permittivity Measurements. *IEEE transactions on Microwave Theory and Techniques* **1999**, 47 (10), 2014–2020; DOI: 10.1109/22.795077.
13. Bergman, R.; Alvarez, F.; Alegıa, A.; Colmenero, J. Dielectric relaxation in PMMA revisited. *J. Non-Cryst. Solids* **1998**, 235, 580–583; DOI: 10.1016/S0022-3093(98)00639-5.

14. Reyes, N.; Casado, F.; Tapia, V.; Jarufe, C.; Finger, R.; Bronfman, L. Complex dielectric permittivity of engineering and 3D-printing polymers at Q-band. *J. Infrared, Millimeter, Tetrahertz Waves* **2018**, *39* (11), 1140–1147; DOI: 10.1007/s10762-018-0528-9.

Chapter 4

Dielectric Spectroscopy at High Electric Fields

Ranko Richert*

School of Molecular Sciences, Arizona State University, Tempe, Arizona 85287, United States

***Email: ranko@asu.edu**

The field of nonlinear dielectric effects started with Debye's work on dielectric saturation. The signature of this effect is a reduction of the dielectric constant, ε, with increasing electric field. The work of Piekara added the so-called "chemical effect", an increase in ε that originates from an equilibrium of two species with different dipole moments. More recently, dielectric hole burning and similar techniques based on high ac fields demonstrated that the energy absorbed from a field can accelerate dynamics, even for isothermal conditions. For strong static (dc) fields, the dynamics of polar liquids can become slower relative to the field free case. Thus, depending of the type of field applied, these reversible field effects can lead to amplitudes increasing or decreasing, and to the loss spectrum being shifted to the left or right along the logarithmic frequency scale. Additionally, crystallization outcomes can be modified by static fields of moderate amplitude.

Introduction

Dielectric relaxation spectroscopy is a powerful tool for characterizing the dynamics of liquids as well as glassy and crystalline solids (*1*). The technique rests on the interaction between mobile charges in the sample and an external electric field, with the mobility of ions or dipoles leading to measurable currents or polarizations. In the majority of applications, one is interested in measuring properties of the sample that are not altered by applying the electric field that is needed to measure dielectric relaxation. The field range for which this is warranted is called the regime of linear response, i.e., where polarization P is linearly related to the electric field E. This linearity is the basis for the constitutive equations that define the material specific properties, susceptibility χ and permittivity ε, via

$$P = \chi\varepsilon_0 E = (\varepsilon - 1)\varepsilon_0 E \,. \tag{1}$$

In the typical case of sufficiently small fields, the proportionality between P and E implies that both χ and $\varepsilon = 1 + \chi$ are independent of the magnitude of the field used to obtain either the dielectric constant ε or the frequency dependent permittivity, $\varepsilon(\omega)$ (*2*). If the entire loss spectrum, $\varepsilon''(\omega)$, remains field invariant, this implies that quantities such as relaxation amplitude $\Delta\varepsilon$ or relaxation time constant τ_D are not affected by the magnitude of the electric field. The high frequency limit of permittivity, ε_∞, is often governed by electronic polarizability and density, and thus not susceptible to changes of considerable magnitude, even at relatively high electric fields. Therefore, a field effect on the static dielectric constant, ε_s, is dominated by the relaxation amplitude, $\Delta\varepsilon = \varepsilon_s - \varepsilon_\infty$.

It is of course interesting to learn at what field amplitude the linear regime ends and at what level nonlinearity will set in. There is no straightforward answer to this question, as the onset of nonlinearity is a matter of the experimental resolution. In other words, there is no threshold separating linear from nonlinear behavior, but rather a continuous increase of nonlinear contributions as the magnitude of the electric field is elevated (3). The field at which this nonlinear effect can be observed will then obviously depend on the resolution regarding permittivity.

The aim of this Chapter is to provide an outline of the various high-field dielectric approaches and to present the different field protocols can be exploited to discriminate among the different sources of nonlinear polarization in glass forming materials. It will also be shown how the nonlinear effects approach equilibrium along the time scale, and how this time dependence relates to structural recovery and physical aging. As far as is known, the origin for each of the effects will be discussed.

Field Effects on Relaxation Amplitude

Dielectric saturation is a nonlinear effect recognized and studied about 100 years ago (5). Saturation refers to the feature that dipole orientation has an upper bound given by the situation in which all dipoles are aligned towards the field, i.e., $\langle\cos\theta\rangle = 1$, where θ is the angle between dipole and field direction. Therefore, polarization due to dipole orientation can not continue to increase linearly with the field, but will remain below the value expected by linearly extrapolating the low field behavior to higher field amplitudes (3). In terms of permittivity, this means that the dielectric constant, ε_s, will appear to become smaller at high fields. A typical metric for this effect is the Piekara coefficient a defined by

$$\Delta\varepsilon_E = -aE^2 \,. \qquad (2)$$

where the dielectric constant decrement $\Delta\varepsilon_E$ is the difference between the values of ε_s observed at a high field E and in the low field limit (6). Based on symmetry considerations, $P(E) = -P(-E)$, the dominant term of $\Delta\varepsilon_E$ is quadratic in the field. Due to the general nature of the origin of saturation, every dipolar liquid is expected to show this effect, and Figure 1 displays an example of a strong effect for a highly polar liquid. Note that the dielectric constant decrement can be very small even when the field approaches the limit given by dielectric breakdown, i.e., near the field-induced sample failure. There is also the possibility of saturation being compensated by another effect discussed below.

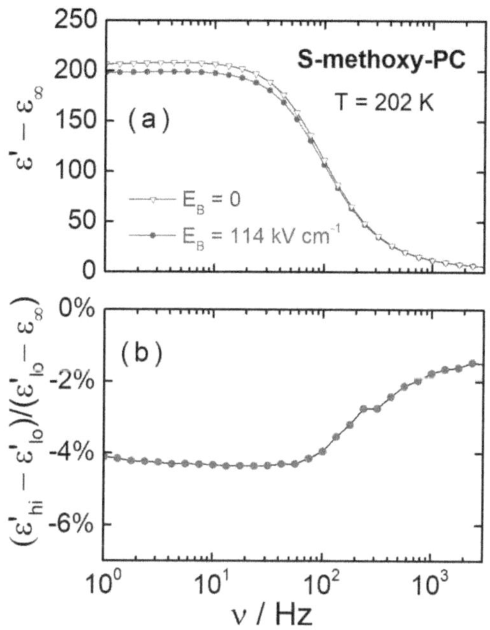

Figure 1. (a) Frequency dependent real part of permittivity, $\varepsilon' - \varepsilon_\infty$, for (S)-(−)-methoxy-propylene carbonate at $T = 202$ K, measured for a high bias field of $E_B = 114$ kV cm^{-1} (solid symbols) and for zero bias (open symbols). In the static limit, the field induced reduction of $\Delta\varepsilon$ amounts to 4.3 %. (b) Field-induced relative change of the quantity $\varepsilon' - \varepsilon_\infty$ versus frequency. Lines serve as guides only. Data from reference (4).

Some materials show the opposite of dielectric saturation, namely an increase of the dielectric constant with the electric field magnitude (*7, 8*). This effect requires samples in which two states with different dipole moments are in thermodynamic equilibrium so that the field can act on the equilibrium constant by lowering further the free energy of the more polar species. This can occur when the field is capable of modifying the charge distribution within the molecule, thus enhancing the molecular dipole moment (*9*). This so called "chemical effect" can be particularly strong for monohydric alcohols in the case of coexisting ring- and chain-like hydrogen-bonded structures (*10*). Because a ring-like structure has zero net dipole moment, shifting the equilibrium more towards chain-like configurations can lead to a considerable increase in the dielectric constant at high fields. The case of 5-methyl-3-heptanol shows this effect in a very pronounced manner, with Figure 2 demonstrating that a moderately high field of 171 kV cm^{-1} increases $\Delta\varepsilon$ by 11% (*7*).

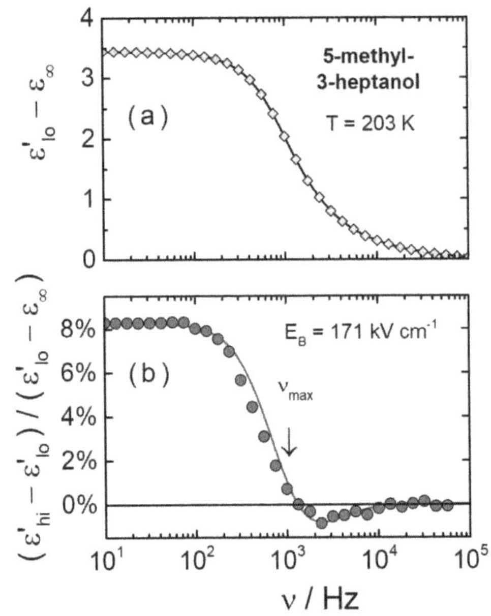

Figure 2. (a) Low-field permittivity ε'_{lo} - ε_∞ for 5-methyl-3-heptanol at T = 203 K. The line serves as guide only. (b) Experimental steady state spectra of the relative change of the real part of permittivity, (ε'_{hi} - ε'_{lo})/(ε'_{lo} - ε_∞), induced by a static field of E_B = 171 kV cm^{-1}. The line is a fit assuming an 11.0% increase in the amplitude ($\Delta\varepsilon$) and a 12.6% increase in the time constant (τ_D). The arrow indicates the frequency position of the low field loss peak, ν_{max}. Panel (a) helps to locate the field-induced changes relative to the linear spectrum. Data from reference (7).

Both amplitude effects, saturation and chemical effects, can be observed experimentally using either a high static (dc) field on top of an alternating field suitable for measuring the static dielectric constant, or a high amplitude alternating (ac) field of suitable frequency (*11*). For a typical polar liquid that displays a loss peak as a result of primary structural relaxation (dipole reorientation), the ac field induced amplitude change may be reduced near or beyond the loss peak frequency ν_{max}. Because such field-induced changes require a concomitant adjustment in the structure of the liquid, it takes a time approximately equal to the structural relaxation time, $\tau_\alpha \approx 1/(2\pi\nu_{max})$, to establish the nonlinear effect (*12*). As a result, the effect will fade for frequencies $\nu > \nu_{max}$.

It should be noted that an apparent rise in permittivity at high fields could originate from electrostriction. This may lead to a reduction in sample thickness due to the coulombic attraction of surface charge, and the resulting increase in capacitance could appear as enhanced permittivity (*11*, *13*).

Faster Dynamics Due to High Fields

It is well established that a dielectric can absorb energy from a time-dependent electric field. For a sinusoidal field, $E(t) = E_0\sin(\omega t)$, the amount absorbed is proportional to the dielectric loss at that frequency, $\varepsilon''(\omega)$, hence the term "loss":

$$q_{eq} = \pi \varepsilon_0 E_0^2 \varepsilon''(\omega)\ . \tag{3}$$

Here, q_{eq} is the energy density absorbed per cycle and ε_0 is the permittivity of vacuum. This is the basis for microwave heating and microwave chemistry (*14*), but it is important to realize that this energy enters the system as work done on the dipoles, not as heat. In a microwave oven using 2.45 GHz, energy enters the sample faster than it is released to the surroundings, and the sample temperature thus increases.

Interestingly, such an energy absorption experiment can be performed at much lower frequencies and under conditions at which a thin sample is in contact with metal electrodes of relatively high heat capacity. In such a dielectric capacitor arrangement, energy uptake from the field is slow compared with the rate of heat transport to an isothermal reservoir, and the high-field effect can be studied without impacting the sample temperature (*15*). For an alternating field of frequency ω, heterogeneous dynamics will lead to only those domains absorbing energy from the field whose loss at ω is considerable, while homogeneous dynamics will result in energy absorption that is uniform across the sample volume (*16*). The technique of dielectric hole burning has exploited this difference and demonstrated spectral selectivity, i.e., the effect that the frequency of the field determines which part of the spectrum is modified by the field (*16, 17*). This spectral selectivity implies that fast and slow modes are relatively independent, which is the hallmark of the heterogeneous nature of dynamics observed as a general feature of supercooled liquids (*18–21*).

The observed field-induced modification resulting from energy absorption follows a model that assumes an increase in local temperature, better described as fictive temperature T_{fic}, which can be determined from the energy absorbed and the heat capacity of the mode in question via

$$T_{fic} = T + \varepsilon_0 E_0^2 \Delta\varepsilon \,/\, (2\rho \Delta C_p)\ . \tag{4}$$

where ρ is the density and ΔC_p the glass-to-liquid heat capacity step. Because this high field effect results in a rise of the fictive temperature, $T_{fic} > T$, it leads to a reduction in time constants and faster dynamics without changing T itself (*16, 17*).

Quantitative models of the result of energy absorption led to the realization that impedance or permittivity spectra at high electric fields will also yield similar effects (*21*). This had subsequently been verified experimentally for numerous supercooled liquids (*22, 23*). The signature of this energy absorption is an increase of the loss for frequencies exceeding the peak value, $\nu > \nu_{max}$, as shown in Figure 3. Consistent with the concept of increased fictive temperatures, it can be shown that the origin of this change in spectral shape is based on shifting relaxation time constants to lower values (i.e., higher frequencies) rather than changing amplitudes (*24*). Note that elevating the (fictive) temperature affects more strongly the time constant than the relaxation amplitude of a Debye type mode, and that the field has no measurable effect on the bath or phonon temperature of the sample.

As will be discussed below, the field-induced changes such as those seen in Figure 3 are not fully developed the instant that the high field is applied. Instead, it requires a time on the order of the structural relaxation time for the sample to establish equilibrium with the high electric field (*25*).

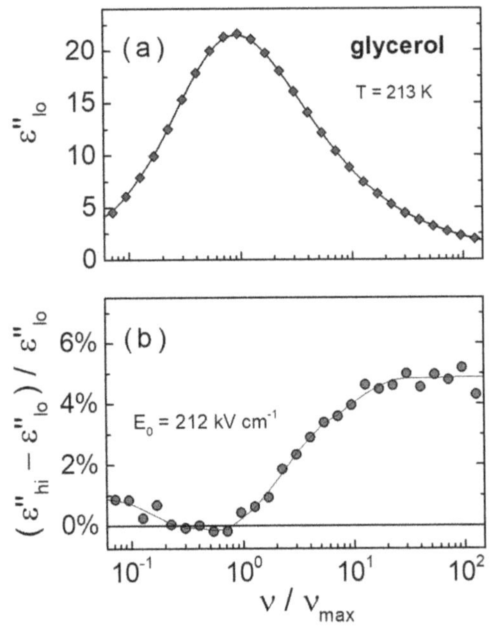

Figure 3. (a) Low-field loss spectrum, $\varepsilon''_{lo}(\omega)$, of glycerol at $T = 213$ K, where the loss peak frequency is $\nu_{max} = 150$ Hz. (b) Effect of energy absorption from an ac field with peak value $E_0 = 212$ kV/cm in terms of the relative change of the dielectric loss on the same frequency scale as the upper panel. Lines serve as guides only. Data from reference (22).

Slower Dynamics Due to High Fields

Moynihan and Lesikar predicted a dependence of T_g on the external static electric field E based on equilibrium thermodynamics, equivalent to a change in the structural relaxation time at constant temperature (26). For the case of a constant volume (before and after relaxation), the change of T_g was assumed to follow

$$\Delta T_g = \varepsilon_0 \, (\Delta[\partial(\chi V)/\partial P]) \, / \, (2V\Delta\alpha) \times E^2, \tag{5}$$

where $\chi = \varepsilon - 1$ is the dielectric susceptibility, V is the volume, P is the pressure, and α the volume expansion coefficient. In this expression, "Δ..." refers to the change from the liquid to the glassy state. For glycerol at $T = 218$ K, equation (5) yields $\chi T_g = +2$mK for a field of $\Delta E = 100$ kV cm^{-1}, which at the time was considered too small to be observable. Nowadays, such changes can be resolved (27).

More recently, Johari presented an approach to a similar field effect on the dynamics, but based on combining the field induced change of thermodynamic entropy with the Adam-Gibbs model (28). According to thermodynamics, the dependence of entropy S on a static electric field E (at constant volume) is given by,

$$\Delta_E S = \upsilon \, (\varepsilon_0/2) \, (\partial \varepsilon_s/\partial T)_V \, E^2 \, , \tag{6}$$

with υ representing the molar volume (29). Because the term $(\partial \varepsilon_s/\partial T)$ is usually negative, the application of a field leads to a reduction of S for the majority of liquids. According to the Adam-Gibbs model (*30*), the super-Arrhenius increase of relaxation times is a matter of exhausting configurational entropy as the temperature is decreased, quantitatively described by

$$\log_{10}(\tau_\alpha / \mathrm{s}) = A + C / (T \times S_{\mathrm{cfg}}(T))\ . \tag{7}$$

Accordingly, a field-induced entropy decrease should lead to increased relaxation times, i.e., slower dynamics and an increase of T_g.

In part, this prediction motivated the search for a change in dynamics brought about by a static electric field. It was found that an electric field does modify relaxation time constants, and in the direction of increasing τ values as predicted (*27, 31*). However, none of the approaches mentioned above provide a quantitative account of the observations (*27, 32*). An example of the signature of this effect is shown in Figure 4, which reveals a field-induced elevation of the loss for frequencies below the peak at ν_{max}, and reduced loss for $\nu > \nu_{max}$, consistent with a peak shift. If only the time constants were altered, one would expect exactly zero change at $\nu = \nu_{max}$. The observation that the relative change vanishes at a frequency just below instead of at ν_{max} is based on the simultaneous effect of dielectric saturation, which reduces the amplitude of $\varepsilon''(\omega)$.

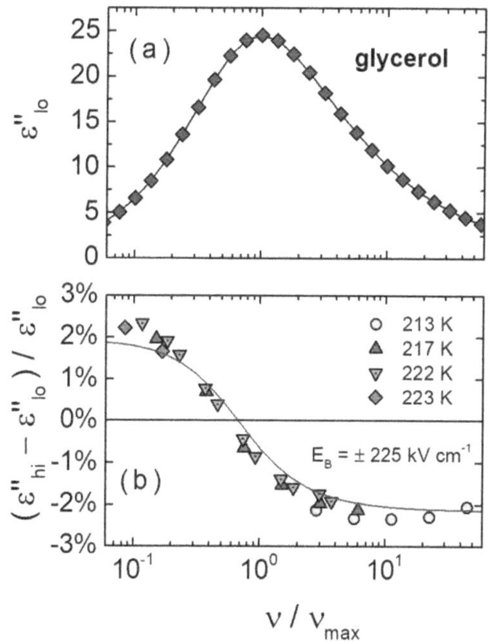

Figure 4. (a) Low-field loss spectrum, $\varepsilon''_{lo}(\omega)$, of glycerol at $T = 213$ K. (b) Effect of entropy reduction and saturation from a dc field with $E_B = 225$ kV cm^{-1} in terms of the relative change of the dielectric loss on the same frequency scale as the upper panel. The results are compiled from different temperatures as indicated. Lines serve as guides only. Data from reference (27).

Time Dependence of Nonlinear Effects

The examples of nonlinear effects shown in the previous Sections all refer to situations in which the sample is in equilibrium with the high electric field. These steady state result are not what would be observed immediately after applying the field, dc or ac. Whether it is dielectric saturation, the chemical effect, or the result of energy uptake or entropy reduction, it requires a time of the order of the structural relaxation time for these steady state levels to be established.

In the case of high amplitude alternating fields with no dc bias, a consequence of this time dependence for the nonlinear effect to fully emerge is that the steady state nonlinear features are diminished at frequencies above the loss peak value, i.e., when the field alternates fast relative to the average relaxation time (*33, 34*). The problem of actually observing this lies in the fact that energy absorption will tend to set in as the main effect at frequencies above ν_{max}, see Figure 3. The nonlinear effect associated with energy absorption differs from the others in that it is not diminished at high frequencies, $\nu > \nu_{max}$. The reason for this effect is that the power associated with a certain field magnitude increases with frequency.

It turns out that observing the time dependence of nonlinear effects can be performed directly using time resolved techniques, where a strong dc field is added to a small ac signal or the amplitude of an ac signal is increased to high levels at time zero (*24, 35*). Subsequent to that field amplitude change, permittivity is recorded and analyzed period-by-period so that ε' or ε'' at a given frequency is known as a function of time with a resolution of one period of the frequency used (*24*). An example of such an experiment is depicted in Figure 5, where a bias field of $E = 174$ kV cm^{-1} is superimposed onto a small field alternating at $\nu = 4$ kHz. Because this frequency is on the high frequency side of the loss peak, the reduction of ε'' observed in Figure 5 is largely the result of the effect associated with entropy reduction. Actually, dielectric saturation also contributes to reducing the loss.

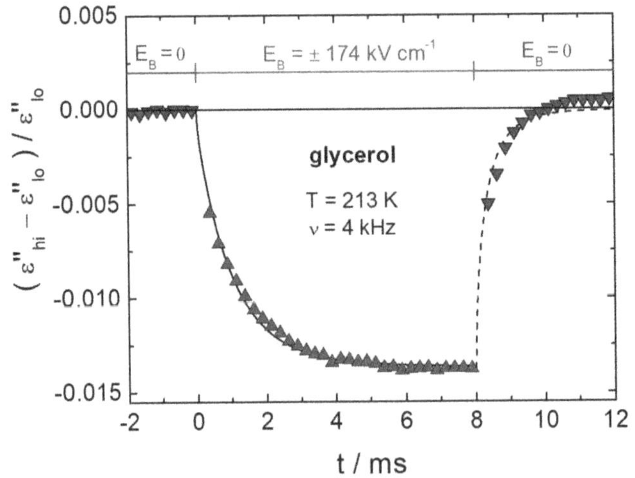

Figure 5. Time resolved relative change of the dielectric loss of glycerol at T = 213 K, where the loss peak frequency is $\nu_{max} = 150$ Hz. Data are measured at a frequency of 4 kHz in response to switching a dc field of $E_B = 174$ kV cm^{-1} on at t = 0 and off at t = 8 ms. Changes reflect the effects of both entropy reduction and saturation. The time dependence is governed by the average structural relaxation time, τ_α, and the rise/fall asymmetry is a matter of the quadratic field dependence. The fit lines for rise and fall curves are based on the same time constant. Data from reference (36).

An unexpected feature of the time resolved change in Figure 5 may be the apparent asymmetry of the response to adding the dc field at $t = 0$ and removing it at $t = 8$ ms. Obviously, the establishment of the nonlinear effect ("rise") beginning at $t = 0$ is slower than the return ("decay") to the low field level of ε'' in the $t = 8$ to 10 ms range. The origin of this feature is the quadratic dependence of the nonlinear effect on the field (*36*). To illustrate this, consider a Debye like or exponential linear response of the form $\phi(t) = exp(-t/\tau)$. The symmetry of the linear responses, $1 - \phi(t)$ and $\phi(t)$, is eliminated by the quadratic dependence, so that the "rise" follows $[1 - \phi(t)]^2$, while the "decay" with $[\phi(t)]^2$ appears as a more rapid change.

This relatively slow approach to a new state is entirely analogous to aging in response to a temperature step. Physical aging is usually observed after a temperature down-jump from an equilibrium state to a temperature at which the system slowly changes its fictive temperature and thus relaxation time constants. The "entropy" effect resulting from applying a static field also leads to changes in the time constants, and these changes can be mapped onto a fictive electric field (*12*). This comparison to aging has been discussed in detail elsewhere (*25, 37*).

Higher Harmonics

In the previous Sections, nonlinear effects have been observed through changes in the permittivity behavior at the fundamental frequency, i.e., at the frequency of the field used to measure the dielectric behavior. When field amplitudes reach levels that drive the system beyond the regime of linear responses, the polarization will include higher harmonics (*38*). For high amplitude alternating fields with no dc bias, these will appear at frequencies 3ω, 5ω, 7ω, and so on. Generally, the polarization response to a field $E(t) = E_0\sin(\omega t)$ can be expressed as a Fourier series with each complex susceptibility, $\hat{\chi}_n^{(k)} = \chi'^{(k)}_n + i\chi''^{(k)}_n$, being associated with a certain harmonic n and power k of the field:

$$P(t) = \sum_{n=1}^{\infty} \sum_{k=n}^{\infty} \left(\chi'^{(k)}_n \sin(n\omega t) + \chi''^{(k)}_n \cos(n\omega t)\right) E_0^k \, . \tag{8}$$

The susceptibility $\hat{\chi}_n$ (without superscript k) is then meant to reflect all contributions at the frequency $n\omega$, regardless of k, and $|\chi_n| = \sqrt{\chi'^2_n + \chi''^2_n}$ is the mathematical modulus of $\hat{\chi}_n$. A typical result for the third harmonic response is shown in Figure 6. It is customary to present such results as the dimensionless quantity $|\chi_3|E^2$, but a limitation of using $|\chi_3|$ is that it is no longer possible to assess whether the χ_3 effect is associated with an increase or decrease of χ_1.

It is a common feature of such $|\chi_3|$ and also $|\chi_5|$ curves to display a low frequency plateau as well as "hump", whose peak is positioned well below the loss peak frequency, ν_{max}, of the linear response permittivity (*4, 40, 41*). As with all nonlinear dielectric effects in polar liquids, this $|\chi_3|$ behavior can not be predicted on the basis of the linear behavior alone, implying that such results bear information about the system that is not contained in the low field counterpart. For systems near a critical point, it has been argued that the qualitative appearance of such a "hump" as seen in Figure 6a is the result of dynamical correlations across considerable length scales that increase with decreasing temperature (*41, 42*), a feature associated with the heterogeneous nature of the dynamics of supercooled liquids (*19, 20*).

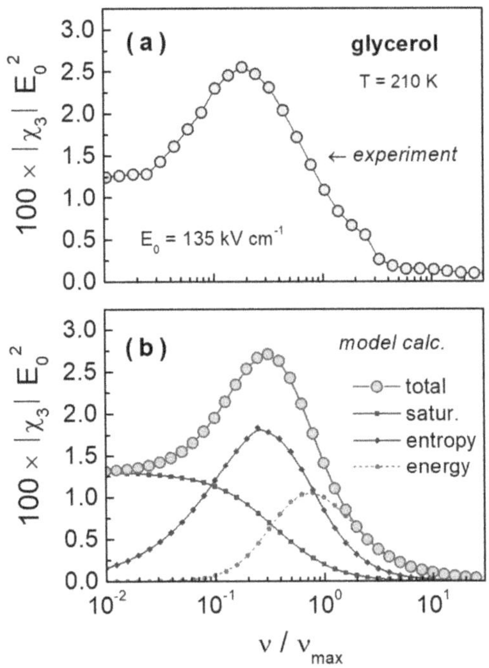

Figure 6. (a) Spectra of the third harmonic susceptibility of glycerol at $T = 210$ K, reported in terms of the quantity $|\chi_3|E_0^2$, based on peak fields of $E_0 = 135$ kV cm^{-1}, versus frequency normalized to the loss peak position, ν_{max}, of the linear response χ''. The line serves as guide only. (b) Model calculation matching the situation of the experiment, showing the total χ_3 effect as well as the individual contributions from saturation, entropy effect, and energy absorption. Data from reference (39).

An alternative and more quantitative approach to this $|\chi_3|$ result is based on the steady state magnitudes of the saturation, energy absorption, and entropy reduction effects, as well as the time dependence discussed in the previous Section. Without adjustable parameters, this model provides a total $|\chi_3|$ spectrum (large symbols in Figure 6b) that is highly reminiscent of the experimental data (39). In principle, analogous calculations are possible for higher harmonics such as χ_5 and χ_7 (39). Moreover, this model facilitates separating the individual contributions, as indicated by the three curves with smaller symbols shown in Figure 6b.

Naturally, the higher harmonics beyond $n = 3$ become progressively smaller in amplitude and are thus more challenging regarding their unambiguous detection. Nevertheless, harmonics up to $n = 5$ have been detected in the permittivity of polar molecular liquids (*41*), and $n = 9$ signals have been reported for the conductivity of ionic materials (*43*).

Crystallization at High Fields

Common to the nonlinear dielectric effects discussed above is the feature that the field-induced modifications disappear within a time range of about the structural relaxation time after removing the external electric field. There are also examples of nonlinear effects that are not reversible in that sense, with dielectric breakdown and the resulting sample failure being an obvious case. An example of relatively persistent field-induced changes without chemical changes is the effect an electric field can have on the crystallization behavior and outcome, even for neat molecular glass-forming materials such as vinyl ethylene carbonate (VEC) (*44*).

For VEC, a polar liquid with static dielectric constant of $\varepsilon_s = 80$ near T_g (*4*), it has been shown that an electric field can accelerate the crystallization rate by more than a factor of 10 (*44*). Interestingly, field-assisted crystallization can lead to a new VEC polymorph, a crystal structure that had not been observed prior to using high fields and that melts at a temperature considerably below the melting temperature of the ordinary crystal. This is demonstrated in Figure 7, which shows isothermal crystallization and the distinct melting events by raising the temperature immediately after the end of the crystallization scans (*45*). Here, a measurement of ε' is used to detect the liquid or crystal volume fraction of the sample, as only the liquid state contributes to the dipole reorientation that brings ε' up from ε_∞ to ε_s. After the sharp rise of ε' at the respective melting points, T_{m2} or T_{m1}, in Figure 7b, the sample remains liquid as the permittivity follows $\varepsilon_s(T)$ indicated by the dashed line.

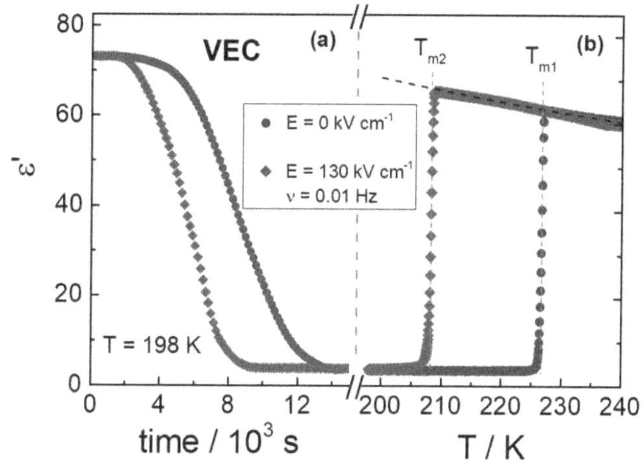

Figure 7. (a) Crystallization kinetics of VEC at T = 198 K with and without the application of a high electric field. In both cases, the sample had been cooled to near T_g in order to nucleate the ordinary crystal. (b) Temperature scan at about 1 K min^{-1} performed immediately after the crystallization shown in the left panel. Overall, the field not only accelerates crystallization but leads to a new polymorph that melts about 20 K below the ordinary (field free) crystal. Data from reference (45).

Concluding Remarks

This survey of nonlinear dielectric effects hopes to convey that there are many facets of how permittivity is affected by high amplitude electric fields, both static and alternating. A high field can decease or increase the dielectric relaxation amplitude, and it can slow down or accelerate the dynamics. None of the effect are very large, even at field near what the material can tolerate in term of its dielectric strength (breakdown field). The reason for this is that the energy μE of a dipole moment μ oriented in a field E usually remains very small compared with $k_B T$, with a typical experimental limit being μE 0.1 $k_B T$. Nonetheless, the resolution of modern dielectric equipment is capable of clearly resolving such small effects.

The nonlinear effects have been discussed separately by the changes that occur. In reality, however, all effects mentioned here may occur simultaneously. Energy absorption can be eliminated by avoiding time-dependent field of high amplitudes, and the chemical effect may be negligible for many systems because it requires the equilibrium of more and less polar species. Therefore, separating the contributions experimentally required selecting the appropriate field (ac versus dc) and analyzing the permittivities at various frequencies.

Generally, non-linear effects provide information on the structure and dynamics that is complimentary to the linear counterparts. Therefore, non-linear dielectric studies have improved fundamental aspects of our understanding of liquids, particularly those near the threshold of becoming a glassy solid. Studying the behavior of materials at high fields will also help to illuminate the precursor states of dielectric breakdown, potentially leading to improvements in high-energy capacitors. Further applications are the possibilities of finding novel polymorphs by crystallization at high electric fields, which is particularly advantageous in the pharmaceutical sciences.

It is beyond the scope of this chapter to discuss the numerous theoretical approaches to nonlinear features in detail. Nonlinear phenomena in more complex systems have been reviewed in a book edited by Rzoska and Zhelezny (*11*). An outline of the state-of-the-art of experimental and theoretical approaches to nonlinear dielectric features has been compiled from various authors in a recent publication (*46*). Other details may also be found in recent reviews on the subject of nonlinear approaches to structure and dynamics of soft materials (*47, 48*). It should also be noted that nonlinear effects have been studied by techniques other than dielectric relaxation, e.g., rheology, mechanical shear, and calorimetry (*46, 48*).

References

1. Kremer, F., Schönhals, A., Eds. *Broadband Dielectric Spectroscopy*; Springer: Berlin, 2002.
2. Richert, R. Supercooled liquids and glasses by dielectric relaxation spectroscopy. *Adv. Chem. Phys.* **2014**, *156*, 101–195.
3. Debye, P. *Polar Molecules*; Chemical Catalog Company: New York, 1929.
4. Young-Gonzales, A. R.; Adrjanowicz, K.; Paluch, M.; Richert, R. Nonlinear dielectric features of highly polar glass formers: Derivatives of propylene carbonate. *J. Chem. Phys.* **2017**, *147*, 224501.
5. Herweg, J. Die elektrischen Dipole in flüssigen Dielektricis. *Z. Phys.* **1920**, *3*, 36–47.
6. Chełkowski, A. *Dielectric Physics*; Elsevier Scientific Pub. Co.: Amsterdam, 1980.
7. Young-Gonzales, A. R.; Richert, R. Field induced changes in the ring/chain equilibrium of hydrogen bonded structures: 5-Methyl-3-heptanol. *J. Chem. Phys.* **2016**, *145*, 074503.
8. Piekara, A.; Piekara, B. Saturation électrique dans les liquides purs et leurs mélanges. *Compt. Rend. Acad. Sci. (Paris)* **1936**, *203*, 852–854.
9. Małecki, J. Non-linear dielectric behaviour and chemical equilibria in liquids. *Electrochim. Acta* **1988**, *33*, 1235–1241.
10. Singh, L. P.; Richert, R. Watching hydrogen bonded structures in an alcohol convert from rings to chains. *Phys. Rev. Lett.* **2012**, *109*, 167802.
11. Rzoska, S. J., Zhelezny, V. P., Eds. *Nonlinear Dielectric Phenomena in Complex Liquids*; Kluwer Academic Publishers: Dordrecht, 2004.
12. Riechers, B.; Richert, R. Structural recovery and fictive variables: The fictive electric field. *Thermochim. Acta* **2019**, *677*, 54–59.
13. Weinstein, S.; Richert, R. Nonlinear features in the dielectric behavior of propylene glycol. *Phys. Rev. B* **2007**, *75*, 064302.
14. Galema, S. A. Microwave chemistry. *Chem. Soc. Rev.* **1997**, *26*, 233–238.

15. Khalife, A.; Pathak, U.; Richert, R. Heating liquid dielectrics by time dependent fields. *Eur. Phys. J. B* **2011**, *83*, 429–435.
16. Schiener, B.; Böhmer, R.; Loidl, A.; Chamberlin, R. V. Nonresonant spectral hole burning in the slow dielectric response of supercooled liquids. *Science* **1996**, *274*, 752–754.
17. Jeffrey, K. R.; Richert, R.; Duvvuri, K. Dielectric hole burning: Signature of dielectric and thermal relaxation time heterogeneity. *J. Chem. Phys.* **2003**, *119*, 6150–6156.
18. Böhmer, R.; Chamberlin, R. V.; Diezemann, G.; Geil, B.; Heuer, A.; Hinze, G.; Kuebler, S. C.; Richert, R.; Schiener, B.; Sillescu, H.; Spiess, H. W.; Tracht, U.; Wilhelm, M. Nature of the non-exponential primary relaxation in structural glass-formers probed by dynamically selective experiments. *J. Non-Cryst. Solids* **1998**, *235-237*, 1–9.
19. Ediger, M. D. Spatially heterogeneous dynamics in supercooled liquids. *Annu. Rev. Phys. Chem.* **2000**, *51*, 99–128.
20. Richert, R. Heterogeneous dynamics in liquids: Fluctuations in space and time. *J. Phys.: Condens. Matter* **2002**, *14*, R703–R738.
21. Weinstein, S.; Richert, R. Heterogeneous thermal excitation and relaxation in supercooled liquids. *J. Chem. Phys.* **2005**, *123*, 224506.
22. Weinstein, S.; Richert, R. Nonlinear dielectric response and thermodynamic heterogeneity in liquids. *Phys. Rev. Lett.* **2006**, *97*, 095703.
23. Bauer, T.; Lunkenheimer, P.; Kastner, S.; Loidl, A. Nonlinear dielectric response at the excess wing of glass-forming liquids. *Phys. Rev. Lett.* **2013**, *110*, 107603.
24. Huang, W.; Richert, R. Dynamics of glass-forming liquids. XIII. Microwave heating in slow motion. *J. Chem. Phys.* **2009**, *130*, 194509.
25. Samanta, S.; Richert, R. Limitations of heterogeneous models of liquid dynamics: Very slow rate exchange in the excess wing. *J. Chem. Phys.* **2014**, *140*, 054503.
26. Moynihan, C. T.; Lesikar, A. V. Comparison and analysis of relaxation processes at the glass transition temperature. *Ann. N. Y. Acad. Sci.* **1981**, *371*, 151–169.
27. Samanta, S.; Richert, R. Dynamics of glass-forming liquids. XVIII. Does entropy control structural relaxation times? *J. Chem. Phys.* **2015**, *142*, 044504.
28. Johari, G. P. Effects of electric field on the entropy, viscosity, relaxation time, and glass-formation. *J. Chem. Phys.* **2013**, *138*, 154503.
29. Fröhlich, H. *Theory of Dielectrics*; Clarendon Press: Oxford, 1958.
30. Adam, G.; Gibbs, J. H. On the temperature dependence of cooperative relaxation properties in glass-forming liquids. *J. Chem. Phys.* **1965**, *43*, 139–146.
31. L'Hôte, D.; Tourbot, R.; Ladieu, F.; Gadige, P. Control parameter for the glass transition of glycerol evidenced by the static-field-induced nonlinear response. *Phys. Rev. B* **2014**, *90*, 104202.
32. Richert, R. Relaxation time and excess entropy in viscous liquids: Electric field versus temperature as control parameter. *J. Chem. Phys.* **2017**, *146*, 064501.
33. Richert, R. Frequency dependence of dielectric saturation. *Phys. Rev. E* **2013**, *88*, 062313.
34. Richert, R. Nonlinear dielectric signatures of entropy changes in liquids subject to time-dependent electric fields. *J. Chem. Phys.* **2016**, *144*, 114501.
35. Pathak, U.; Richert, R. Dielectric loss of poly(vinylacetate) at electric fields of 400 kV/cm. *Colloid Polym. Sci.* **2014**, *292*, 1905–1911.

36. Young-Gonzales, A. R.; Samanta, S.; Richert, R. Dynamics of glass-forming liquids. XIX. Rise and decay of field induced anisotropy in the non-linear regime. *J. Chem. Phys.* **2015**, *143*, 104504.
37. Riechers, B.; Richert, R. Rate exchange rather than relaxation controls structural recovery. *Phys. Chem. Chem. Phys.* **2019**, *21*, 32–37.
38. Déjardin, J. L.; Kalmykov, Yu. P. Nonlinear dielectric relaxation of polar molecules in a strong ac electric field: Steady state response. *Phys. Rev. E* **2000**, *61*, 1211–1217.
39. Kim, P.; Young-Gonzales, A. R.; Richert, R. Dynamics of glass-forming liquids. XX. Third harmonic experiments of non-linear dielectric effects versus a phenomenological model. *J. Chem. Phys.* **2016**, *145*, 064510.
40. Bauer, Th.; Lunkenheimer, P.; Loidl, A. Cooperativity and the freezing of molecular motion at the glass transition. *Phys. Rev. Lett.* **2013**, *111*, 225702.
41. Albert, S.; Bauer, Th.; Michl, M.; Biroli, G.; Bouchaud, J.-P.; Loidl, A.; Lunkenheimer, P.; Tourbot, R.; Wiertel-Gasquet, C.; Ladieu, F. Fifth-order susceptibility unveils growth of thermodynamic amorphous order in glass-formers. *Science* **2016**, *352*, 1308–1311.
42. Berthier, L.; Biroli, G.; Bouchaud, J.-P.; Cipelletti, L.; El Masri, D.; L'Hôte, D.; Ladieu, F.; Pierno, M. Direct experimental evidence of a growing length scale accompanying the glass transition. *Science* **2005**, *310*, 1797–1800.
43. Patro, L. N.; Burghaus, O.; Roling, B. Anomalous Wien effects in supercooled ionic liquids. *Phys. Rev. Lett.* **2016**, *116*, 185901.
44. Adrjanowicz, K.; Paluch, M.; Richert, R. Formation of new polymorphs and control of crystallization in molecular glass-formers by electric field. *Phys. Chem. Chem. Phys.* **2018**, *20*, 925–931.
45. Duarte, D. M.; Richert, R.; Adrjanowicz, K. Exploring the effect of electric fields on the crystallization behavior of a molecular liquid: Vinyl ethylene carbonate. *Cryst. Growth Des.*, submitted.
46. Richert, R., Ed. *Nonlinear Dielectric Spectroscopy*; Springer: Cham, 2018.
47. Richert, R. Nonlinear dielectric effects in liquids: A guided tour. *J. Phys.: Condens. Matter* **2017**, *29*, 363001.
48. Richert, R. Perspective: Nonlinear approaches to structure and dynamics of soft materials. *J. Chem. Phys.* **2018**, *149*, 240901.

Chapter 5

Dielectric Probes: A Versatile Tool for the Study of Molecular Dynamics in Polymers

Michael Wübbenhorst*

Laboratory for Soft Matter and Biophysics, Department of Physics and Astronomy, KU Leuven, Celestijnenlaan 200D, Leuven 3001, Belgium

*Email: wubbenhorst@kuleuven.be

Due to its extraordinary wide dynamic range, covering more than 15 decades in time or frequency, broadband dielectric spectroscopy (BDS) has been widely used to characterize the molecular dynamics in polymers, organic liquids and solids. Since dielectric relaxation processes sense fluctuations of molecular dipoles driven by molecular motions on various length scales, BDS can provide deep insights in molecular dynamic processes involving simple bond rotations, segmental dynamics (the dynamic glass transition) or even whole molecule reorientations. A prerequisite for the application of BDS is the presence of molecular dipoles, which provides the link between molecular motions and the interactions with an external electrical field. Though most of the polymer systems contain at least weakly polar groups in either the main or the side chains, a substantial class of nonpolar polymers, for example polyolefins, does not. A successful strategy to make molecular relaxation "visible" to BDS is the addition of rigid molecules acting as "dielectric probes" in analogy to the well-established fluorescent probe techniques. This approach, developed and refined in the past 15 years, relies on the fact that the strength of a dielectric relaxation $\Delta\varepsilon$ scales according to $\Delta\varepsilon \propto n\mu^2$, i.e. depends on the concentration n and the square of the molecular dipole moment μ^2. Because of this scaling, a substantial enhancement of the dielectric α-relaxation already at "doping" levels less than 1wt% using the highly polar probe molecule *E-4,4'-N,N-dibutylamino-nitro-stilbene* (DBANS) could be achieved as demonstrated for isotactic polypropylene (i-PP), polyethylene (LDPE), atactic polystyrene (a-PS) and polypropylene (a-PP). This chapter will give an introduction into probe-enhanced BDS and will then focus on various specific applications of dielectric probes: i) the study of glass transition dynamics in nonpolar polymers, ii) probe response in relation to dynamic heterogeneities, iii) spatially resolved dynamic information in ultrathin polymer films, and iv) tracer diffusion studied by dielectric probes in multi-layer systems.

Introduction

In the past decades, considerable progress in electronic instrumentation has pushed broadband dielectric relaxation spectroscopy (DRS) forward to a readily available technique for the study of molecular dynamics in many liquids and polymers in an unmatched dynamic range spanning more than 15 decades in frequency (*1*). By virtue of probing orientation fluctuations of molecular dipoles, DRS allows one to study molecular motions on various length scales ranging from individual bond rotations, the cooperative glass transition, and up to relaxation processes involving the contour of the entire polymer coil (normal process). Since DRS covers a wide frequency range and can easily be performed at different temperatures, this technique yields a wealth of information on molecular dynamics in the solid and liquid state.

Typical examples of polymers studied by DRS include traditional amorphous (PMMA, PC) and semi-crystalline polymers (PET, PEN, PEO) (*2–4*), complex, (multi-phase) liquid-crystalline polymers (*5, 6*) as well as nano-composites or supra-molecular polymers (*7*). A prerequisite for the application of dielectric relaxation spectroscopy is the presence of molecular dipoles, which provide the link between specific molecular motions and an external electrical (probing) field. Though most polymer systems contain at least weakly polar groups, there is a substantial class of non-polar polymers like polyolefins, which do not contain dipolar moieties. To overcome this problem, various techniques that induce polar groups by chemical modification like oxidation or chlorination have been proposed (*8, 9*).

An interesting approach for enhancing the dielectric response of a weakly polar system relies on the addition of small quantities of a polar diluent to form an asymmetric binary mixture. Early dielectric studies on such binary mixtures date back to the late 60's and 70's (Williams and co-workers (*10, 11*), Johari and Goldstein (*12*)) and are of persisting interest as proven by recent studies (*13–15*). Williams and Hains studied the dielectric relaxations of dilute solutions of phthalic anhydride and four other guest molecules in ortho-terphenyl (OTP) at diluent concentration between 2.1 and 8.5% (*10*), observing a strong, non-Debye-type α-peak in the supercooled state of OTP. Dielectric relaxations of mixtures of polystyrene and di-n-butyl phthalate (DBP, $9.6 < c_{DBP} < 100\%$) and two other plasticizers (TTP and DOP) were reported by the same authors (*11*). Mixtures of various rigid molecules such as chlorobenzene in cis-decalin were investigated by Johari and Goldstein in the diluent concentration range from 8 – 60 mol% (*12*). Other studies on binary mixtures involved poly(vinyl acetate)/p-nitroaniline (*p*NA) mixtures (*13*), poly(n-alkyl-methacrylate)s/*p*NA (*14*) or various rigid probe molecules (e.g. picoline, tert-butyl-pyridine, cyanobenzene) in apolar styrene oligomers (*15*).

The majority of the previously mentioned work on mixtures of rigid polar molecules and an apolar glassy or viscous matrix was predominantly motivated in the context of understanding the Johari-Goldstein (JG)-β-relaxation and correlations with the α-process as e.g. predicted by the Coupling Model (CM) (*16*). For this purpose, probe concentrations of typically 10% and more, in a few cases down to 2% were considered and revealed valuable insights in the dynamics of supercooled liquids involving effects like probe induced spectral broadening of the α-relaxation, partial relaxation of the rigid probes and many others.

The approach by van den Berg and co-workers, using the polar, rigid rod-type probe N,N-dibutylamino-E-nitro-stilbene (DBANS, cf. Figure 1) as a probe for segmental dynamics in apolar polymers at "doping" level concentrations below 1 mol% (*17*) was inspired by a different idea, namely to utilize dielectric probes as a *tool* like fluorescent labels or any other type of markers in condensed matter. For this purpose, probe molecules were added at concentrations where probe-induced effects on the matrix dynamics can be neglected – this in contrast of most work on mixtures (see previous section). In their first paper (*17*) the authors demonstrated and validated their approach by using DBANS at concentrations between 0.1 and 1.0 wt% in combination with the three apolar polymers low density polyethylene (LDPE), isotactic polypropylene (i-PP) and atactic polystyrene (a-PS). Since the rigid core of DBANS lacks any internal degree of freedom that could give rise to intrinsic probe relaxations, no dielectric activity of the probe was found in the glassy state of the polymers. Above the glass transition temperature, T_g, the cooperative segmental motions, i.e., the dynamic glass transition, cause large scale fluctuations of the polar axis of the rigid probe molecules, while other, local processes and the dynamics involving the crystalline phase do not affect the dielectric probe response. In polystyrene, the dielectric probe was found to "amplify" solely the intrinsic relaxation behaviour of the matrix polymer by specifically enhancing the dielectric glass transition (α-)process. The increase in dielectric strength was shown to follow linearly the probe concentration up to about 1.0%. The relaxation time of the probe coincided with the mean relaxation time of undoped polystyrene within one order of magnitude in frequency.

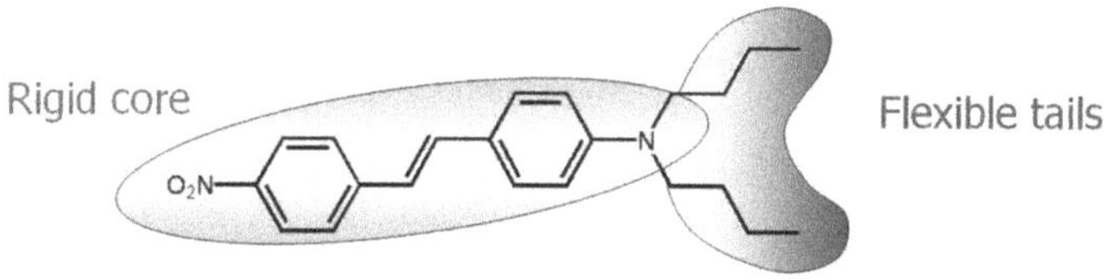

Figure 1. Chemical structure of 4,4'-N,N-dibutylamino-nitro-stilbene. The shaded areas are an artistic impression of the hydrodynamic volume of the rigid (blue) and flexible (red) part of the molecule.

For an optimal performance, a dielectric probe molecule has to meet a number of requirements: (i) first, the probe molecule must be rigid to eliminate dielectric relaxations associated to intramolecular motions, (ii) it should have a large permanent dipole moment and (iii) it must possess a sufficient solubility in an apolar environment.

According to the above three criteria, a rigid-rod type, conjugated molecular design seems to be the natural choice for an effective dielectric probe molecule. A high dipole moment is the second prerequisite because of the favourable scaling law for the probe related dielectric relaxation strength $\Delta\varepsilon$ according to

$$\Delta\varepsilon \propto n\mu^2 \qquad (1)$$

Where n is the probe concentration and μ^2 represents the squared molecular dipole moment. Because of this μ^2 dependence, probe molecules with a high dipole moment (> 5 D) are able to generate a considerable dielectric response already at "doping" levels less than 1wt% as we will discuss in detail in the following sections. Using probes at such low concentrations also ensures a good solubility and thus good probe dispersion even in case of unfavorable probe/polymer interactions as the miscibility of highly asymmetric binary mixtures is governed by the entropy of mixing.

This book chapter will give an introduction into probe-enhanced BDS and will then focus on various specific applications of dielectric probes: i) the study of glass transition dynamics in nonpolar systems, ii) probe response in relation to dynamic heterogeneities, iii) Spatially resolved dynamic information in ultrathin polymer films, and iv) tracer diffusion studied by dielectric probes in multi-layer systems.

Experimental

Sample Preparation

In this section, the general strategy of "doping" polymeric samples with a dielectric probe will be described briefly for the example a-PS, while specific details regarding sample preparation will be mentioned later and can be found in (*17–21*). Starting from a commercial grade of polystyrene, mixtures of a-PS with probe were prepared by dissolving in spectroscopic grade dichloromethane and solution casting on a glass plate. The plates were left to dry for 24 h at room temperature, then 48 h at 50 °C and 20 mmHg. The dry sheets were cut with a razor blade and a drop of water was added to release the sheets from the glass substrate. Squared pieces of 1 cm^2 were then cut and pressed together with 100 μm glass fiber spacers between circular brass electrodes (Ø = 2 cm) resulting in samples with a well-defined geometry and dimensional stability for DRS experiments.

Ultrathin films were prepared by spin-coating on a glass or mica-substrate instead of casting (*21*). For labelled thin films (*22, 23*), no mixing was required since the label concentration was given by the co-polymer composition.

Dielectric Relaxation Spectroscopy

Dielectric experiments were performed using a high resolution dielectric analyzer (ALPHA AN, Novocontrol Technology) in combination with a Novocontrol Quatro temperature system providing control of the sample temperature with an accuracy better than 50 mK. Dielectric measurements were performed at typical frequencies between 0.1 Hz and 10^7 Hz. Temperature dependent experiments were realized by consecutive isothermal frequency sweeps in the temperature range from +200 °C to -120 °C in steps of 5 K, which resulted in an effective cooling rate of about 0.5 K/m.

Spectra were further analysed employing usual fit procedures based on one or more generic relaxation functions such as the Havriliak–Negami (*24*) relaxation function (Eq. 2) using the Levenberg–Marquardt (*25, 26*) algorithm:

$$\varepsilon'' = -Im\left\{\frac{\Delta\varepsilon}{(1+(i\omega\tau)^a)^b}\right\} + \frac{\sigma}{\varepsilon_0\omega} \qquad (2)$$

Here, $\Delta\varepsilon$ and τ denote the relaxation strength and the mean relaxation time of the relaxation process. The two shape parameters a and b, which determine the logarithmic slope of the low frequency loss tail a and the high frequency loss tail $-a{\cdot}b$, are related to the underlying distribution in relaxation times. The second term in Eq. 2 accounts for ohmic conduction. In case of strongly overlapping peaks, derivative techniques or 2-dimensional fitting based on simultaneous fitting in the frequency- and temperature domain were applied (*27*).

Results and Discussion

Studying the Glass Transition Dynamics in Weakly Polar Polymers

The effect of DBANS in the low-concentration limit on the dielectric permittivity of a-PS at f=13 kHz is demonstrated in Figure 2. In the range from -120 °C (not fully shown in Figure 2) up to T_g + 25K, all curve collapse notwithstanding differences in "doping" level and reveal the volumetric glass transition as a kink in the slope $d\varepsilon'/dT$. Between 120 and 150 °C, the dielectric α-process manifests as a step in $\varepsilon'(T)$, the height of which increases monotonically with increasing probe concentration. As seen from Figure 2, a probe content of just about 0.2 wt% (interpolated) results in a doubling of the strength $\Delta\varepsilon_\alpha$ indicating that DBANS acts as a real probe that "amplifies" the intrinsic segmental dynamics.

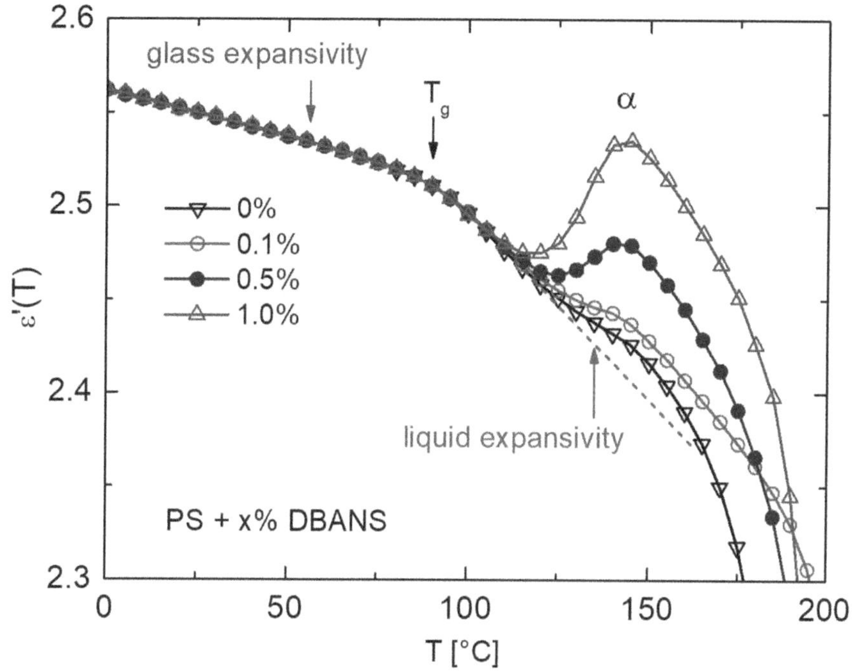

Figure 2. Dielectric constant ε' of undoped PS and PS/DBANS blends at f = 13 kHz as a function of temperature. Reproduced with permission from reference (17). Copyright 2004 American Chemical Society.

A second example taken from van den Berg *et al.* displays isochronal dielectric loss curves $\varepsilon''(T)$ for LDPE as function of DBANS content (*17*). Here the effect on the strength $\Delta\varepsilon_\alpha$ is even more pronounced since neat LDPE exhibits no noticeable dielectric activity. Moreover, Figure 3 also reveals that the probe response exclusively couples to the segmental dynamics (α-process), while both the local γ-relaxation and the high-temperature α_c-process related to defect dynamics within the crystalline phase are not sensed by the probe molecule. This unambiguous identification of the α-process around 0°C in Figure 3 to the dynamic glass transition of LDPE, together with the paper by Frübing and co-workers from 2001 (*9*), can be seen as the end of a long lasting controversial debate regarding the actual glass transition temperature of PE that was either claimed to be located below -100°C (based on mechanical relaxation experiments, thermal expansivity of TSDC (*28*)) or around -35°C (this work). Figure 4 summarizes, in an Arrhenius representation, all relaxation time data for the structural relaxation for undoped and DBANS-doped PS, i-PP, and LDPE samples. For each polymer, the relaxation times nearly coincide independent from the probe content with the little exception LDPE (see discussion in (*17*)), all systems confirming clear VFT-behaviour as the signature of the dynamic glass transition process. Another striking feature of Figure 4 is the nearness of the τ-data for i-PP and LDPE. This proximity of both glass transitions is actually expected when considering the strong analogy regarding chain flexibility and the common lack of specific intermolecular interactions (such as H-bonds, dipole-dipole interactions) supporting once more the assignment of the α-process of LPDE to the polymer segmental mobility.

The exclusive sensitivity of DBANS for the dynamic glass transition of polyolefins and PS can be applied to study blends of these polymers. PE, PP and PS are immiscible like most polymer/polymer combinations in symmetric or not too asymmetric blends, which leads to phase separation on the micrometre scale. This phenomenon has been investigated by Sengers et al. (*18*), who were able to study the glass transition processes in PE/PP, PE/PS and PS/PP blends that were doped with 0.5 wt% DBANS. By virtue of detecting and analysing the two, well separated α-processes, the authors could verify full phase separation and even quantitatively recover the blend composition from the α-relaxation strengths.

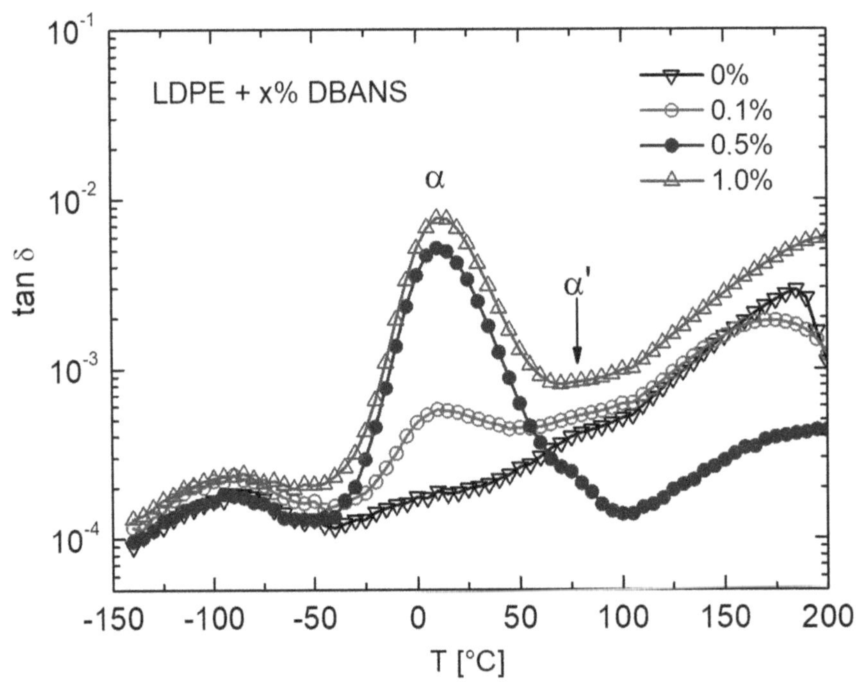

Figure 3. Dielectric loss tangent as function of temperature at f = 13 kHz for undoped LDPE and three LDPE/DBANS blends. Reproduced with permission from reference (17). Copyright 2004 American Chemical Society.

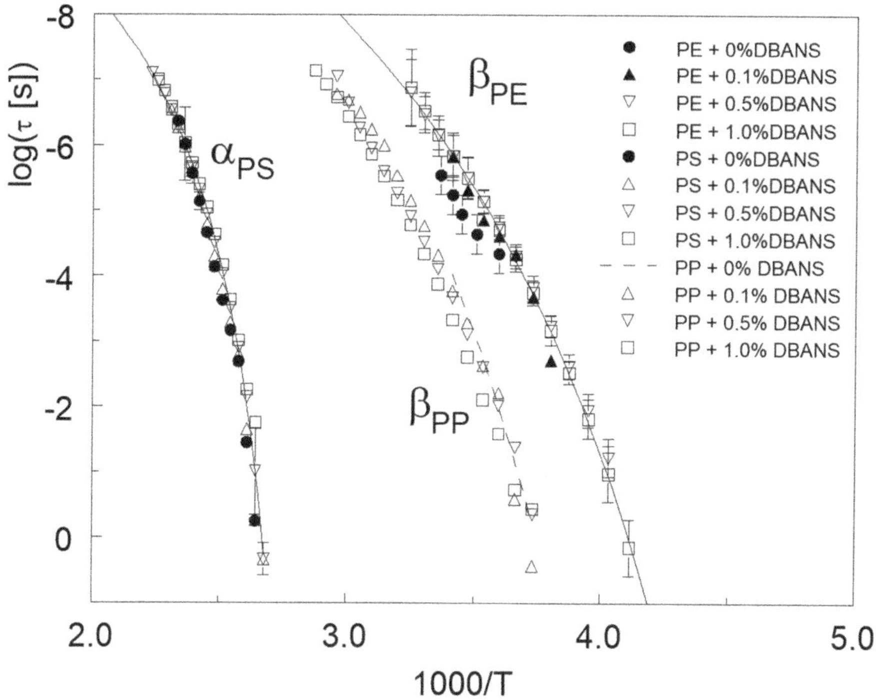

Figure 4. Arrhenius representation of the structural relaxation time for undoped and DBANS-doped PS, i-PP, and LDPE samples. Full details about the fit procedure used are given in (17). Reproduced with permission from reference (17). Copyright 2004 American Chemical Society.

Linearity between Probe Concentration and Probe Response

A crucial check whether a polar molecule behaves as a dielectric probe is the relation between probe concentration and the strength of the probe response. For the probe molecule DBANS, this relation has been investigated in detail for amorphous polymers such as a-PS (*17–19*, *29*), a-PP (*20*) and the semi-crystalline polymers i-PP and LDPE (*17*, *18*). In Figure 5, the concentration dependence of the relaxation strength is displayed for three polymers, i-PP, LDPE and a-PS, which were doped by DBANS at concentrations between 0% and 1% m/m. Figure 5 essentially confirms a linear relation between the probe concentration and the intensity of the primary relaxation process for LDPE and PS. For i-PP, the same trend was found up to a probe concentration of 0.5%; however, at 1% m/m, the relaxation strength levels off. This anomaly was discussed as a possible effect of the semi-crystalline nature of i-PP that leads to locally enhanced probe concentrations up to 1.67 % m/m for i-PP due to its high degree of crystallinity (*17*). From fluorescence spectroscopy it is known that at probe well above 1% m/m concentrations aggregation of probe molecules starts to occur that might result in a fraction of inactive probe molecules and hence a deviation from the linearity in $\Delta\varepsilon(c)$.

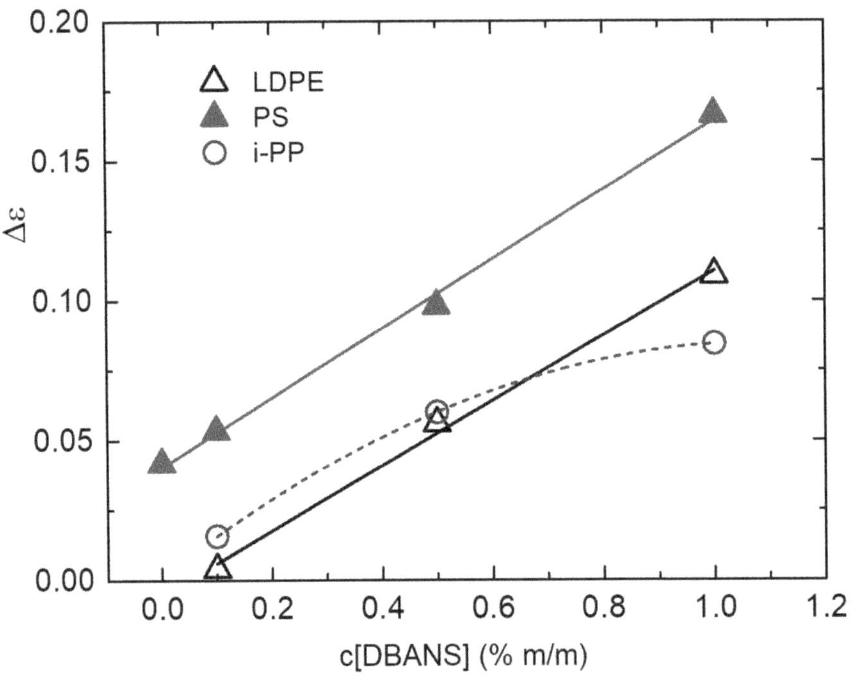

Figure 5. Relaxation strength $\Delta\varepsilon_\alpha$ of the dielectric glass transition process vs probe concentration for LDPE (T = 70°C), PS (T = 135 °C), and i-PP (T = 40 °C). The solid lines for LDPE and PS correspond to actual linear fits to the $\Delta\varepsilon_\alpha$ (c) data, whereas the dashed line in the case of i-PP is just a guide to the eye. Reproduced with permission from reference (17). Copyright 2004 American Chemical Society.

Effect of Probe Size on the Coupling to Specific Dynamic Processes

In the pioneering work by van den Berg et al. (*17*), the choice of a nitro stilbene based probe molecule (DBANS) was originally motivated by its strong fluorescence response along with a large dipole moment. Compared to other fluorescent probes such as pyrene (*30*) or rubrene, DBANS is a rather small probe molecule with a rigid rod shape and a large dipole moment like e.g. disperse red 1 (DR1) (23).

In this section we will discuss the impact of the probe length and concentration on the dielectric response in the model system atactic polystyrene (PS). The weak polarity of the polymer, due to the small dipole moment of the phenyl ethylene moieties, allows for a direct comparison between the intrinsic dielectric relaxation processes of the pristine polymer with those of the doped polymer. Three types of probe molecules, having different lengths of the rigid core l_c of 0.64, 1.3 and 2.0 nm (cf. Figure 6), were studied in combination with PS as the polymer matrix. These molecules all possess relative large dipole moments (The dipole moments of A, B and C were calculated using molecular dynamics software B3LYP/cc-pV02. For A a dipole moment of 8.1 D was calculated, for B this was 10.8 D and for C 12.1 D), while the aliphatic tails ensure a good solubility of these molecules in non-polar media.

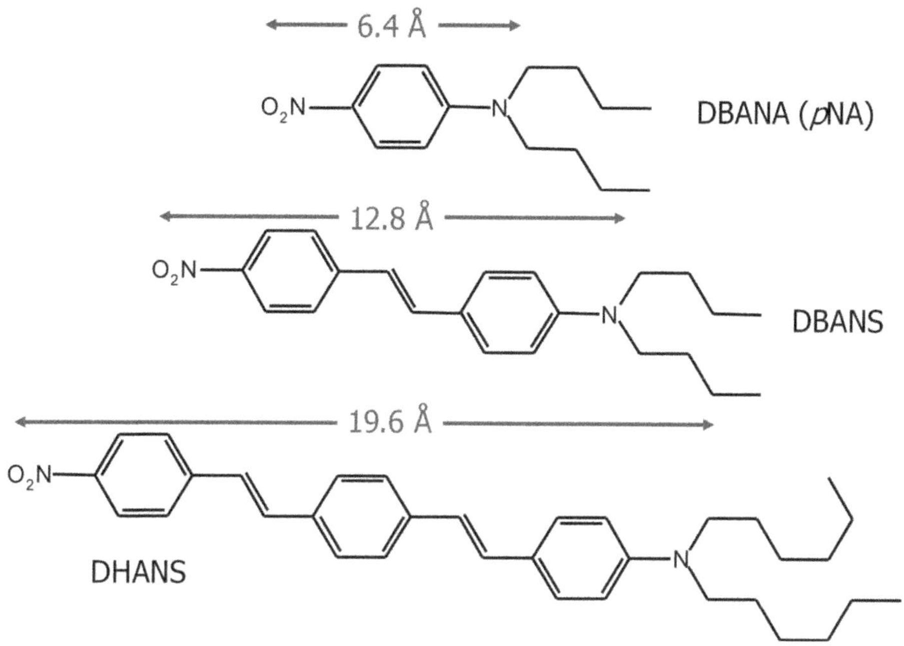

Figure 6. Chemical structures and size indications of the three dielectric probes. Reproduced with permission from reference (19). Copyright 2005 Elsevier.

The essential results from the study by Van den Berg and co-workers (*19*) can be summarized as follows:

i) For all three probes a linear relation between probe concentration and the gain in dielectric strength $\Delta\varepsilon_\alpha$ of the α-relaxation was found.
ii) Compared to the reference probe DBANS (B), the longest probe DHANS (C) revealed similar behaviour regarding amplification of the α-relaxation. Minor shifts in the glass transition temperature were observed and will be discussed in the following in terms of (anti)plasticization.
iii) The short probe, DBANA (A), did couple to the segmental dynamics as well, however, the dielectric spectra revealed an unexpected sub-T_g relaxation process, β*, that was not reported before as an intrinsic dielectric mode for a-PS (cf. Figure 7a).

Further analysis of the two relaxation processes yielded the relaxation strength $\Delta\varepsilon_{\beta^*}$ and $\Delta\varepsilon_\alpha$ for different probe concentrations, which are plotted in Figure 7b. Here, the strength of both relaxation processes obeys a clear linear dependence on the concentration of DBANA and shows the same slope. The latter implies that the probe contribution to both relaxation mode is nearly equal, i.e. the effective, mean square dipole moment $\langle\mu^2\rangle_{\beta^*}$ is about half of the maximum value $\langle\mu^2\rangle_{max}$ corresponding to unrestricted angular displacements of the probe molecule. The offset in $\Delta\varepsilon_\alpha$ for the pure PS sample corresponds to the intrinsic strength of the α-process of the undoped material.

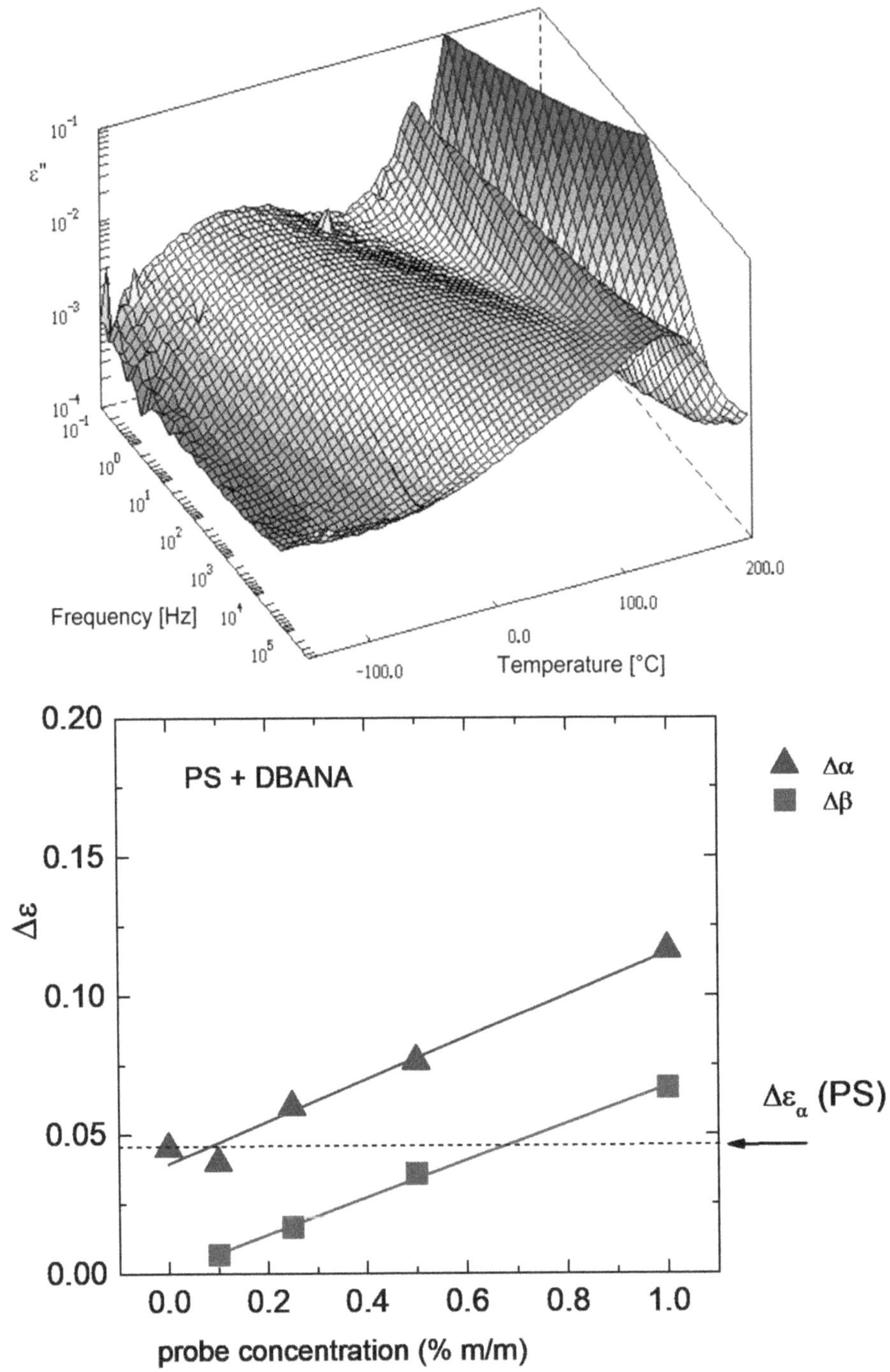

Figure 7. Upper panel: Dielectric loss of DBANA as function of frequency and temperature $\varepsilon''(f,T)$ in a 3D representation. Lower panel: Relaxation strength $\Delta\varepsilon$ of the α- and β^*-process vs. probe concentration for PS doped with A (T = 135 °C). Reproduced with permission from reference (19). Copyright 2005 Elsevier.

The peculiar feature of the shortest probe (A) is its remarkable rotational mobility (β*-process) even in the glassy state of the polymer matrix, where environmental fluctuations on the length scale of the segmental dynamics are frozen in. Since rotational diffusion in a viscous matrix can be ruled out as the mechanism behind the dielectric β*-process, a discrete jump scenario involving large jump angles (*31*) is more likely as supported by two observations: i) In a rotational jump scenario, assuming a rigid molecule being captured in a non-spherical "free volume" cavity, not all angular positions are accessible. Hence the dielectric relaxation strength $\Delta\varepsilon_{\beta}$* does not represent the total dielectric activity of the probe A, as confirmed by the additional probe contribution to the α-process (cf. discussion in previous section). This reasoning is consistent with the earlier observation of two relaxation modes upon addition of a polar, rigid probe to an apolar liquid by Hains and Williams (*11*), who explained the low-temperature mode in their DBP/a-PS mixture as a partial relaxation of the DBP's dipole moment. (ii) There is an abrupt change in the activation energy of the β*-process at about 80 °C, which separates a low temperature Arrhenius part (E_a = 50–65 kJ/mol) from a high temperature branch with an activation energy of approximately 135 kJ/mol (cf. Figure 8). This discontinuity, which was independent from the probe concentration is located slightly below the volumetric glass transition and hence implies that β* is sensitive to the thermal expansion of free volume cavities that changes drastically at T_g as earlier revealed by PALS experiments on PS networks. In other words, opposed to the probe B and C, the short probe has the unique property to sense free volume in the glassy PS matrix as well as to contribute to the "amplification" of the α-relaxation in the polymer melt.

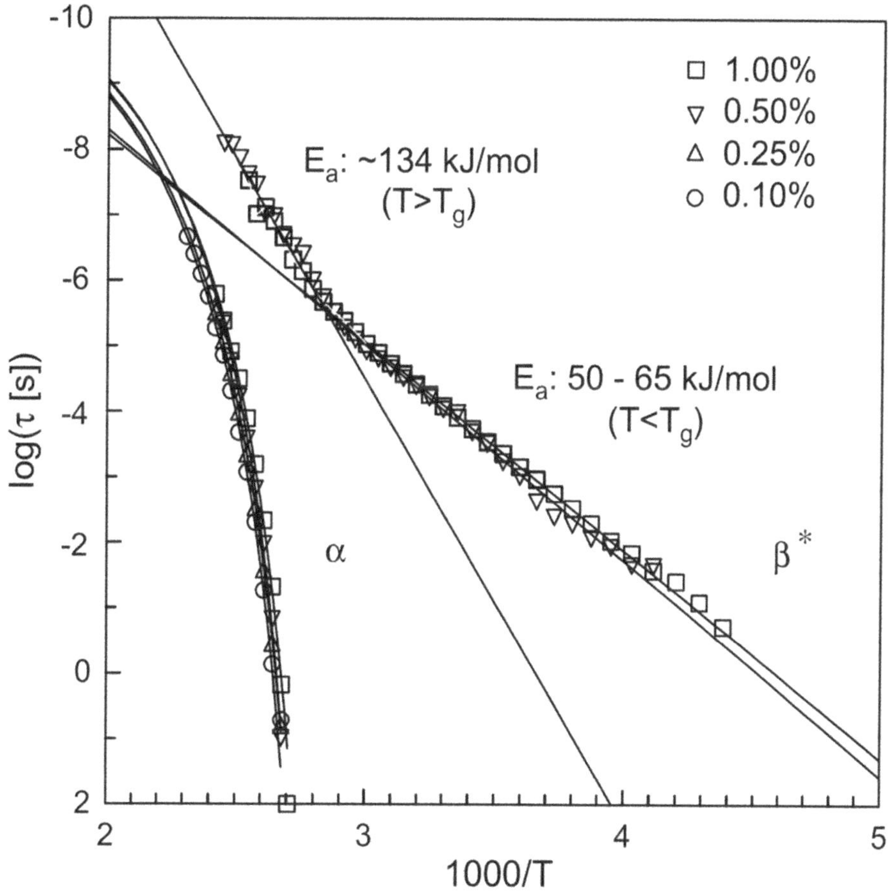

Figure 8. Arrhenius diagram for the α and β-processes of PS doped with DBANA. Reproduced with permission from reference (19). Copyright 2005 Elsevier.*

Figure 9 summarizes and illustrates the various probe regimes for the short vs. long probe in the glassy environment compared to the melt state above T_g.

	long probe	short probe
$T > T_g$	large angular fluctuations in "viscous" matrix → amplification of α-process	large angular fluctuations → contribution to α-process
$T < T_g$	small angular fluctuations → no dielectric response	large angular fluctuations, restricted → β*-process → **free-volume probe**

Figure 9. Interpretation schema for the short- and long-probe response in relation to motional processes above and below the glass transition temperature.

Rotational Diffusion Times as Function of the Rigid Probe Length Well above Tg

So far we have implicitly assumed, based on the efficient coupling of the probe dipole moment to the intrinsic segmental mobility, that the rigid probe A – C behave like a true rotational probe that samples the micro-viscosity of a liquid in a correct way. This point was further investigated by considering the high-temperature/high frequency response of probe A, B and C in relation to their rigid core size (*19*).

Assuming quasi-homogeneous dynamics for high frequencies we can relate the probe relaxation time (cf. Figure 10) to predictions for the rotational diffusion time as a function of the probe length. According to Broersma (32), the rotational diffusion time for Brownian rigid rods, in the dilute regime, obeys the relation $\tau \propto l^3$. Consequently, the relaxation time of the probe-related α-process is expected to slow-down progressively with increasing probe length.

This trend is clearly seen in Figure 10b showing the relaxation time dependencies $\tau_\alpha(T)$ for all probes in PS. At high temperatures, a fairly parallel shift of the τ_α (T) curves towards longer relaxation times from $A \rightarrow B \rightarrow C$ is manifested. To prove the validity of the $\tau \propto l^3$ relation we have evaluated τ/l^3 for relaxation time data at 180 °C. The results are listed in Table 1 and show a pretty universal value for τ/l^3 around 30 ns/nm³ and hence qualify the dielectric probes A, B and C as true rotational diffusion probes obeying Brownian rigid rod dynamics.

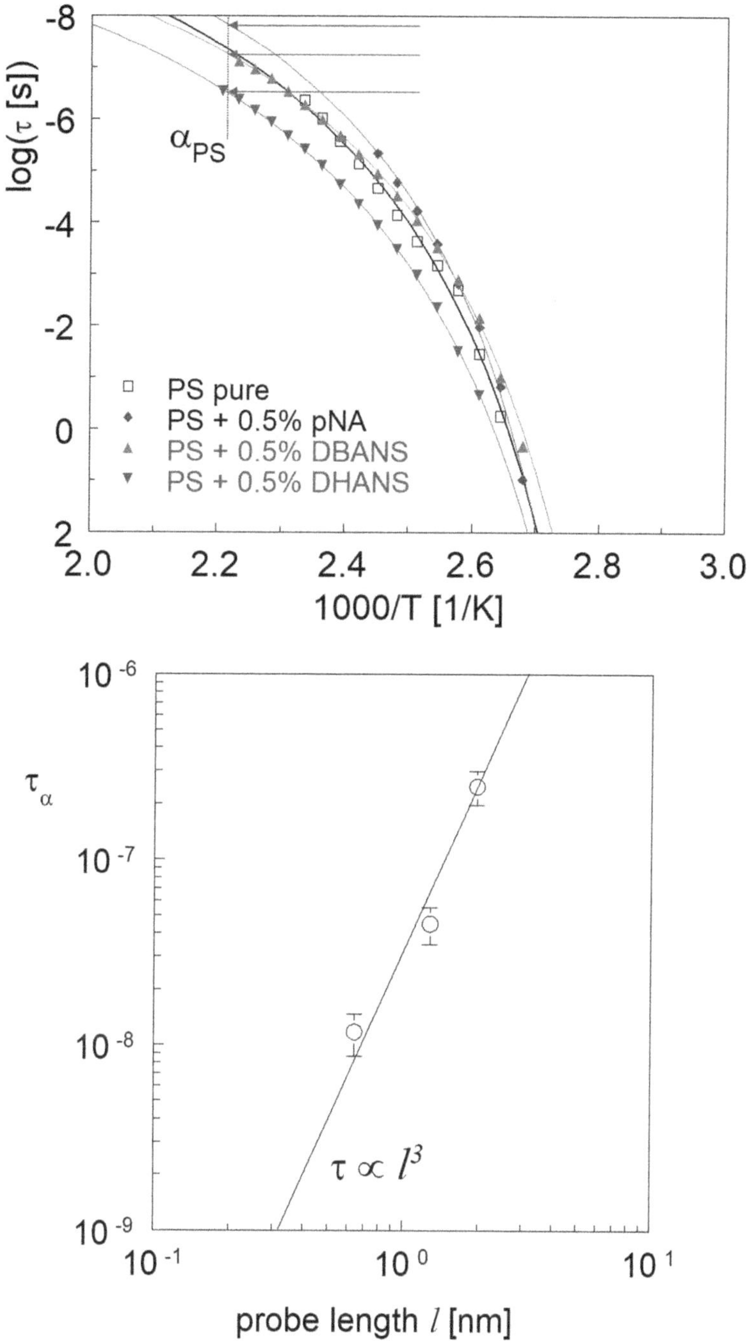

Figure 10. Upper panel (10a): Relaxation time $log(\tau_\alpha)$ vs. 1/T for PS doped with 0.5% for probe A, B and C. Lower panel (10b): $log(\tau_\alpha)$ at T = 180 °C as function of the logarithmic probe length. The solid line corresponds to the slope $dlog(\tau_\alpha)/dlog(l) = 3$. Reproduced with permission from reference (19). Copyright 2005 Elsevier.

Table 1. Dielectric relaxation time τ_α at 180 °C, probe length and reduced probe length for all dielectric probe types[a]

Probe	τ *[ns]*	*l [nm]*	l^3 *[nm³]*	τ/l^3
A	246	1.96	7.50	32
B	44.8	1.28	2.10	21
C	11.7	0.64	0.26	45

[a] Reproduced with permission from reference (*19*). Copyright 2005 Elsevier.

Plasticizing and Anti-plasticizing Effects

In an ideal probe scenario, the probe molecules satisfy three main criteria: i) the probe size matches the length scale of the dynamic process to be sensed (cf. previous section), ii) the probe's molecular interactions with its environment are very similar to those of the monomer-monomer interactions. i.e. the probe possesses highest compatibility with the polymer matrix, and iii) the probe concentration is low enough to neglect probe-probe interactions. In a realistic scenario, the probe will always affect its local environment due to some mismatch regarding shape, size and energetic interactions, which will result in a probe relaxation time that depends on its size and concentration.

In previous work we have shown that there is indeed a small yet significant effect of the probe concentration on both the calorimetric "dielectric" glass transition temperature that was attributed to plasticization of the polymer matrix by the aliphatic tail fraction of the probe molecules (*17*, *19*, *20*). This concentration dependence was particularly investigated and discussed in (*19*) for the three probes of different length, DBANA, DBANS and DHAHS. While the shortest probes revealed a monotonic T_g-down shift with increasing probe concentration, no such monotonic behaviour was found for the longest probe DHANS. Instead, this longest probe showed the smallest T_g-reduction among all three probe versions, which was interpreted as a counterbalance between plasticization by the flexible tail and anti-plasticization caused by very long aromatic core. To account for the different ratio between rigid and flexible part–a consideration that reminds on the basic assumption of the McMillan theory for the smectic A phase (33)–we have plotted the dielectric T_g as function of the probe-specific aliphatic volume fraction in Figure 11. Since all curves show the same slope dT_g/c_{aliph} within the error margin, we may conclude that the flexible probe part is indeed the relevant structural feature governing the probe-induced deviation from the bulk-T_g.

A further discussion of the effect of probe-induced (anti)plasticization considering intrinsic length scale of the cooperative dynamics in relation with mean probe-probe distances can be found in (*19*).

Probe Response in Relation to Dynamic Heterogeneities

Atactic polypropylene (a-PP) is an amorphous polymer with a similar glass transition temperature to its crystallizable isotactic counterpart (i-PP). This polymer exhibits a sub-T_g, β-relaxation with a loss peak maximum around 150K and an α-process around 280K at $f = 13$kHz. The aim of the study by Kessairi et al. (*20*) was to reveal the complete relaxation map of a-PP by means of dielectric spectroscopy and particularly utilizing the dielectric probe DBANS to enhance the response of the α-process.

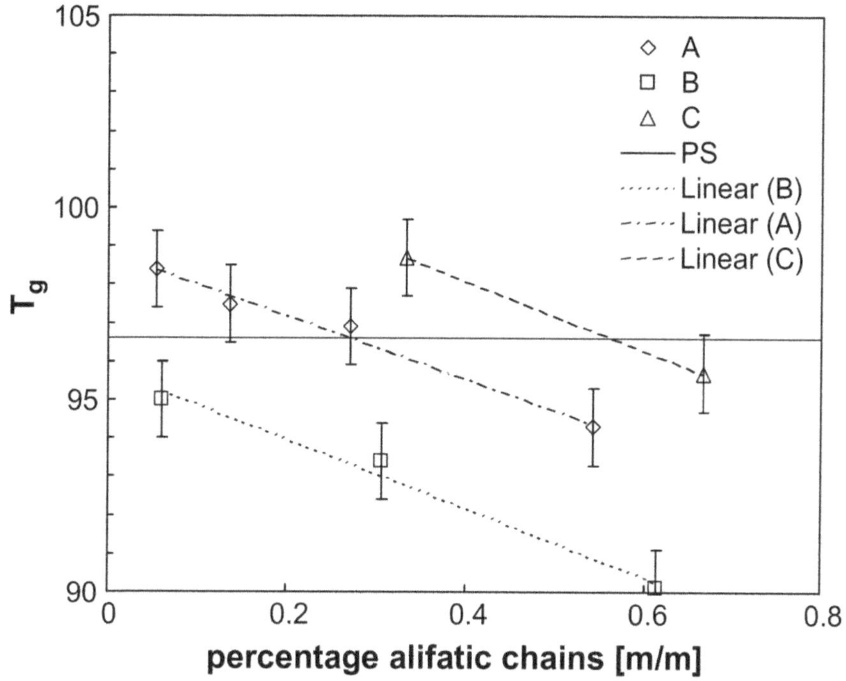

Figure 11. "Dielectric" glass transition temperatures as function of the probe-related aliphatic volume fraction for PS doped with A, B and C. Reproduced with permission from reference (19). Copyright 2005 Elsevier.

Atactic poly(propylene) (a-PP, $M_w = 12\ 000$ g/mol, $M_w/M_n = 2.33$, $T_{g,\ DSC} = 249.2$ K) was purchased from Scientific Polymer Products and used as received. Mixtures of PP and DBANS were prepared by simultaneously dissolving exact amounts of the two components in o-xylene (Aldrich) and subsequently mixing to obtain a homogeneous solution. Polymer films were prepared by casting at T = 110 °C on circular brass electrodes and annealed for 1 h at 150 °C under a continuous nitrogen flow.

For probe concentrations between 0.2% and 1%, a linear trend between probe concentration and the increase in $\Delta\varepsilon_\alpha$ was found, while the dielectric β-process was not affected at all upon addition of DBANS. On the other hand, little plasticizer effects were noticed, which are discussed in detail in the original paper (*20*).

Here we will focus on the spectral shape of the probe-enhanced α-process that is displayed in Figure 12 for four different concentrations of DBANS. Note that the shape parameter a_{HN} represents the symmetric broadening of the dielectric loss peak and coincides with the low-frequency slope of the loss in log-log representation (cf. Equation 2). For pure a-PP, the shape parameter increases monotonously from ~0.4 to 0.5, a trend that is line with known scaling trends for polymeric glass formers (*34*) though the absolute values are on the low side. Adding 0.2% DBANS, the trend is maintained but shifts to the range from 0.5 to 0.6. At this probe concentration, the actual peak shape is influenced by both the intrinsic response and the probe response. At even higher concentrations of DBANS (0.5 and 1.0 wt%), the scenario changes completely: while, in the vicinity of T_g, the shape parameter a_{HN} stays around 0.5 for all probe concentrations, there is a continuous and non-linear increase of a_{HN} towards 0.9 indicating an almost Debye-type probe response in the probe-dominated regime.

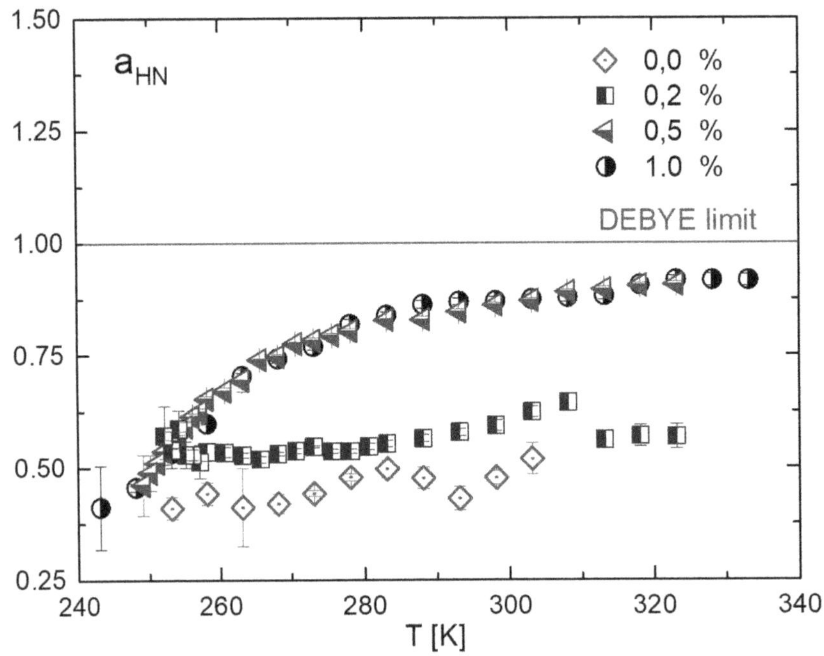

Figure 12. Temperature dependent Havriliak-Negami shape parameter a_{HN} of the structural α-relaxation for neat and DBANS-doped a-PP. Reproduced with permission from reference (20). Copyright 2007 American Chemical Society.

This scaling can be rationalized by comparing the probe size with the (temperature dependent) length scale of dynamic heterogeneities, which is comparable with the intrinsic size cooperativity length ξ(T) in supercooled liquids. Two limiting cases are illustrated in Figure 13. Close to T_g, the size of cooperatively rearranging regions (CRR), which is reported around 2 nm at T_g (35, 36), is larger than the rigid probe length, $L = 1.5$nm in the case of DBANS. In this condition, the probe molecule will sense the local mobility around the probe in a heterogeneous scenario and its individual relaxation time will reflect the local (heterogeneous) mobility.

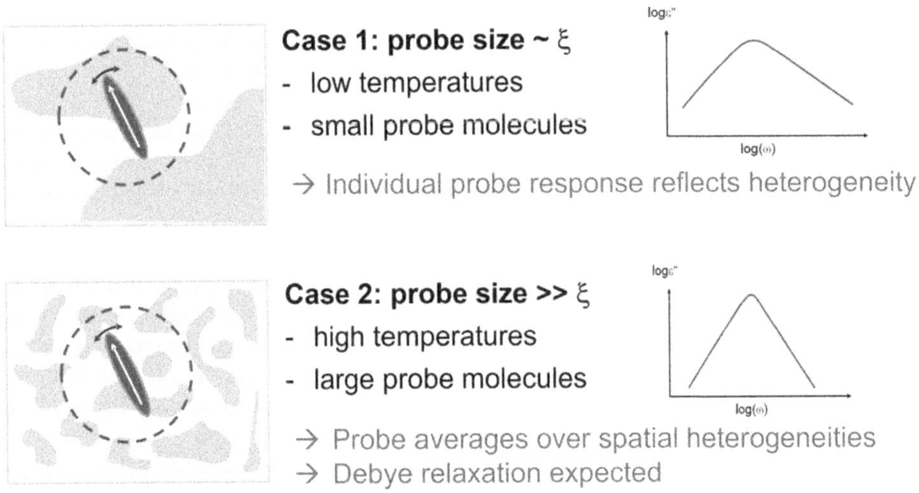

Figure 13. Low-temperature vs. high-temperature scenario for a dielectric probe exposed to a supercooled liquid assuming dynamic heterogeneities, the characteristic size of which varies with temperature.

On the other hand, at higher temperatures and higher frequencies, the probe dynamics is significantly slower than the α-process of the PP matrix. This effect is an inherent feature of a rotational probe as discussed before. In this regime, its hydrodynamic extension (larger than L, the size of DBANS probe ~1.5 nm (*17*, *19*)) will by far exceed the size ξ(T) that gets reduced to maybe 0.5 nm at temperatures well above T_g. Consequently, the probe will average over the mobility of several CRR's (for $L \gg \xi(T \gg T_g)$), between which no dynamic correlation exists. For these reasons, in these conditions, the probe relaxation is not affected by the heterogeneous nature of the glass transition dynamics and enters into a hydrodynamic regime (*37*, *38*).

Spatially Resolved Dynamic Information in Ultrathin Polymer Films

The extensive study in the past decades of polymer layers exposed to interfaces or confined in nanometer-sized geometries has unambiguously revealed the impact of surface interactions, size restrictions and preparative conditions on physical properties like the glass transition temperature (*39*) and the related segmental mobility (*40–43*), flow behaviour and entanglement density (*44*), small molecule diffusion (*45*), crystallization kinetics (*46*, *47*), and physical aging (*48*). While in bulk local chain mobility can be investigated by several experimental techniques, a method probing the segmental dynamics in ultrathin polymer films, particularly in the freely standing geometry, is not readily available for obvious technical reasons.

A successful approach to achieve depth resolution for the glass transition temperature in ultrathin polymer films was proposed and utilized by Torkelson and co-workers using fluorescence techniques (*30*, *49*). By placing a (fluorescent) dye-labeled polymer layer at a specific position within a stack of unlabeled polymer layers, the authors managed to determine the volumetric glass transition temperature averaged over a depth region of typically 15 nm (*30*).

In our experiment, we adopted the multilayer approach (cf. Figure 14) by taking advantage of the sensitivity of dielectric spectroscopy to determine the profile of the structural relaxation time-resolved along the thickness. In fact, this technique, in contrast to dilatometric measurements, is able to investigate the relaxation processes and thus the relaxation time distribution over a broad frequency range for temperatures extending from above to below the bulk glass transition. In addition to the information achieved on a depth profile of T_g, the investigation of the structural relaxation allowed us to map the molecular mobility also by means of the dielectric strength, a quantity proportional to the amount of mobile molecules relaxing on the time scale and the length scale of the dynamic glass transition.

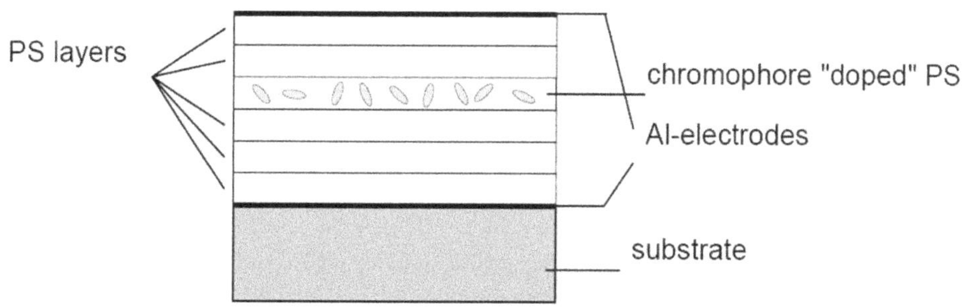

Figure 14. Multilayer sample arrangement containing dielectric probes or a labelled polymer layer at a specific depth position to gain localized dynamic information by DRS.

Depth Specific Segmental Dynamics Probed by Dye-Labelled a-PS

Although the multilayer approach sounds appealing, the experimental realization of a "proper" experiment is challenging, a reason why the number of published studies is still very sparse. One successful application of this principle was demonstrated by Rotella et al. (22) who prepared asymmetric bi-layers and a symmetric tri-layer of a-PS with the purpose to obtain selective dynamic information from the metal/polymer interface in comparison with expected bulk-dynamics in case the labelled film was placed in the centre of the tri-layer arrangement. To maintain sharp probe profiles during extended measurements, a high molecular mass polystyrene-based random copolymer was synthesised that was functionalized with the chromophore DR1 at a mass fraction of about 3% (w/w) (22).

Thin film samples were prepared following the usual procedure for single ultrathin polymer films as described in e.g. (*50*) or (*51*). Bi-layer and tri-layer films were prepared by spin-coating individual layers on mica sheets and subsequently floating them from a water reservoir onto the blank substrate or onto a previously deposited and annealed polymer layer. Water occasionally trapped between consecutive layers was allowed to evaporate at room temperature for a couple of days before the deposition of the next layer. After assembling, the multilayer films were again annealed at 110 °C for 1 h.

The main results of this experiment are seen in Figure 15 that displays the isochronal loss curves in heating and subsequent cooling at 1 kHz of labelled two-layer and tri-layer films of a-PS. For a given frequency, the position of the loss peak is a good measure for the dynamic glass transition temperature. For the tri-layer film, this response mainly originates from the labelled PS-layer in the middle of the stack showing only a marginal upshift in cooling after heating to 150°C, while the strength of the process remains unaltered. In contrast, both asymmetric bi-layers reveal a drastic drop in the intensity $\Delta\varepsilon_\alpha$ to about 50% of its original value, which is accompanied by a significant up-shift in $T_{\alpha,max}$. Since no polar group can escape from the bi-layer films, the only way to rationalize this drop in strength is the immobilization of a significant fraction of polar DR1 dipoles by adsorption of DR1 side groups on the aluminium (oxide) surface. This means that polar labels had to diffuse over the distance of several micrometres to reach the polymer/substrate interface to get subsequently adsorbed on the time scale of the experiment. This finding was the first observation of label adsorption and gave rise to a series of systematic studies on irreversible polymer adsorption on solid substrates as documented in e.g. (*22, 23, 52*) and other papers.

Monitoring Probe Diffusion in a Freely-Standing Film of DBANS-Doped a-PS

As mentioned before, the measurement of dielectric relaxations in ultrathin polymer films in the freely-standing geometry are extremely challenging from the technical point of view. This problem was successfully solved by the use of interdigitated comb electrodes (IDE), on which a freely-suspended film of DBANS-doped a-PS was transferred (*21*). Using the same preparation procedure, another PS-sample (M_W = 160 kg/mol) was studied by repeated heating and cooling cycles with a step-wise increase of the maximum temperature in steps of 10 K. The evolution of the dielectric permittivity $\varepsilon'(T)$ and the dielectric loss $\varepsilon''(T)$ during this temperature cycling is shown in Figure 16.

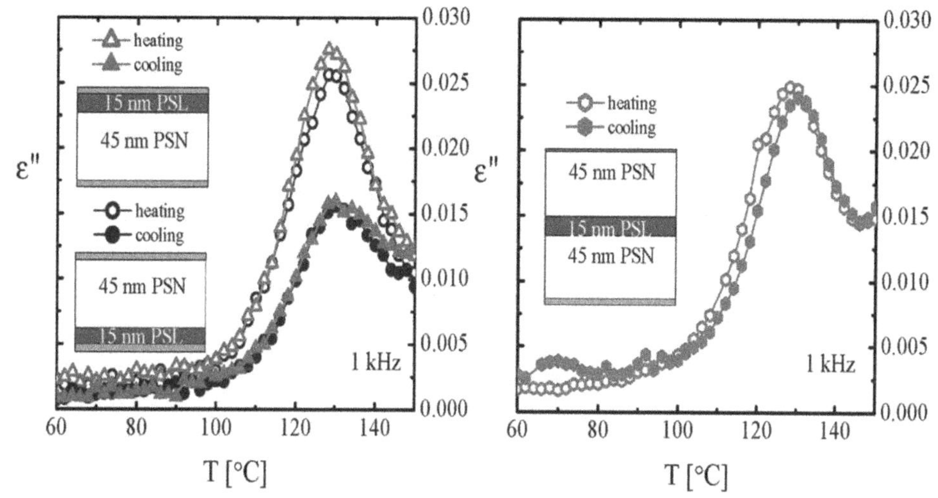

Figure 15. Isochronal loss curves in heating and subsequent cooling at 1 kHz of labelled two-layer (left panel) and tri-layer (right panel) films of a-PS. Reproduced with permission from reference (22). Copyright 2010 American Chemical Society.

Inspecting the permittivity, we see that heating and cooling the sample below 100°C does not alter the traces substantially as also manifested by an invariant peak position in the corresponding loss curve. In contrast, subsequent heating to 110°C, 115°C and 120°C results in a gradual drop of ε′(T) and finally in a stabilization of the response on a minimum level in both the permittivity and the loss. Little discrepancies between the cooling and heating parts belonging to the same cycle are due to a little temperature hysteresis that were due to a poor heat transfer between the freely-standing sample and the special sample cell (ΔT < 2K). The shift of the peak temperature of the α-process T_{max} and as well as the height of the dielectric loss peak ε''_{max} are shown in Figure 17.

This particular experiment demonstrates both the strength and the weakness of the use of small molecule dielectric probes (DBANS) that are not bound to the polymer structure (labels) as discussed in the previous section. The continuous reduction of the dielectric response clearly proves that probe molecules evaporate at elevated temperatures $(T \gtrsim T_g)$, which is a noticeable effect because of the very low film thickness in combination with the translational diffusion coefficient D_t(T) of the probe molecules. Assuming a steady concentration profile of DBANS over the film thickness L reaching from $c = c_{max}$ in the film core to $c = 0$ at both surfaces (cf. Figure 18), one can expect, at a given temperature, an exponential mass loss at the time scale of the diffusion time according to

$$t = \frac{L^2}{D_t} \qquad (3)$$

Combining Eq. 3 with the idea of a thermally activated translational diffusion coefficient that obeys either an Arrhenius-law or a VFT-type dependence, the temperature dependent $D_t(T)$ can be obtained by fitting the "excess" permittivity data

$$\Delta\varepsilon_{ex} = \varepsilon'(T) - \varepsilon'_{\infty}(T) \propto \langle c_{probe} \rangle. \quad (4)$$

Here $\varepsilon'_{\infty}(T)$ denotes the final permittivity trace for a zero probe concentration after extended film annealing. The mean probe concentration (indicated by Figure 18) is proportional to the total probe mass, which represents the physical quantity that can be treated using proven formalisms known from thermally stimulated discharge (TSDC) (53) or thermally stimulated luminescence experiments (54).

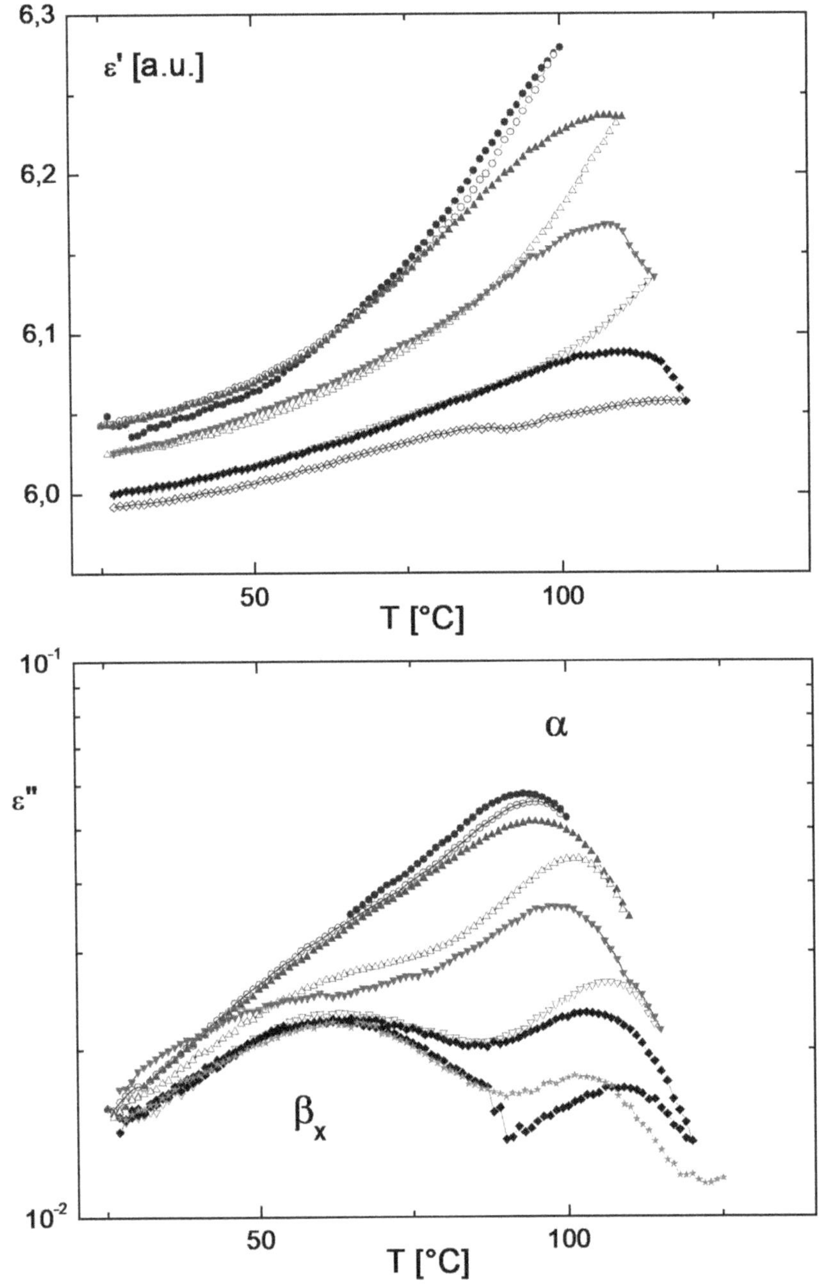

Figure 16. Dielectric permittivity ε'(T) (upper panel, Figure 16a) and dielectric loss ε''(T) (lower panel, Figure 16b) at f = 1kHz for DBANS-doped freely standing film of a-PS. For experimental details refer to (21).

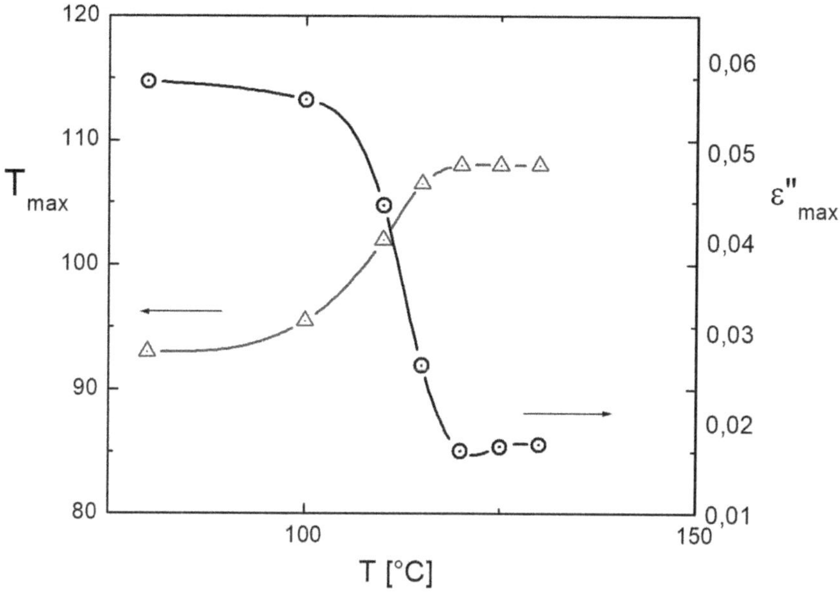

Figure 17. Evolution of the peak temperature of the α-process T_{max} (blue symbols) and with the height of the dielectric loss peak ε''_{max} as function of the maximum heating temperature.

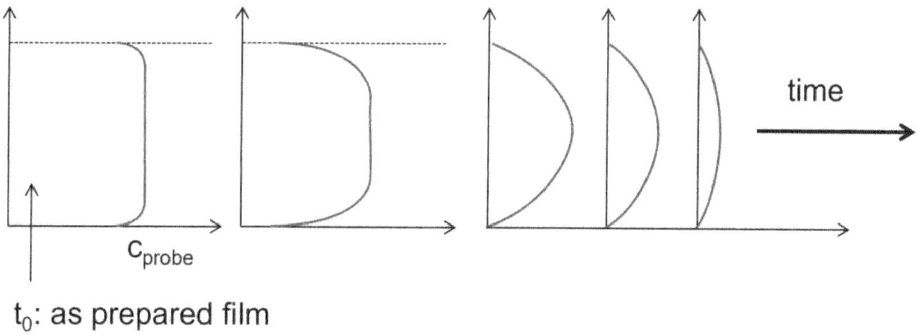

Figure 18. Schematic evolution of the probe distribution over the film thickness as function of time in a dye-doped freely-standing polymer film.

Conclusions

This chapter has presented some history, the basic principles, preparative details and several applications of dielectric probes for the study of dielectric relaxations in nonpolar or weakly polar materials. While asymmetric mixtures of rigid probe molecules and nonpolar solvents have been studied since the 60's, the demonstration that highly polar rigid probes like DBANS are suitable probes for segmental dynamics in nonpolar polymers at "doping" level concentrations below 1 mol% has opened a wide field of applications for the dielectric probe techniques.

We have shown that, at very low concentrations, the response of dielectric probes is linear to their concentration. A particular strength of dielectric probes is their exclusive coupling to segmental dynamics, a feature that can be used to discriminate between cooperative or local dynamics as demonstrated for LDPE.

An essential feature is the probe size: Very short probes (DBANA) were found to act as free volume probe, while the systematic variation of the probe size confirmed that DBANS and longer probe molecules behave as a true Brownian rigid rod.

Analysis of the spectral shape as a function of the temperature and probe content in DBANS doped atactic PP provided strong evidence for the existence and length scale of dynamic heterogeneities confirming other approaches.

Finally, possibilities for the use of dielectric probes for the study of ultrathin polymer films have been explored. Chemically attached probes (labels) in multilayer films were shown to monitor probe-adsorption at polymer/substrate interfaces including its kinetics. Moreover, dielectric (rotational) probes might also be applied for the study of tracer diffusion in ultrathin polymer films as demonstrated for a DBANS-doped freely standing PS film.

References

1. Kremer, F., Schönhals, A., Eds. *Broadband Dielectric Spectroscopy*; Springer-Verlag: Berlin, 2003.
2. Runt, J. P., Fitzgerald, J. J., Eds. *Dielectric Spectroscopy of Polymeric Materials: Fundamentals and Applications*; American Chemical Society: Washington, DC, 1997.
3. Wübbenhorst, M.; de Rooij, A. L.; van Turnhout, J.; Tacx, J.; Mathot, V. New dielectric relaxation process reveals mesomorphic ordering in rapidly cooled poly(ethylene naphthalate). *Colloid Polym. Sci.* **2001**, *279*, 525–531.
4. McCrum, N. G.; Read, B. E.; Williams, G. *Anelastic and Dielectric Effects in Polymeric Solids*; Dover Publications: New York, 1991.
5. Boersma, A.; van Turnhout, J.; Wübbenhorst, M. Dielectric Characterization of a Thermotropic Liquid Crystalline Copolyesteramide: 1 Relaxation Peak Assignment. *Macromolecules* **1998**, *31*, 7453–7460.
6. Simon, G. P.. Dielectric Properties of Polymeric Liquid Crystals. In *Dielectric Spectroscopy of Polymeric Materials: Fundamentals and Applications*; Runt, J. P., Fitzgerald, J. J., Eds.; ACS Professional Reference Series; American Chemical Society: Washington, DC, 1997.
7. Wübbenhorst, M.; Folmer, B. J. B.; van Turnhout, J.; Sijbesma, R. P.; Meijer, E. W. Complex Dynamics of Hydrogen Bonded Self-assembling Polymers. *IEEE Trans. Dielectr. Electr. Insul.* **2001**, *8*, 365–372.
8. Ashcraft, C. R.; R.H., B. A dielectric study of molecular relaxation in oxidized and chlorinated polyethylenes. *J. Polym. Sci., Polym. Phys. Ed.* **1976**, *14*, 2153–2193.
9. Frübing, P.; Blischke, D.; Gerhard-Multhaupt, R.; Khalil, M. S. Complete relaxation map of polyethylene: filler-induced chemical modifications as dielectric probes. *J. Phys. D: Appl. Phys.* **2001**, *34*, 3051–3057.
10. Williams, G.; Hains, P. J. Dielectric study of the motion of molecules in the supercooled liquid state. *Faraday Symp. Chem. Soc.* **1972**, *6*, 14–22.
11. Hains, P. J.; Williams, G. Molecular motion in polystyrene-plasticizer systems as studied by dielectric relaxation. *Polymer (Guildf).* **1975**, *16* (10), 725–729.
12. Johari, C. P.; Goldstein, M. Viscous liquids and the glass transition. II. Secondary relaxations in glasses of rigid molecules. *J. Chem. Phys.* **1970**, 53 (6), 2372–2388.
13. Dionísio, M. S.; Moura-Ramos, J. J.; Williams, G. Molecular motions in poly(vinyl acetate) and in poly(vinyl acetate) p-nitroaniline mixtures. *Polymer (Guildf).* **1993**, *34* (19), 4105–4110.

14. Dionísio, M. S.; Moura-Ramos, J. J.; Williams, G. Dielectric relaxation in poly(n-alkyl methacrylate)s and their mixtures with p-nitroaniline. *Polymer (Guildf).* **1994**, *35* (8), 1705–1713.
15. Capaccioli, S.; Kessairi, K.; Thayyil, M. S.; Prevosto, D.; Lucchesi, M. The Johari-Goldstein β-relaxation of glass-forming binary mixtures. *J. Non-Cryst. Solids* **2011**, *357* (2), 251–257.
16. Ngai, K. L. An extended coupling model description of the evolution of dynamics with time in supercooled liquids and ionic conductors. *J. Phys. Condens. Matter* **2003**, *15* (11), S1107.
17. van den Berg, O.; Sengers, W. G. F.; Jager, W. F.; Picken, S. J.; Wübbenhorst, M. Dielectric and Fluorescent Probes To Investigate Glass Transition, Melt, Crystallization in Polyolefins. *Macromolecules* **2004**, *37* (7), 2460–2470.
18. Sengers, W. G. F.; van den Berg, O.; Wübbenhorst, M.; Gotsis, A. D. Dielectric spectroscopy using dielectric probes: a new approach to study glass transition dynamics in immiscible apolar polymer blends. *Polymer (Guildf).* **2005**, *46*, 6064–6074.
19. van den Berg, O.; Wübbenhorst, M.; Picken, S. J.; Jager, W. F. Characteristic Size of Molecular Dynamics in Polymers Probed by Dielectric Probes of Variable Length. *J. Non.-Cryst. Solids* **2005**, *351*, 2694–2702.
20. Kessairi, K.; Napolitano, S.; Capaccioli, S.; Rolla, P. A.; Wübbenhorst, M. Molecular dynamics of atactic poly(propylene) investigated by broadband dielectric spectroscopy. *Macromolecules* **2007**, *40* (6), 1786–1788.
21. Rotella, C.; Napolitano, S.; Wübbenhorst, M. Segmental mobility and glass transition temperature of freely standing ultrathin polymer films. *Macromolecules* **2009**, *42* (5), 1415–1417.
22. Rotella, C.; Napolitano, S.; De Cremer, L.; Koeckelberghs, G.; Wübbenhorst, M. Distribution of segmental mobility in ultrathin polymer films. *Macromolecules* **2010**, *43* (20), 8686–8691.
23. Rotella, C.; Napolitano, S.; Vandendriessche, S.; Valev, V. K.; Verbiest, T.; Larkowska, M.; Kucharski, S.; Wübbenhorst, M. Adsorption Kinetics of Ultrathin Polymer Films in the Melt Probed by Dielectric Spectroscopy and Second-Harmonic Generation. *Langmuir* **2011**, *27*, 13533–13538.
24. Havriliak, S.; Negami, S. A complex plane representation of dielectric and mechanical relaxation processes in some polymers. *Polymer* **1967**, *8*, 161–210.
25. Levenberg, K. A method for the solution of certain non-linear problems in least squares. *Q. Appl. Math.* **1944**, 2 (2), 164–168.
26. Marquardt, D. W.; Dupont, E.; R, B.; Burrell, G. An algorithm for least-squares estimation of nonlinear parameters. *J. Soc. Indust. Appl. Math.* **1963**, *11*, 431–412004.
27. Wübbenhorst, M.; van Turnhout, J. Analysis of complex dielectric spectra. I. One-dimensional derivative techniques and three-dimensional modelling. *J. Non.-Cryst. Solids* **2002**, *305* (1–3), 40–49.
28. Laredo, E.; Suarez, N.; Bello, A.; Marquez, L. The glass transition in linear low density polyethylene determined by thermally stimulated depolarization currents. *J. Polym. Sci., Part B: Polym. Phys.* **1996**, *34* (4), 641–648.
29. Sengers, W. G. F.; Wübbenhorst, M.; Picken, S. J.; Gotsis, A. D. Distribution of oil in olefinic thermoplastic elastomer blends. *Polymer* **2005**, *46*, 6391–6401.

30. Ellison, C. J.; Torkelson, J. M. The distribution of glass-transition temperatures in nanoscopically confined glass formers. *Nat. Mater.* **2003**, *2* (10), 695–700.
31. Döß, A.; Paluch, M.; Sillescu, H.; Hinze, G. Dynamics in supercooled polyalcohols: Primary and secondary relaxation. *J. Chem. Phys.* **2002**, *117* (14), 6582–6589.
32. Broersma, S. Rotational diffusion constant of a cylindrical particle. *J. Chem. Phys.* **1960**, *32* (6), 1626–1631.
33. McMillan, W. L. Simple molecular model for the smectic A phase of liquid crystals. *Phys. Rev. A* **1971**, *4* (3), 1238–1246.
34. Schönhals, A.; Kremer, F.; Schlosser, E. Scaling of the relaxation in low-molecular-weight glass-forming liquids and polymers. *Phys. Rev. Lett.* **1991**, *67* (8), 999–1002.
35. Donth, E. *The Glass Transition. Relaxation Dynamics in Liquids and Disordered Materials*; Springer: Berlin Heidelberg, 2001; Vol. 48.
36. Ediger, M. D. Spatially heterogeneous dynamics in supercooled liquids. *Annu. Rev. Phys. Chem.* **2000**, *51*, 99–128.
37. Wang, L. M.; Richert, R. Exponential probe rotation in glass-forming liquids. *J. Chem. Phys.* **2004**, *120* (23), 11082–11089.
38. Huang, W.; Richert, R. Dynamics of glass-forming liquids. XI. Fluctuating environments by dielectric spectroscopy. *J. Chem. Phys.* **2006**, *124* (16), 164510.
39. Forrest, J.; Dalnoki-Veress, K.; Stevens, J.; Dutcher, J. Effect of Free Surfaces on the Glass Transition Temperature of Thin Polymer Films. *Phys. Rev. Lett.* **1996**, *77* (10), 2002–2005.
40. Fukao, K.; Miyamoto, Y. Glass transitions and dynamics in thin polymer films: Dielectric relaxation of thin films of polystyrene. *Phys. Rev. E - Stat. Physics, Plasmas, Fluids, Relat. Interdiscip. Top.* **2000**, *61* (2), 1743–1754.
41. Fakhraai, Z.; Forrest, J. A. Measuring the surface dynamics of glassy polymers. *Science* **2008**, *319* (5863), 600–604.
42. Napolitano, S.; Lupaşcu, V.; Wübbenhorst, M. Temperature dependence of the deviations from bulk behavior in ultrathin polymer films. *Macromolecules* **2008**, *41* (4), 1061–1063.
43. Serghei, A.; Kremer, F. Confinement-induced relaxation process in thin films of cis-polyisoprene. *Phys. Rev. Lett.* **2003**, *91* (16), 165702.
44. Shin, K.; Obukhov, S.; Chen, J.-T.; Huh, J.; Hwang, Y.; Mok, S.; Dobriyal, P.; Thiyagarajan, P.; Russell, T. P. Enhanced mobility of confined polymers. *Nat. Mater.* **2007**, *6* (12), 961–965.
45. Pu, Y.; White, H.; Rafailovich, M. H.; Sokolov, J.; Patel, A.; White, C.; Wu, W. L.; Zaitsev, V.; Schwarz, S. A. Probe diffusion in thin PS free-standing films. *Macromolecules* **2001**, *34* (24), 8518–8522.
46. Capitán, M. J.; Rueda, D. R.; Ezquerra, T. A. Inhibition of the crystallization in nanofilms of poly(3-hydroxybutyrate). *Macromolecules* **2004**, *37* (15), 5653–5659.
47. Napolitano, S.; Wübbenhorst, M. Slowing down of the crystallization kinetics in ultrathin polymer films: A size or an interface effect? *Macromolecules* **2006**, *39* (18), 5967–5970.
48. Priestley, R. D.; Ellison, C. J.; Broadbelt, L. J.; Torkelson, J. M. Structural relaxation of polymer glasses at surfaces, interfaces, in between. *Science* **2005**, *309* (5733), 456–459.
49. Kim, S.; Roth, C. B.; Torkelson, J. M. Effect of nanoscale confinement on the glass transition temperature of free-standing polymer films: Novel, self-referencing fluorescence method. *J. Polym. Sci., Part B: Polym. Phys.* **2008**, *46* (24), 2754–2764.

50. Lupaşcu, V.; Huth, H.; Schick, C.; Wübbenhorst, M. Specific heat and dielectric relaxations in ultra-thin polystyrene layers. *Thermochim. Acta* **2005**, *432* (2), 222–228.
51. Lupaşcu, V.; Picken, S. J.; Wübbenhorst, M. Cooperative and non-cooperative dynamics in ultra-thin films of polystyrene studied by dielectric spectroscopy and capacitive dilatometry. *J. Non. Cryst. Solids* **2006**, *352* (52–54), 5594–5600.
52. Napolitano, S.; Wübbenhorst, M. The lifetime of the deviations from bulk behaviour in polymers confined at the nanoscale. *Nat. Commun.* **2011**, *2* (1), 1–7.
53. van Turnhout, J. *Thermally stimulated discharge of electrets*; Springer: Berlin, Heidelberg, 1980; pp 81–215.
54. Głowacki, I.; Ulański, J. Simultaneous measurements of thermoluminescence and thermally stimulated currents in poly(N-vinylcarbazole)/polycarbonate blends. *J. Appl. Phys.* **1995**, *78* (2), 1019–1025.

Applied Dielectric Spectroscopy of Polymer Systems

Chapter 6

Vitrification and Physical Aging in Polymer Glasses by Broadband Dielectric Spectroscopy

Daniele Cangialosi[1,2,*]

[1]Centro de Física de Materiales (CSIC–UPV/EHU), Paseo Manuel de Lardizabal 5, 20018 San Sebastián, Spain

[2]Donostia International Physics Center, Paseo Manuel de Lardizabal 4, 20018 San Sebastián, Spain

[*]E-mail: daniele.cangialosi@ehu.eus

Physical aging is an ubiquitous phenomenon in glasses. It takes place on liquids previously cooled down below their melting temperature and subsequently further cooled below their glass transition temperature, T_g, a phenomenon known as vitrification. Here, we present the strategies employed to study vitrification and physical aging in polymers and other glasses by dielectric spectroscopy. Specifically, the effect on the main α relaxation, associated to the glass transition, and on sub-T_g relaxations, in particular the β relaxation, is scrutinized. Finally, we discuss some recent new developments on the phenomenology of physical aging, in particular the complex behavior of physical aging and its acceleration in confined polymers, and the possible contribution that can be provided by dielectric spectroscopy.

Introduction

Glasses are a class of materials of utmost importance as they are used since the origin of civilization in a wide variety of applications, ranging from the most basic to the most advanced technologies. Furthermore glasses are ubiquitous in nature and constitute a significant portion of the solid state matter in the universe (*1*). A sub-class of glassy materials is represented by glass-forming polymers (*2*), whose importance has gained increasing importance since the discovery of the first synthetic polymers at the beginning of the last century. The precondition to obtain a glass is that during the process of its formation crystallization is avoided. Cooling from the liquid state is by far the most common procedure followed to form a glass, a phenomenon known as vitrification or glass transition. Forming a glass by cooling implies that the liquid must be supercooled below the melting temperature, T_M, fast enough to hinder crystallization. While for the vast majority of liquids fast cooling is required to achieve glass formation, there exists a wide range of polymers that do not crystallize under any condition. The main reason is that, due to chain connectivity, polymer crystallization entails chain folding, a process whose complexity bears large activation barriers (*3*). Furthermore, conformational disorder, for instance in atactic polymers, constitutes an additional factor inhibiting crystallization, thereby favoring glass formation.

Once they are formed, glasses are in non-equilibrium and, therefore, their thermodynamic state changes over time, a phenomenon known as structural recovery (*4, 5*) or physical aging (*6–8*). While for some kinds of glasses, including polymers amenable to crystallization, the evolution of a non-equilibrium glass may also entail the transformation into the crystalline state, for non-crystallizing polymers physical aging is the only possible evolution. The important consequence of the change of state resulting from physical aging is that virtually any property sensitive to the glassy thermodynamic state changes with time. Hence, apart from the important implications of the presence of technological properties changing with time, the kinetics of physical aging can be monitored by following the time evolution of these properties. This chapter is specifically devoted to how physical aging impacts all properties associated to the presence of a dipole moment in the glass. Though polymeric glasses will be the main focus of this chapter, the physical aging behavior by dielectric spectroscopy of other kinds of glasses will be considered. This kind of approach relies on the universality of the phenomenology of physical aging, resulting in a common behavior among glasses of different nature.

Starting from these premises, the present chapter is organized as follows: i) we will first provide an overview of physical aging and its connection to vitrification; ii) Secondly, the way dielectric spectroscopy can be fruitfully exploited to provide a signature of the glass transition and to gain insights on physical aging will be presented; iii) Finally, perspectives on what kind of new insights can be delivered by dielectric spectroscopy will be discussed.

Vitrification and Physical Aging

The process of vitrification or glass transition allows transforming a supercooled liquid into a glass. From a thermodynamic viewpoint, when this is done on cooling a supercooled liquid–previously obtained by avoiding crystallization (see scheme in Figure 1, right panel)–vitrification occurs with a kink in a first order thermodynamic property, for instance the specific volume, the enthalpy or the entropy. The temperature at which this kink is observed is commonly referred to as the glass transition temperature, T_g (see scheme in Figure 1, right panel). The presence of a kink in a first order thermodynamic property may be interpreted as a second order thermodynamic transition within the Erhenfest classification (*1*). However, several observations rather indicate that the glass transition is a kinetic phenomenon and not a true thermodynamic transition. Among them, one of the most emblematic is that showing the T_g dependence on the applied cooling rate, β_c, which defines the time scale of the experiment. Specifically, the slower the cooling rate, the lower will also be T_g. This outcome is understood considering the time scale of molecular motions responsible for vitrification, whose time scale becomes increasingly larger on decreasing the temperature. As a result, on lowering the temperature the supercooled liquid requires larger time scales to rearrange and maintain equilibrium and, therefore, lower cooling rates must be applied to this aim.

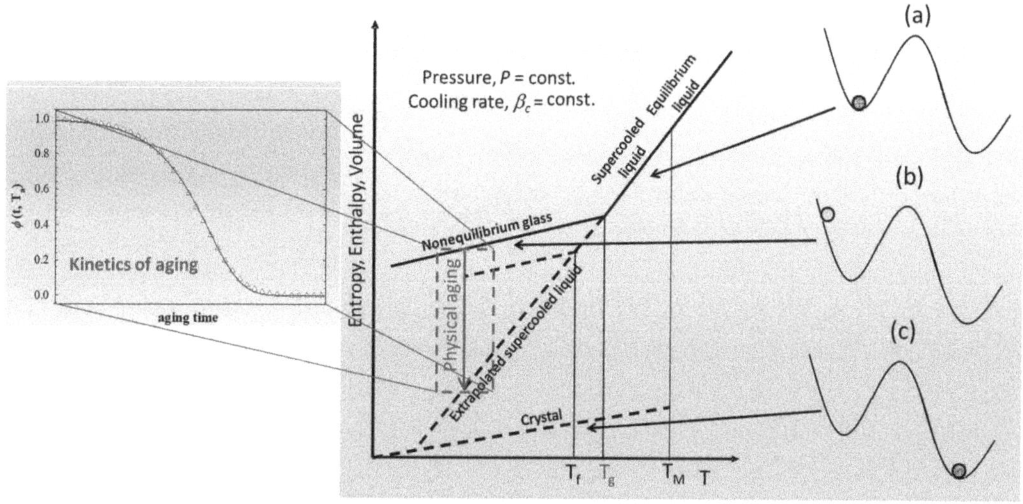

Figure 1. (Right panel) Schematic thermodynamic stability plot of a system that may crystallize or supercooled below the melting temperature, T_M, and form a glass at a given cooling rate, β_c, below T_g. The scheme shows the kinetic path underlying the physical aging of a glass formed at β_c and left evolving isothermally. The T_f of a glass with the thermodynamic state corresponding to a certain degree of aging is shown. The right part of the panel schematically shows the free energy location of a supercooled liquid (a), a glass (b) and a crystal (c). (Left panel) Typical recovery function of any property sensitive to physical aging as a function of aging time.

Due to the mentioned kinetic nature of vitrification, from a thermodynamic viewpoint a glass is a non-equilibrium system, not only with respect to the crystalline state–located at the bottom of the free energy landscape–but also with respect to the supercooled liquid (see scheme in Figure 1, right panel). Hence, one of the possible evolutions of the non-equilibrium glass thermodynamic state is the spontaneous approach towards the closest equilibrium, that is, the supercooled equilibrium state (see scheme in Figure 1). This phenomenon is generally addressed as structural relaxation (*4*) or physical aging (*6, 7, 9*). The way equilibrium is approached, that is the kinetics of physical aging, generally exhibits a sigmoid-shaped behavior as exemplified in the left panel of Figure 1. The position in the thermodynamic landscape of a glass with such a given state is often defined using the concept of fictive temperature, T_f, long ago introduced by Tool (*10*) and subsequently implemented by Moynihan (*11*), who introduced a method to estimate T_f. This is based on the definition of T_f, that is the intercept of the a glassy line, traced from the thermodynamic state of a given glass, with the extrapolated supercooled liquid line (see scheme in Figure 1).

The behavior shown in the left panel of Figure 1 originates from the role of molecular motions involved in the kinetic evolution towards equilibrium. Generally aging carried out close to T_g is uniquely connected to the main so-called α relaxation process, detectable by standard dynamic scattering (*12*) and spectroscopic techniques (*13–15*), including dielectric spectroscopy (*13*). This introduces to one of the main signatures of physical aging, that is, its non-exponential nature, which can be described by the Kohlrausch-Williams-Watts (KWW) equation:

$$\phi(t) = \exp\left[-(t/\tau)^{\beta_{KWW}}\right] \quad (1)$$

Where $\phi(t) = [P(t) - P_0)]/[P_\infty - P_0]$, varying between 0 and 1 is the relaxation function, being P a generic property evolving during physical aging—e.g. in the thermodynamic scheme of Figure 1 the specific volume (or equivalently the density), the enthalpy or the entropy—assuming the value $P(t)$, P_0 and P_∞ after a given aging time t, at zero and infinite aging time, respectively. $\beta_{KWW} < 1$ is the stretching exponent and τ a characteristic time of physical aging, which can be associated to the typical time scale of the α process.

The fact that the approach to equilibrium entails a density increase of the glass introduces to another main signature of physical aging, its non-linear nature (*10, 11, 16, 17*). Specifically the density increase results in an increase of the relaxation time, τ in eq. 1. As a consequence, for aging conducted after cooling through T_g, non-linearity induces an additional stretching beyond that induced by $\beta_{KWW} < 1$, which makes the kinetics of physical aging encompassing many decades in aging time. The non-exponential and non-linear nature of physical aging are incorporated in different phenomenological models, such as the Kovacs-Hutchinson-Aklonis-Ramos (KAHR) (*18*), and the Tool-Narayanaswamy (*16*). models. The latter has been implemented either expliciting the dependence of τ on the glass thermodynamic state in the so-called TNM model (*11*), or keeping it implicit, as in the single parameter aging model (*19, 20*).

General Features of Dielectric Spectroscopy

Dielectric spectroscopy deals with the detection of the response to the application of an electric field (*21*). If this is done in the linear regime, that is applying small fields (*22, 23*), the intrinsic response of the material is obtained. The perturbed property in dielectric spectroscopy is the polarization vector, whose existence relies on the presence of a dipole moment (*21*). In the case of polymers, the vector associated to it can either be parallel or perpendicular to the polymer main chain. Polymers with the former configuration are named Type-A, whereas those with perpendicular configuration belong to the Type-B class. In the latter, fluctuations of the dipole moment are related to the cooperative motion responsible for the α relaxation, as well as more localized secondary motions, generally addressed as β, γ, etc. processes. As will be discussed, vitrification and physical aging are associated to these processes and, therefore, these can be fruitfully used for their study. In Type-A polymers, dipoles parallel to the chain add up together and the resulting dipole moment extends along the whole chain. Its associated dielectric relaxation predominantly conveys information on the slow dynamics the entire chain, observed above T_g. Therefore the dielectric response associated to these dipoles is not relevant for the study of vitrification and physical aging. Still, in Type-A polymers, the dielectric response associated to the α process and secondary relaxations can be fruitfully exploited to study vitrification kinetics and physical aging.

The most common method to carry out dielectric spectroscopy relies on the response to a sinusoidal electric field in a broad frequency range. In this case, a frequency sweep delivers, among other observables, the complex dielectric permittivity, that is the equivalent of the compliance in mechanical spectroscopy: $\epsilon^* = \epsilon' - i\epsilon''$, or equivalently its inverse, the dielectric complex modulus: $M^* = M' + iM''$. ϵ' and M' are the storage part, also known as dispersion, of the dielectric permittivity and modulus, respectively; whereas ϵ'' andM'' are the loss parts. If the experiment is conducted at frequencies lower than those characteristic of all molecular motions in the system, the storage part of the permittivity will assume its completely relaxed value, known as static permittivity, ϵ_s. This is the largest value of ϵ', excluding phenomena due to charge transport in medium (*21*), due to the fact that all dipoles in the material are oriented in the direction of the applied electric field. At the other extreme, if an electric field at frequencies exceeding those of all kinds of dipolar reorientations is applied, the only response to such perturbation will be that of atomic and electronic polarization, which delivers the unrelaxed value of the real part of the permittivity, ϵ_∞. This is connected to the density, $\rho = 1/V$, via the Clausius-Mossotti equation:

$$\frac{\epsilon_\infty - \epsilon_0}{\epsilon_\infty + 2\epsilon_0} = \frac{4\pi N_A \alpha}{3} \frac{\rho}{M} \tag{2}$$

where ϵ_0 is the vacuum permittivity, N_A the Avogadro number, α the polarization, and M the molar mass.

In the frequency range between the low and high limits at which ϵ' equals ϵ_s and ϵ_∞, respectively, this magnitude exhibits several step decays, each corresponding to the relaxation of a given molecular motion (α, β etc.). Furthermore, via the Kramer-Kronig relations (*21*), the decay in ϵ' is accompanied by a peak in ϵ'', since (*24*):

$$\epsilon'' \approx -\frac{2}{\pi} \frac{\partial \epsilon'}{\partial \ln \omega} \tag{3}$$

where ω is the angular frequency in rad/s. To sum up, the signature of a relaxational process in a certain frequency range and temperature is the presence of decay in ϵ' and a peak in ϵ''.

A typical dielectric response after a single frequency sweep at fixed temperature is shown in Figure 2, where the α, β processes can be observed. The latter process is located at relatively high frequencies. Regarding the α process, in the typical conditions at which physical aging is conducted, that is, with the system in the glassy state, only its high frequency flank is visible, whereas to detect the low frequency flank, time consuming experiments should be performed to access this low frequency range. This would be detrimental for the real time monitoring of the evolution of the dielectric spectrum during aging.

Vitrification and Physical Aging by Dielectric Spectroscopy

Vitrification by Capacitive Dilatometry

As discussed, once the decay of the β process is complete, provided that no other secondary processes are present in the glass former, ϵ_∞ is measured and, therefore, information on the density of the system is conveyed via the Clausius-Mossotti equation (eq. 2). Hence, if ϵ_∞ is obtained over a wide temperature range, a kink will be detected at T_g, in ways analogous to a standard volumetric measurement (*4*). This method was first introduced by Meingast et al. (*25*) for studying the glass transition of an oriententional glass, and later extended to polymeric glass formers by different groups (*26–29*). In this way, vitrification at a given cooling rate has been successfully characterized by dielectric spectroscopy and these pioneering studies later provided impetus for an intense activity aiming to provide insights on vitrification kinetics by capacitive dilatometry of polymer glasses of different nature, including nanocomposites (*30–32*), thin films (*33–38*), nanospheres (*39*), polymers with varying molecular weight (*40*), as well as non-polymeric glass formers (*41*).

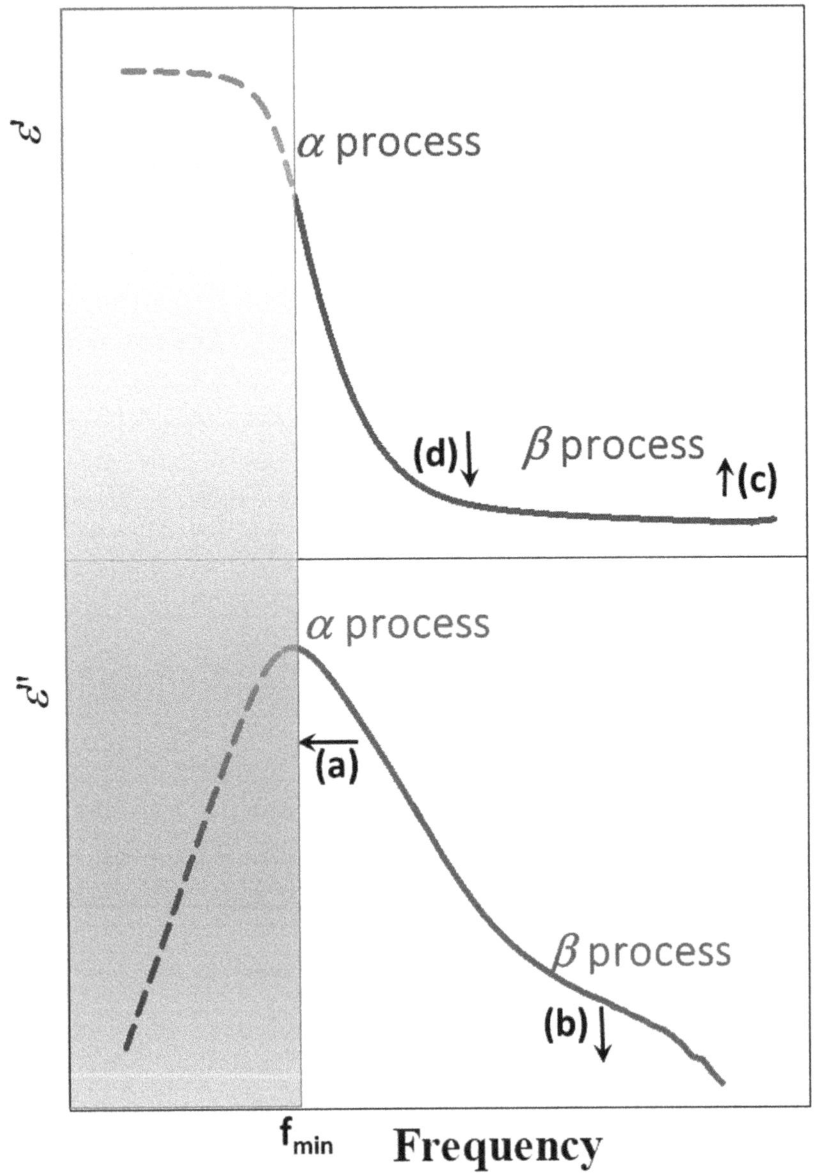

Figure 2. Typical frequency dependent dielectric spectrum in the aging regime in terms of ϵ' (upper panel) and ϵ" (lower panel). The letters indicates the evolution during physical aging of: (a) the shift of ϵ" to lower frequency due to the slow-down of the α relaxation; (b) the decrease of $\Delta\epsilon_\beta$ implying an overall decrease of the ϵ" in the region of the β relaxation; (c) The increase of ϵ' at high frequency, that is, ϵ'_∞; and (d) the decrease of ϵ' at frequencies between the α and β relaxations. The shaded area represents the frequency range inaccessible in experimental time scales; and f_{min} is the minimum frequency accessed in a typical dielectric experiment, typically located in the sub-Hz range.

As an example, Figure 3 shows the temperature dependence of the dielectric capacitance, *C*, proportional to the dielectric permittivity via geometrical factors: $C = \epsilon_0 \epsilon_\infty A/h$, with *A* and *h* the area and the thickness of the capacitor, respectively—at different frequencies. Here, it is worth of mention that, via the Clausius-Mossotti equation (eq. 2), the capacitance is directly related to the density: $dC/dT = d\rho^{-1}/dT \epsilon_0 A\rho/3h(\epsilon_\infty^2 + \epsilon - 2)$. A kink in *C*, underlying the glass transition, can be observed at all frequencies. However, it is worth pointing out that lowering the frequency implies a progressive reduction of the temperature at which relaxational processes contribute to ϵ' and, therefore, this magnitude cannot be used anymore as a reporter of the density/volume of the glass former.

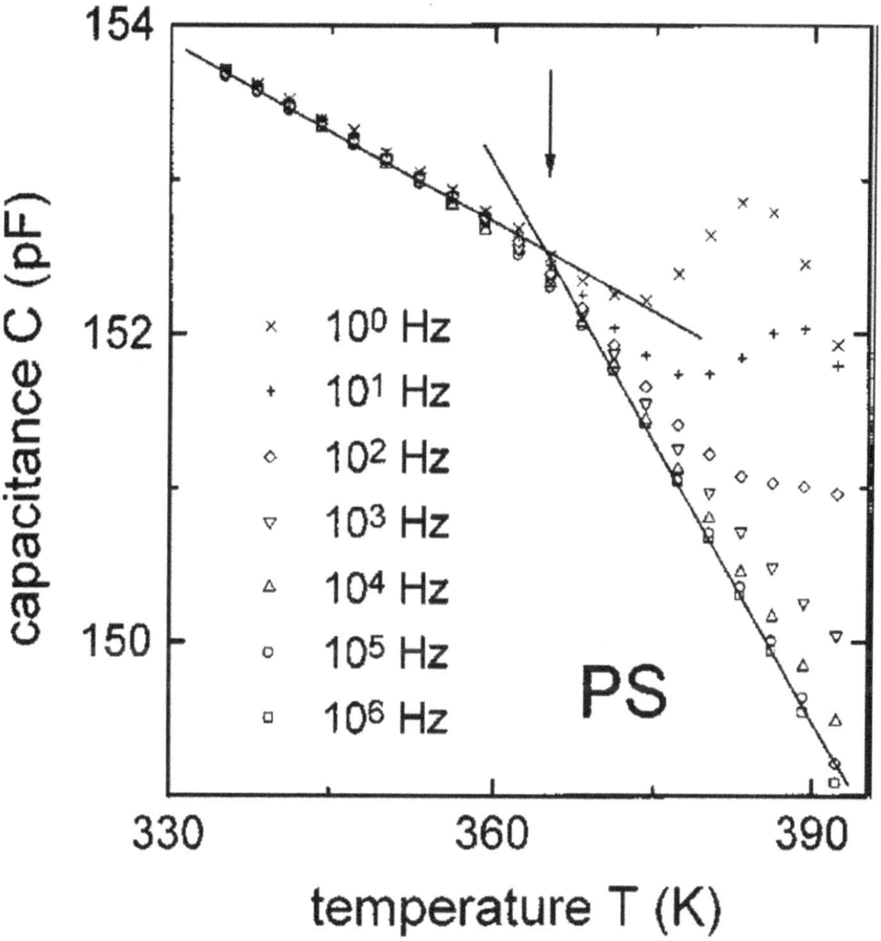

Figure 3. Capacitance, proportional via geometric factors to ϵ', as a function of temperature and at different frequencies for PS. Reproduced with permission from reference (28). Copyright 2000 American Physical Society.

Methods to Study Physical Aging by Dielectric Spectroscopy

The fact that only the high frequency flank of the α process is visible in a spectrum as that of Figure 2 means that the typical frequencies of the slow parts of this process are smaller than the inverse of the time scale of the experiment; in this case several minutes. As a result, a system exhibiting a spectrum as that of Figure 2 is a non-equilibrium glass. The thermodyamic evolution taking place in such a system generally modifies the pattern of the dielectric spectrum. Figure 2 shows how various flanks of the dielectric spectrum can be modified by physical aging and, therefore, offers an overview on the different approaches employed to study this phenomenon. These can be summarized considering the label indicated in Figure 2: (a) Shift of the α relaxation to lower frequencies; (b) Decrease, and possibly also shift, of the β relaxation; (c) Increase of ϵ' at high frequencies, that is ϵ_∞; and (d) decrease of ϵ' at frequencies between that of the α and β processes.

Modification of the α Relaxation

The approach labelled as (a) in Figure 2 relies on the fact that during physical aging the α process undergoes a slow-down due to densification and, therefore, the dielectric spectrum shifts to lower frequency. The modification of the high frequency flank of the α process may in principle originate also from a change in the shape of the α relaxation, becoming narrower, and/or to a decrease in the dielectric strength $\Delta\epsilon_\alpha$, that is, the area underlying the ϵ'' peak, or equivalently the step in ϵ'. However, none of the two effects are of relevance. Regarding the former, densification on of the glass during physical aging induces an increase of the dielectric strength, $\Delta\epsilon_\alpha$, as a results of an increase in the Kirkwood factor (*21*), underlying an enhancement of the orientational order, which is among other factors related to a density increase (*42*). This effect would produce an increase of ϵ'' at the low frequencies relevant for the α process, whereas in experiments the opposite is observed. Regarding the possible narrowing of the α relaxation, this would entail a modification of the dynamical heterogeneity of the glass. Experiments where aging was monitored after the application of non-linear electric field-which in ways analogous to a temperature jump are able to draw glass forming liquids out of equilibrium (*43*)—showed that at least the high frequency modes of the α relaxation age in a homogeneous fashion (*44, 45*), thus ruling out the scenario that a modification of the high frequency flank of the α process would be caused by its narrowing during aging.

The first experiments where the α relaxation was monitored after prolonged physical aging on an archetypal glassy polymer, poly(vinyl acetate) (PVAc), dates back the late 1980s (*46*). Subsequently, Schönhals and Schossler carried a systematic monitoring of the evolution of the α relaxation on the same polymer (*47*). Since these pioneering studies, several efforts have been devoted to the study of physical aging by this method in both polymeric (*48*) and small molecules (*49–51*) glasses.

An example in this sense is showed in Figure 4, where the effect of physical aging on the high frequency flank of the α relaxation at the indicated aging temperature, T_a, and different aging times, t_a is shown for PVAc (*48*). As discussed at the beginning of this section, aging provokes an overall shift of the spectrum to lower frequencies, whereas a modification of $\Delta\epsilon_\alpha$ has negligible effects on the spectrum. The evolution of the typical α relaxation time, obtained from data of Figure 4 and shown in the insert of this figure, indicates the typical evolution of an observable sensitive to aging, namely an initial logarithmic change followed by the approach to a plateau. Other studies monitored ϵ''at a fixed frequency below T_g (*24, 29*). Though the dielectric spectrum at the investigated frequency is not explicitly shown, the change in ϵ''must be associated to the slow-down of the α relaxation, at least in T_a values close to T_g.

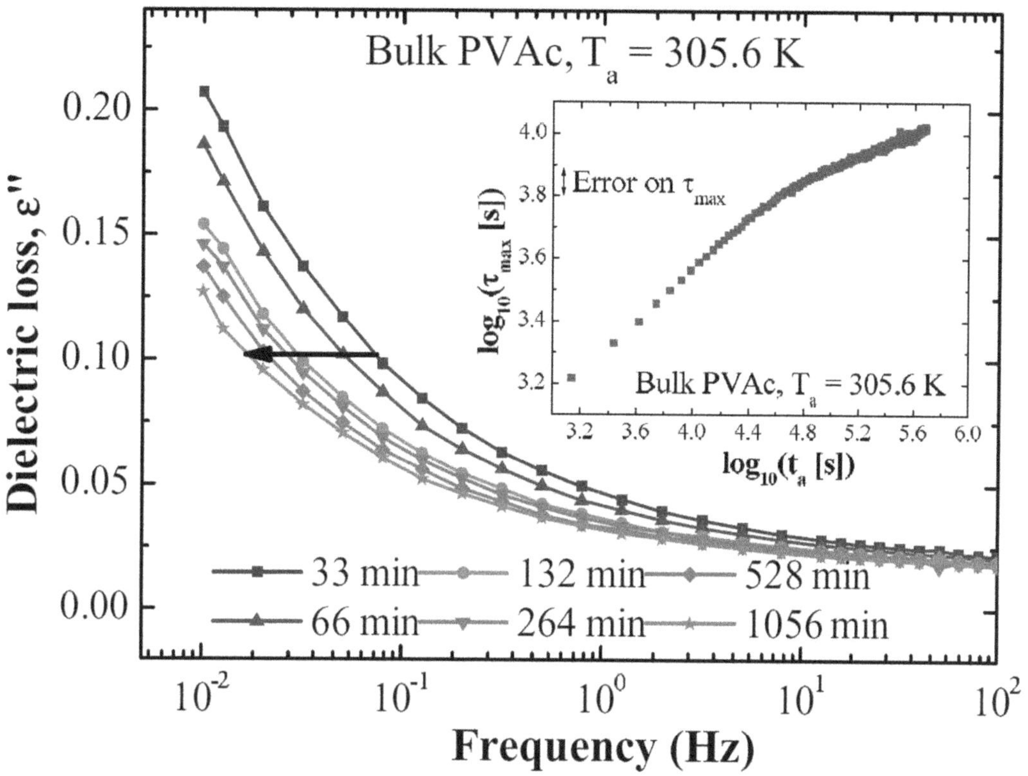

Figure 4. Evolution of ϵ″ spectrum with aging time, t_a, for bulk PVAc at the indicated aging temperature, T_a. The insert shows the corresponding aging time evolution of the most probable relaxation time, τ_{max}, obtained as $\tau_{max} = 1/(2\pi f_{max})$, where f_{max} is the frequency of the maximum of ϵ″. The latter is obtained fitting the α relaxation via the Havriliak-Negami empirical equation. Lines are the fits to this equation. Reproduced with permission from reference (48). Copyright 2012 American Physical Society.

Among the many experiments aiming to monitor the modification of the α relaxation, one group is represented by those where the shift to higher temperatures of this process at fixed frequency is followed (52, 53). In this case, the shift in T_g is equivalent to that of τ_α to lower frequencies in isothermal conditions. The drawback of this method is that on approaching the glass transition the thermodynamic state at some point rapidly evolves towards equilibrium due to the increase of the rate of physical aging.

Apart from measurements in the frequency domain, the evolution of the α relaxation during physical aging has also been monitored by dielectric spectroscopy techniques operating in the time domain, in particular thermally stimulated depolarization current (TSDC), which can be operated in essentially two different modes. One method, operating in isothermal conditions, is based on polarizing the glass with an electric field in the linear regime, that is, a small enough field to avoid modifying the glass properties, and subsequently observing the depolarization current resulting from switching-off of the field (54–56). The depolarization is carried out at various aging times and for durations short enough that no significant modification of the thermodynamic state takes place. To this aim the procedure originally proposed by Struik (6) in mechanical analysis during physical aging is generally adopted, which consists of following depolarization for a time duration shorter than 10

% of the aging time. In such a way a snapshot in quasi-invariant thermodynamic conditions of the dielectric response is obtained. Alternatively, the non-isothermal method is adopted consisting of polarizing th sample above T_g, cooling it down to the aging temperature and, after isothermal aging, observing the depolarization of the sample on heating (*48, 57–65*). If aging is carried out in proximity to T_g, both procedures allow attaining information on the evolution of the α relaxation in the glassy state during physical aging. Details regarding the formalism required to obtain τ_α are reported by Alegria et al. (*55*) for the isothermal method and by Boucher et al. (*48*) and Alegria et al. (*58*) for the non-isothermal one. An example of the application of the non-isothermal method is shown in Figure 5, where the physical aging of a nanocomposite of PVAc with silica is presented (*48*). The depolarization current undergoes a shift to higher temperature during the course of aging (see Figure 5a), which, after using the appropriate formalism (*58*), is translated into a temperature and aging time-dependent relaxation time of the α process (see Figure 5b). In such a way, the evolution of the α relaxation time at any temperature, including that of aging, can be obtained as shown in the insert of Figure 5b.

The drawback of using ϵ'' to follow physical aging is that only its high frequency tail can be monitored and therefore the peak frequency is hardly accessible, except close to T_g. To overcome this problem, other methods are based on monitoring the evolution of the α relaxation during aging by dielectric techniques rely on monitoring the complex dielectric modulus $M^* = 1/\epsilon^*$ (*66*). The features of the relaxation spectrum in terms of M^* is shifted to frequencies higher than those at which ϵ^* exhibits a maximum (*13*). For an exponential response, $\beta_{KWW} = 1$, normally addressed as Debye response in dielectric spectroscopy (*21*), $\tau_M = \tau_\epsilon \epsilon_\infty/\epsilon_s$, where τ_M is the relaxation time obtained from the frequency of the maximum ofM'' and τ_ϵ that from the maximum of ϵ''. For a non-exponential KWW behavior, as that of the α relaxation, the reduction of τ_M with respect to τ_ϵ becomes even more significant (*66*). As a result, the peak frequency of M'' can be visualized even if the glass is cooled significantly below its T_g.

A method of indirectly probing the change of the α relaxation and also relying on the dielectric modulus is used for glasses exhibiting large ionic conductivity. In this case ϵ'' only exhibits a power law decrease with frequency whose intensity masks the α relaxation: $\epsilon'' = \sigma/\epsilon_0\omega$. However, concomitantly the loss part of the dielectric modulus M'' is characterized by a peak (*21*). This method has been employed to study physical aging in non-polymeric highly conductive glass-formers (*66–70*). For these kinds of glass formers, methods directly relying on the modification of the ionic conductivity have received attention (*68, 69, 71, 72*). Furthermore, it been highlighted how this method—considering the particular features of a dielectric spectrum associated with ionic conduction may—lead to different conclusions regarding the nature of physical aging (*71*). Hence, these methods could be exploited in the future to study physical aging of highly conductive polymers, for which the α relaxation is masked.

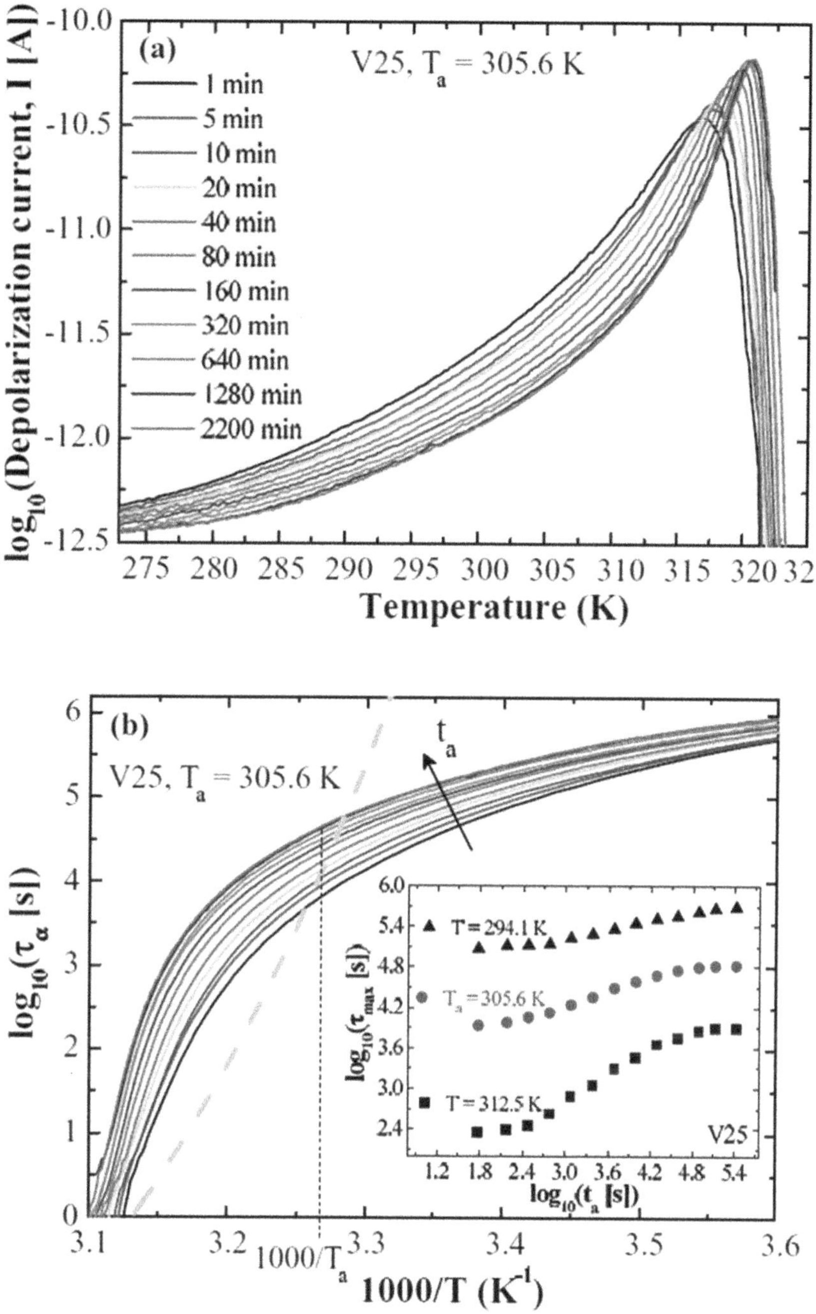

Figure 5. Evolution of the depolarization current (upper panel) and the relaxation time (lower panel) with aging time at the indicated aging temperature as a function of temperature for a nanocomposite of PVAc with silica (labeled V25). (lower panel). The insert of the lower panel shows the evolution with aging time of the relaxation time at different temperatures after aging T 305.6 K. Reproduced with permission from reference (48). Copyright 2012 American Physical Society.

Modification of the β Relaxation

The glass densification taking place during physical aging impacts secondary relaxations, provided that these exhibit some degree of coupling with volume changes (73). All secondary relaxations sensitive to physical aging could in principle be considered in the following treatment. However, for the rest of the section, we will focus on the β relaxation, that is by far the secondary relaxation most monitored to study physical aging. Generally speaking, glass densification results in the decrease of the dielectric strength of the β relaxation, $\Delta\epsilon_\beta$ (74). There exist two leading interpretations for such decrease. One relies on the heterogeneous localization of the β relaxation in islands of mobility, which are depleted when the glass former is densified (74). Another one, on the base of nuclear magnetic resonance experiments, rather suggests that the β relaxation is homogeneously distributed throughout the glass former (75). As a consequence, the reduction in $\Delta\epsilon_\beta$ upon densification originates from reduced amplitude of dipole fluctuations.

Whatever the origin of the coupling between $\Delta\epsilon_\beta$ and the density of the glass former, the former displays a kink in its temperature dependence at the glass transition in ways analogous to the density, and therefore $\Delta\epsilon_\beta$ can be used as an indicator of vitrification (76–79). For polymers this has been done in poly(methyl methacrylate) (PMMA), and its nanocomposites with silica (78), and another acrylate polymer, poly(thiomethyl methacrylate) (PTMA) (80), that is, polymers whose $\Delta\epsilon_\beta$ is especially large. The presence of a well defined and separated β relaxation significantly above and below T_g is a precondition to reliably monitor vitrification kinetics by the temperature dependence of $\Delta\epsilon_\beta$. This is a drawback of the method and, therefore, examples of it are relatively scarce. Alternatively, $\Delta\epsilon_\beta$ can be obtained after vitrifying at a different rates (74, 81). In such a way the magnitude of $\Delta\epsilon_\beta$ can be used to assess the thermodynamic state in terms of fictive temperature, T_f, thus attaining information on vitrification kinetics. The mentioned studies positively showed how $\Delta\epsilon_\beta$ depends on the age of the glass in terms of its T_f.

The mentioned limitations of studying vitrification via the β relaxation convert into advantages when physical aging is monitored by following $\Delta\epsilon_\beta$. Indeed, in a wide variety of cases the β relaxation is well visible in a wide range of temperatures below T_g, above all if dielectric spectroscopy over a broad frequency range is employed. Hence, examples where the β relaxation is monitored during physical aging are numerous and, generally speaking and as a result of densification, they all show a decrease of $\Delta\epsilon_\beta$ with aging time. Preliminary reports on the physical aging of glassy polymers were presented by dynamic-mechanical analysis (DMA), that is, the mechanical equivalent of dielectric spectroscopy (82, 83). Regarding dielectric spectroscopy, the first study reporting the change of the β relaxation with aging is that of Haddad and Goldstein (84) on several small molecules glass formers. That study raised the issue of the actual origin of the decrease of ϵ'' in the region of the β relaxation. This may be a serious concern in all those cases for which the β relaxation is in close proximity of the α relaxation. In that case, the apparent reduction of $\Delta\epsilon_\beta$ may be at least in part due to the shift of α relaxation to lower frequencies. Generally speaking, it is safe to attribute the decrease of ϵ'' in the region of the β relaxation exsclusively to this process only when no overlap with the α relaxation exists, that is to say, when $\epsilon''_\alpha \sim 0$.

Since the work of Haddad and Goldstein (84), the physical aging of glasses following the evolution of $\Delta\epsilon_\beta$ has been widely investigated. A wealth of effort has been pursued by Johari and co-workers on small molecules glass formers (79, 85–87) and in one case also a polymeric glass former

(*88*), in an attempt to provide a microscopic picture of the β relaxation in the glassy state. In these studies it was suggested that the decrease of $\Delta\epsilon_\beta$ would be due to the collapse of islands of mobility. Whatever the interpretation, the decrease of $\Delta\epsilon_\beta$ has been observed by dielectric spectroscopy in numerous studies on both polymers (*78, 89–91*) and small molecules (*63, 92–95*). An example of the decrease of $\Delta\epsilon_\beta$ during physical aging on PMMA and its nanocomposites with silica is shown in Figure 6, taken from Boucher et al. (*96*) As discussed, an alternative explanation to the mobility island theory is a decrease of the amplitude of fluctuations of homogeneously distributed dipoles (*75*). The suppression of the β relaxation upon aging may refute this theory in several glass formers. Specifically, it was shown that after fast cooling these glasses the β relaxation can be visible, whereas slow cooling to attainment of low T_f led to its disappearance (*92*). The reduction in $\Delta\epsilon_\beta$ is the predominant feature in aging experiments. However, there exists at least two examples—the aging of amorphous poly(vinyl ethylene) (PVE) (*89*) and xylitol (*93*)—showing also an increase in the frequency of the maximum of ϵ'', which has been explained by accounting for changes in the distribution of the local density (*89*). The proximity of the α relaxation, which shifts to lower frequencies upon aging, may also provide an explanation for this effect.

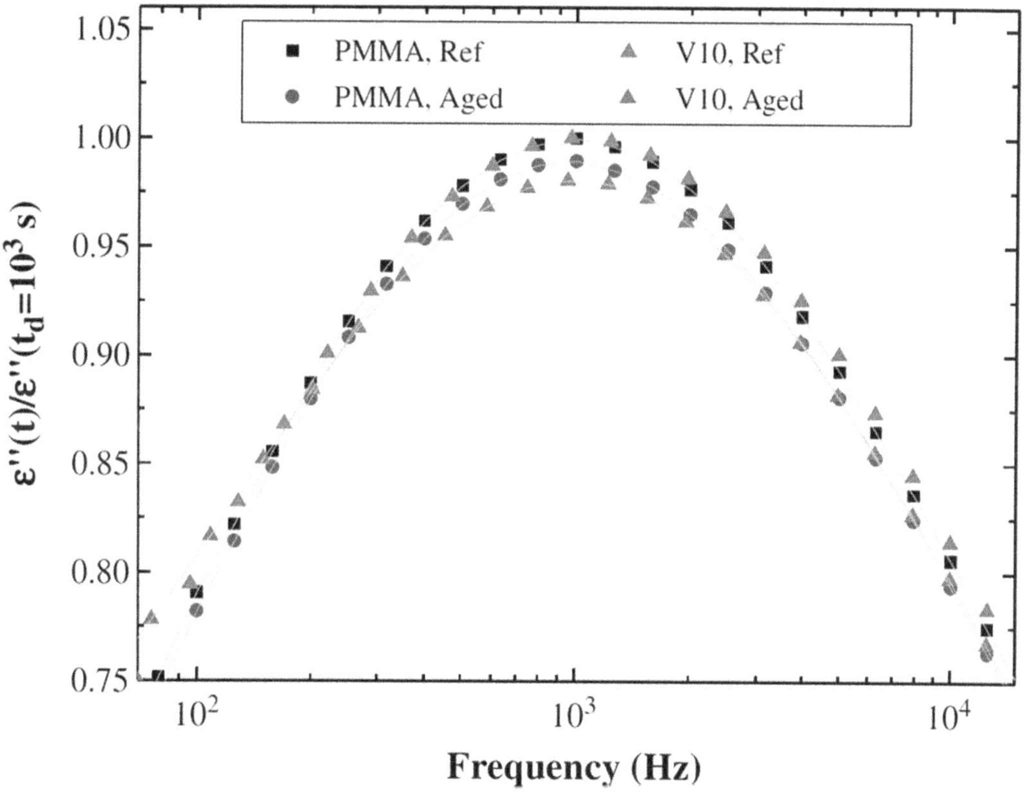

Figure 6. Normalized dielectric losses of bulk PMMA and its nanocomposite with 10 % wt. silica versus frequency at the initial aging time of 10^3 s (black squares for PMMA, light grey triangles for the nanocompisite with silica), and after an aging time t_a of $1.1x10^4$ s (circles for PMMA, dark grey triangles for the nanocompisite with silica), at an aging temperature of 353 K. Continuous lines are the fits through the Havriliak-Negami function. Reproduced with permission from reference (85). Copyright 2011 Elsevier.

As mentioned, in some glasses the frequency/temperature separation between the α and β relaxation can entail a large range and fall in the conditions at which physical aging is monitored. This event is of significant importance in polymers, where the aging time dependence of the dielectric spectrum is indirectly affected by the decrease, at higher frequencies, of $\Delta\epsilon_\beta$. Hence, in cases where two relaxations are separate, the indirect effect on ϵ^* of the modification of the β relaxation can be used to assess the kinetics of equilibrium recovery. This method has been applied to study physical aging of polymers such as poly(ethylene therephtalate) (PET) (*97*) and polycarbonate (PC) (*98*). In both cases, the reduction of ϵ' is observed, which, as discussed, is an indirect consequence of the concomitant reduction of $\Delta\epsilon_\beta$. Although studies by this method are so far relatively scarce (*97, 98*), these can be considered as a sub-class of the much more numerous studies where a direct monitoring of the β relaxation is pursued. Hence, they can be fruitfully exploited in those cases where the β relaxation is located at frequencies higher than those accessible by a given dielectric spectrometer.

Insights by Dielectric Spectroscopy on Multiple Equilibration Mechanisms and Acceleration of Aging in Nanostructured Glasses

Dielectric spectroscopy offers the advantage that physical aging can be monitored following in-situ evolution of a dielectric observable. If this is done concomitant with methods allowing precise determination of tiny changes, dielectric spectroscopy can deliver unique information in the comprehension of physical aging. One aspect that is currently receiving considerable attention is related to newly discovered features beyond conventional behavior based on a single decay toward equilibrium (see Figure 1, left panel). Recent dielectric experiments on a small molecule over prolonged times have shown that, besides the standard decay in ϵ_∞, a subtle long-term tail exists at very long aging times (*99*). These recent findings are supported by activity over the last decade, mainly based on calorimetric measurements (*100*). Such studies show multiple decays towards equilibrium in the physical aging regime in glassy polymers (*101*), indicating the existence of different mechanisms of equilibration (*102*). This outcome is general for different kinds of glasses (*103–105*), a result warranting a universal description.

Specifically, the presence of different mechanisms of equilibrium recovery can be detected by calorimetry—a technique delivering information on the enthalpic state of the glass—by carrying out aging experiments both in isothermal and isochronal conditions. In the former case, aging experiments for times as long as more than one year have been carried out in different polymers at temperatures ranging from close to significantly below T_g (*101, 106–108*). Beside the standard behavior with a single decay of the enthalpy towards equilibrium, observed close to T_g, these experiments show that farther from T_g approach to equilibrium takes place in two steps. This kind of behavior appears to be universal as it has been observed also for chalcogenide (*109, 110*), metallic glasses (*104, 111*) and a low molecular weight glass former (*112*). If physical aging is conducted way far from T_g, only the first decay is observed (*113–115*) even for time scales as large as 30 years (*102*). In this case, the presence of a mechanism beyond that associated to the α relaxation is reflected by the presence of a broad excess endotherm well below T_g (*102, 116–120*). Isochronal determinations of the amount of recovered enthalpy, also provide compelling arguments for the existence of multiple mechanisms of equilibrium recovery. These experiments are based on sampling the effect of aging in a wide range from T_g to much lower temperatures at fixed aging times. While aging experiments conducted close to T_g show the standard behavior with a rather narrow endothermic overshoot in the glass transition region, those carried out far below T_g show the previously mentioned broad en dotherm far below T_g indicating the response to the fast mechanism of equilibrium recovery. The general feature of these experiments is the presence of two maxima in the amount of recovered enthalpy and, equivalently, two minima in T_f. These results are found in relatively short time scales for polymer glasses exhibiting large amount of free interface (*106*, *107*, *121*). The reason for this is that aging is generally accelerated in these conditions with respect to the bulk, as discussed in details later in this section. However, a similar outcome is obtained in bulk polymer glasses, provided that sufficiently long aging times are considered (*102*, *122*).

The employment of dielectric spectroscopy involving devices whose stability allows precise determination of the dielectric response over very large time scales may provide insights of the utmost importance on these relatively new features of physical aging. An intriguing result, which may go in this direction, is the outcome showing that physical aging measured by following the evolution of ϵ_∞, which as discussed is coupled to the density, exhibits a normalized initial value smaller than that of the high frequency mechanical modulus $G\infty$ and that only the latter is coupled to the α relaxation (*123*). We speculate that at times shorter than the initial recorded value a fast decay takes place for ϵ_∞, in line with the existence of a fast mechanism of equilibrium recovery (*101*). These ideas can be further explored by means of devices allowing for fast thermal equilibration. Hence, the identification of complex physical aging behavior relies on covering a broad interval of time scales, ranging from sub-seconds to months or even years, thus covering 7-8 decades.

An aspect of vitrification and physical aging that has received great attention in recent years is how geometrical confinement interferes with these processes (*124–126*). It has been shown that enlarging the amount of free interfacial area of a polymer (*42*, *127–129*) results in T_g depression and acceleration of physical aging. The two phenomena have a common origin and result from a more efficient ability of maintaining equilibrium when cooling through the T_g or recovering equilibrium in the physical aging regime. Importantly, this can take place in polymers under geometrical confinement where the molecular mobility is essentially unaltered (*124*, *125*, *130*). Dielectric spectroscopy has contributed in this relatively new topic with a wealth of results. Regarding vitrification, several studies detecting the T_g by capacitive dilatometry are for thin polymer films (*37*, *38*, *131*) and polymer nanospheres (*39*). The important outcome of these works is that the observed T_g depression is decoupled from the molecular mobility, which, unlike the T_g, does not exhibit any size dependence and is bulk-like, at least in the conditions at which T_g exhibits large depression.

Regarding the acceleration of physical aging in glassy polymers exhibiting large amount of free interface, most work has been devoted to polymer nanocomposites. Studies on PMMA nanocomposites monitoring the aging time evolution of the dielectric strength of the β re laxation (method (b) of Figure 2) showed that increasing the nanofillers content induces an increase of the rate of equilibrium recovery (*78*). An analogous result was found by dielectric spectroscopy on PVAc (*48*) (see Figures 4 and 5) and PS (*32*) silica nanocomposites. In the former case, the kinetics of physical aging was followed by monitoring the evolution of the α relaxation, whereas for PS/silica nanocomposites, given the low dipolar moment of PS, it was convenient to follow physical aging by monitoring ϵ_∞. Importantly, similar results have been found by the mechanical analogous of dielectric spectroscopy, namely DMA, in PC/silica nanocomposites (*132*). It is worth mentioning that all these results concerns nanocomposites whose nanofillers have been previously treated to make them compatible with the polymers. Contrariwise, if bare nanofillers are employed, polymer adsorption on the nanofiller surface may completely erase (*129*, *133*) or even revert (*134*) the efficiency to maintain or recover equilibrium.

An emblematic example of how dielectric spectroscopy can provide insights of extraordinary importance on the nature of physical aging in confinement is shown in Figure 7, where the evolution of the macroscopic volume recovery in PS/silica nanocomposites is compared to that of the microscopic volume recovery, taken from the increase of ϵ_∞ (32). As can be observed, the former shows no influence of the presence of nanofillers and their concentration. In contrast, the microscopic evolution of the specific volume, that is the inverse of the density directly related to ϵ_∞, indicates that increasing of the nanofillers' content induces acceleration of physical aging. These intriguing results indicate a decoupling between macroscopic and microscopic volume contraction during physical aging. To understand this outcome, Cangialosi et al. (32) hypothesize that the shrinkage of the polymer matrix, observed microscopically, involves its cavitation from the silica nanofiller.

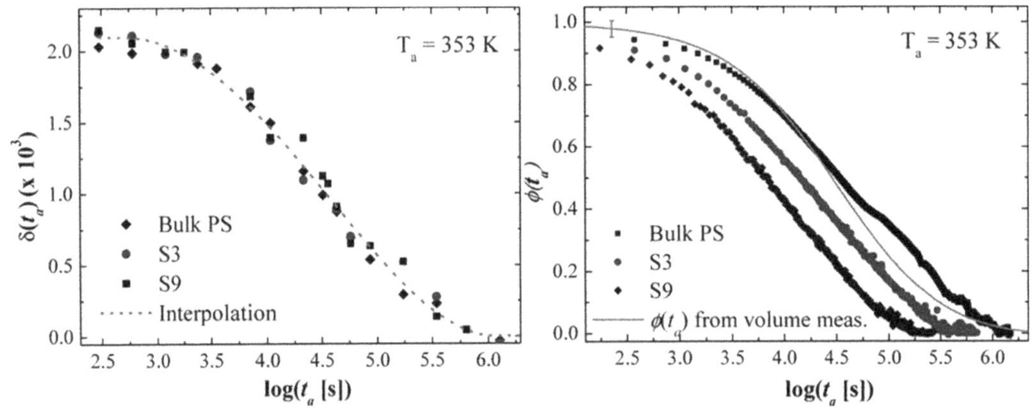

Figure 7. Volume recovery of bulk PS and PS in nanocomposites with 3 % (S3) and 9 % (S9) wt. silica from dilatometric measurements (left panel) and dielectric measurements (right panel). Reproduced with permission from reference (32). Copyright 2013 Wiley.

Conclusions

This work aims to provide an overview on how vitrification and the recovery of equilibrium of glasses, so-called physical aging, can be studied by dielectric spectroscopy. After introducing the main aspects of the phenomenology of vitrification and physical aging, the different strategies that can be pursued to study this phenomenon are presented and the activity carried out following these strategies are reviewed. Specifically, how the dielectrically active α and β relaxations and the high frequency permittivity, directly related to the density via the Clausius-Mossotti equation, can be monitored by dielectric techniques is discussed. We emphasize how each strategy can be adapted to the kind of glass, in particular polymeric ones.

Finally, we discuss how dielectric spectroscopy can be exploited to provide new insights on some new features of physical aging object of the last decade activity. In particular, emphasis is devoted to the complex behavior of physical aging in some specific conditions of time and temperature and the acceleration observed in polymers exhibiting large amounts of free surface, that is, polymers under nanoscale confinement.

Acknowledgments

The author thanks financial support from the project PGC2018-094548-B-I00 (MICINN–Spain and FEDER–UE) and the project IT-1175-19 (Basque Government).

References

1. Schmelzer, J. W. P.; Gutzow, I. S. *Glasses and the glass transition*; Wiley-VCH: Weinheim, 2011.
2. Cangialosi, D. Dynamics and thermodynamics of polymer glasses. *J. Phys.: Condens. Matter* **2014**, *26*, 153101.
3. Strobl, G. R. *The physics of polymers*; Springer: 1997; Vol. 2.
4. Kovacs, A. J. Glass transition in amorphous polymers: a phenomenological study. *Fortschr. Hochpolym.-Forsch.* **1963**, *3*, 394–508.
5. McKenna, G. B. In *Long-Term Durability of Polymeric Matrix Composites*; Pochiraju, K. V., Tandon, G. P., Schoeppner, G. A., Eds.; Springer US: Boston, MA, 2012; pp 237–309.
6. Struik, L. C. E. *Physical aging in amorphous polymers and other materials*; Technische Hogeschool Delft: 1977.
7. Cangialosi, D.; Boucher, V. M.; Alegria, A.; Colmenero, J. Physical aging in polymers and polymer nanocomposites: recent results and open questions. *Soft Matter* **2013**, *9*, 8619–8630.
8. Cangialosi, D. *Encyclopedia of Polymer Science and Technology*; American Cancer Society: 2018; pp 1–36.
9. Hutchinson, J. M. Physical aging of polymers. *Prog. Polym. Sci.* **1995**, *20*, 703–760.
10. Tool, A. Relation between inelastic deformability and thermal expansion of glass in its annealing range. *J. Am. Ceram. Soc.* **1946**, *29*, 240–253.
11. Moynihan, C. T.; Macedo, P. B.; Montrose, C. J.; Gupta, P. K.; De Bolt, M. A.; Dill, J. F.; Dom, B. E.; Drake, P. W.; Eastel, A. J.; Elterman, P. B.; Moeller, R. P.; Sasabe, H.; Wilder, J. A. Structural Relaxation in Vitreous Materials. *Ann. N. Y. Acad. Sci.* **1976**, *279*, 15–35.
12. Colmenero, J.; Arbe, A. Recent progress on polymer dynamics by neutron scattering: From simple polymers to complex materials. *J. Polym. Sci., Part B: Polym. Phys.* **2013**, *51*, 87–113.
13. McCrum, N.; Read, B.; Williams, G. *Anelastic and dielectric effects in polymeric solids*; John Wiley, 1967.
14. Böhmer, R.; Diezemann, G.; Hinze, G.; Rössler, E. Dynamics of supercooled liquids and glassy solids. *Prog. Nucl. Magn. Reson. Spectrosc.* **2001**, *3*, 191–267.
15. Reading, M.; Hourston, D. J. *Modulated Temperature Differential Scanning Calorimetry. Theoretical and Practical Applications in Polymer Characterisation*; Springer: Dordrecht, 2006.
16. Narayanaswamy, O. A model of structural relaxation in glass. *J. Am. Ceram. Soc.* **1971**, *54*, 491–498.
17. Dyre, J. C. Narayanaswamy's 1971 aging theory and material time. *J. Chem. Phys.* **2015**, *143*, 114507.

18. Kovacs, A. J.; Aklonis, J. J.; Hutchinson, J. M.; Ramos, A. R. Isobaric volume and enthalpy recovery of glasses. II. A transparent multiparameter theory. *J. Polym. Sci., Part B: Polym. Phys.* **1979**, *17*, 1097–1162.
19. Hecksher, T.; Olsen, N. B.; Dyre, J. C. Communication: Direct tests of single-parameter aging. *J. Chem. Phys.* **2015**, *142*, 241103.
20. Roed, L. A.; Hecksher, T.; Dyre, J. C.; Niss, K. Generalized single-parameter aging tests and their application to glycerol. *J. Chem. Phys.* **2019**, *150*, 044501.
21. Kremer, F.; Schönhals, A. *Broadband dielectric spectroscopy*; Springer Science & Business Media: 2002.
22. Callen, H.; Greene, R. on a Theorem of Irreversible Thermodynamics. *Phys. Rev.* **1952**, *86*, 702–710.
23. Kubo, R. Statistical-Mechanical Theory of Irreversible Processes. I. General Theory and Simple Applications to Magnetic and Conduction Problems. *J. Phys. Soc. Jpn.* **1957**, *12*, 570–586.
24. Wübbenhorst, M.; Van Turnhout, J. Analysis of complex dielectric spectra. I. One-dimensional derivative techniques and three-dimensional modelling. *J. Non-Cryst. Solids* **2002**, *305*, 40–49.
25. Meingast, C.; Haluska, M.; Kuzmany, H. Non-linear and non-exponential volume relaxation at the orientational 'glass' transition in solid C60. *J. Non-Cryst. Solids* **1996**, *201*, 167–171.
26. Fukao, K.; Miyamoto, Y. Glass transition temperature and dynamics of α-process in thin polymer films. *Europhys. Lett.* **1999**, *46*, 649.
27. Fukao, K.; Miyamoto, Y. Glass transitions and dynamics in thin polymer films: dielectric relaxation of thin films of polystyrene. *Phys. Rev. E* **2000**, *61*, 1743.
28. Bauer, C.; Böhmer, R.; Moreno-Flores, S.; Richert, R.; Sillescu, H.; Neher, D. Capacitive scanning dilatometry and frequency-dependent thermal expansion of polymer films. *Phys. Rev. E* **2000**, *61*, 1755.
29. Fukao, K.; Tahara, D. Aging dynamics in the polymer glass of poly(2-chlorostyrene): Dielectric susceptibility and volume. *Phys. Rev. E* **2009**, *80*, 051802.
30. Mizuno, M.; Nakamura, K.; Konishi, T.; Fukao, K. Glass transition and thermal expansivity in silica-polystyrene nanocomposites. *J. Non-Cryst. Solids* **2011**, *357*, 594–597, 6th International Discussion Meeting on Relaxation in Complex Systems.
31. Oh, H.; Green, P. F. Polymer chain dynamics and glass transition in athermal polymer/nanoparticle mixtures. *Nat. Mater.* **2009**, *8*, 139–143.
32. Cangialosi, D.; Boucher, V. M.; Alegría, A.; Colmenero, J. Volume recovery of polystyrene/silica nanocomposites. *J. Polym. Sci., Part B: Polym. Phys.* **2013**, *51*, 847–853.
33. Priestley, R. D.; Broadbelt, L. J.; Torkelson, J. M.; Fukao, K. Glass transition and α-relaxation dynamics of thin films of labeled polystyrene. *Phys. Rev. E* **2007**, *75*, 061806.
34. Napolitano, S.; Wübbenhorst, M. Dielectric signature of a dead layer in ultrathin films of a nonpolar polymer. *J. Phys. Chem. B* **2007**, *111*, 9197–9199.
35. Napolitano, S.; Pilleri, A.; Rolla, P.; Wübbenhorst, M. Unusual deviations from bulk behavior in ultrathin films of poly (tert-butylstyrene): Can dead layers induce a reduction of T g? *ACS Nano* **2010**, *4*, 841–848.
36. Napolitano, S.; Wübbenhorst, M. The lifetime of the deviations from bulk behaviour in polymers confined at the nanoscale. *Nat. Commun.* **2011**, *2*, 260.

37. Boucher, V. M.; Cangialosi, D.; Yin, H.; Schönhals, A.; Alegria, A.; Colmenero, J. T g depression and invariant segmental dynamics in polystyrene thin films. *Soft Matter* **2012**, *8*, 5119–5122.
38. Yin, H.; Cangialosi, D.; Schönhals, A. Glass transition and segmental dynamics in thin supported polystyrene films: The role of molecular weight and annealing. *Thermochim. Acta* **2013**, *566*, 186–192.
39. Zhang, C.; Boucher, V. M.; Cangialosi, D.; Priestley, R. D. Mobility and glass transition temperature of polymer nanospheres. *Polymer* **2013**, *54*, 230–235.
40. Bogoslovov, R.; Hogan, T.; Roland, C. Clarifying the molecular weight dependence of the segmental dynamics of polybutadiene. *Macromolecules* **2010**, *43*, 2904–2909.
41. Niss, K.; Gundermann, D.; Christensen, T.; Dyre, J. C. Dynamic thermal expansivity of liquids near the glass transition. *Phys. Rev. E* **2012**, *85*, 041501.
42. Capponi, S.; Napolitano, S.; Wübbenhorst, M. Supercooled liquids with enhanced orientational order. *Nat. Commun.* **2012**, 3, 1–7.
43. Richert, R. Perspective: Nonlinear approaches to structure and dynamics of soft materials. *J. Chem. Phys.* **2018**, *149*, 240901.
44. Richert, R. Physical aging and heterogeneous dynamics. *Phys. Rev. Lett.* **2010**, *104*, 085702.
45. Riechers, B.; Richert, R. Rate exchange rather than relaxation controls structural recovery. *Phys. Chem. Chem. Phys.* **2019**, *21*, 32–37.
46. Nozaki, R.; Mashimo, S. Dielectric relaxation measurements of poly(vinyl acetate) in glassy state in the frequency range 10^{-6} 10^{6} Hz. *J. Chem. Phys.* **1987**, *87*, 2271–2277.
47. Schlosser, E.; Schönhals, A. Dielectric relaxation during physical ageing. *Polymer* **1991**, *32*, 2135–2140.
48. Boucher, V. M.; Cangialosi, D.; Alegría, A.; Colmenero, J. Time dependence of the segmental relaxation time of poly(vinyl acetate)-silica nanocomposites. *Phys. Rev. E* **2012**, *86*, 041501.
49. Leheny, R. L.; Nagel, S. R. Frequency-domain study of physical aging in a simple liquid. *Phys. Rev. B* **1998**, *57*, 5154.
50. Lunkenheimer, P.; Wehn, R.; Schneider, U.; Loidl, A. Glassy Aging Dynamics. *Phys. Rev. Lett.* **2005**, 95, 055702.
51. Bhardwaj, S. P.; Suryanarayanan, R. Use of dielectric spectroscopy to monitor molecular mobility in glassy and supercooled trehalose. *J. Phys. Chem. B* **2012**, *116*, 11728–11736.
52. Wypych, A.; Duval, E.; Boiteux, G.; Ulanski, J.; David, L.; Mermet, A. Effect of physical aging on nano- and macroscopic properties of poly(methyl methacrylate) glass. *Polymer* **2005**, *46*, 12523–12531.
53. Grigoriadi, K.; Putzeys, T.; Wübbenhorst, M.; van Breemen, L. C.; Anderson, P. D.; Hütter, M. Effect of low-temperature physical aging on the dynamic transitions of atactic polystyrene in the glassy state. *J. Polym. Sci., Part B: Polym. Phys.* **2019**, *57*, 1394–1401.
54. Alegria, A.; Guerrica-Echevarria, E.; Goitiandia, L.; Telleria, I.; Colmenero, J. Alpha-Relaxation in the Glass-Transition Range of Amorphous Polymers. 1. Temperature Behavior Across the Glass-Transition. *Macromolecules* **1995**, *28*, 1516–1527.
55. Alegria, A.; Goitiandia, L.; Telleria, I.; Colmenero, J. alpha-Relaxation in the glass-transition range of amorphous polymers. 2. Influence of physical aging on the dielectric relaxation. *Macromolecules* **1997**, *30*, 3881–3887.

56. Zhao, J.; McKenna, G. B. Temperature divergence of the dynamics of a poly(vinyl acetate) glass: Dielectric vs. mechanical behaviors. *J. Chem. Phys.* **2012**, *136*, 154901.
57. Schrader, S.; Schönhals, A. *Relaxation in Polymers*; Springer: 1989; pp 93–100.
58. Alegría, A.; Goitiandía, L.; Colmenero, J. On the interpretation of the TSDC results in the study of the -relaxation of amorphous polymers. *Polymer* **1996**, *37*, 2915–2923.
59. Goitiandia, L.; Alegria, A. Physical aging of poly (vinyl acetate). A thermally stimulated depolarization current investigation. *J. Non-Cryst. Solids* **2001**, *287*, 237–241.
60. Descamps, M.; Moura Ramos, J. J.; Correia, N. T. Ageing exploration of the energy landscape of a glass by the TSDC technique. *Mol. Phys.* **2002**, *100*, 2669–2677.
61. Saiter, J.; Dargent, E.; Kattan, M.; Cabot, C.; Grenet, J. Fragility index of drawn or annealed poly (ethylene terephthalate) films studied by thermally stimulated depolarisation currents. *Polymer* **2003**, *44*, 3995–4001.
62. Benrekaa, N.; Gourari, A.; Bendaoud, M.; Saoud, R.; Guerbi, C. TSDC study of structural relaxation on polyethylene terephthalate. *J. Non-Cryst. Solids* **2006**, *352*, 4804–4808.
63. Pinto, S. S.; Ramos, J. J. M.; Diogo, H. P. The slow molecular mobility in poly (vinyl acetate) revisited: New contributions from thermally stimulated currents. *Eur. Polym. J.* **2009**, *45*, 2644–2652.
64. Diogo, H. P.; Ramos, J. J. M. Slow molecular mobility in the amorphous thermoplastic polysulfone. *J. Therm. Anal. Calorim.* **2013**, *111*, 773–779.
65. Arrese-Igor, S.; Alegría, A.; Colmenero, J. Chain dynamics on crossing the glass transition: nonequilibrium effects and recovery of the temperature dependence of the structural relaxation. *ACS Macro Lett.* **2014**, *3*, 1215–1219.
66. Wagner, H.; Richert, R. Dielectric relaxation of the electric field in poly(vinyl acetate): a time domain study in the range 10-3–106 s. *Polymer* **1997**, *38*, 255–261.
67. Richert, R. Comment on "Temperature divergence of the dynamics of a poly(vinyl acetate) glass: Dielectric vs. mechanical behaviors" [J. Chem. Phys.136, 154901 (2012)]. *J. Chem. Phys.* **2013**, *139*, 137101.
68. Wojnarowska, Z.; Roland, C.; Kolodziejczyk, K.; Swiety-Pospiech, A.; Grzybowska, K.; Paluch, M. Quantifying the structural dynamics of pharmaceuticals in the glassy state. *J. Phys. Chem. Lett.* **2012**, *3*, 1238–1241.
69. Paluch, M.; Wojnarowska, Z.; Hensel-Bielowka, S. Heterogeneous dynamics of prototypical ionic glass CKN monitored by physical aging. *Phys. Rev. Lett.* **2013**, *110*, 015702.
70. Wojnarowska, Z.; Ngai, K.; Paluch, M. Deducting the temperature dependence of the structural relaxation time in equilibrium far below the nominal T g by aging the decoupled conductivity relaxation to equilibrium. *J. Chem. Phys.* **2014**, *140*, 174502.
71. Dyre, J. C. Aging of CKN: Modulus versus conductivity analysis. *Phys. Rev. Lett.* **2013**, *110*, 245901.
72. Bragatto, C.; Cassar, D.; Peitl, O.; Souquet, J.-L.; Rodrigues, A. Structural relaxation in AgPO3 glass followed by in situ ionic conductivity measurements. *J. Non-Cryst. Solids* **2016**, *437*, 43–47.
73. Ngai, K.; Paluch, M. Classification of secondary relaxation in glass-formers based on dynamic properties. *J. Chem. Phys.* **2004**, *120*, 857–873.
74. Johari, G. P. Intrinsic mobility of molecular glasses. *J. Chem. Phys.* **1973**, *58*, 1766–1770.

75. Vogel, M.; Rössler, E. Slow β process in simple organic glass formers studied by one- and two-dimensional 2 H nuclear magnetic resonance. *J. Chem. Phys.* **2001**, *114*, 5802–5815.
76. Johari, G.; Power, G.; Vij, J. Localized relaxation's strength and its mimicry of glass-softening thermodynamics. *J. Chem. Phys.* **2002**, *116*, 5908–5909.
77. Power, G.; Johari, G.; Vij, J. Relaxation strength of localized motions in D-sorbitol and mimicry of glass-softening thermodynamics. *J. Chem. Phys.* **2003**, *119*, 435–442.
78. Boucher, V. M.; Cangialosi, D.; Alegría, A.; Colmenero, J.; González-Irun, J.; Liz-Marzan, L. M. Accelerated physical aging in PMMA/silica nanocomposites. *Soft Matter* **2010**, *6*, 3306–3317.
79. Vij, J.; Power, G. Physical ageing and the Johari–Goldstein relaxation in molecular glasses. *J. Non-Cryst. Solids* **2011**, *357*, 783–792.
80. Casalini, R.; Snow, A.; Roland, C. Temperature dependence of the Johari–Goldstein relaxation in poly(methyl methacrylate) and poly(thiomethyl methacrylate). *Macromolecules* **2013**, *46*, 330–334.
81. Johari, G. P.; Goldstein, M. Viscous liquids and the glass transition. II. Secondary relaxations in glasses of rigid molecules. *J. Chem. Phys.* **1970**, *53*, 2372–2388.
82. Frank, W.; Stuart, H. Umwandlungserscheinungen in amorphen Hochpolymeren beim tempern unterhalb der Glastemperatur. *Kolloid-Z.* **1968**, *225*, 1–6.
83. Petrie, S. The effect of excess thermodynamic properties versus structure formation on the physical properties of glassy polymers. *J. Macromol. Sci., Part B: Phys.* **1976**, *12*, 225–247.
84. Haddad, J.; Goldstein, M. Viscous liquids and the glass transition. VIII. Effect of fictive temperature on dielectric relaxation in the glassy state. *J. Non-Cryst. Solids* **1978**, *30*, 1–22.
85. Johari, G. Effect of annealing on the secondary relaxations in glasses. *J. Chem. Phys.* **1982**, *77*, 4619–4626.
86. Power, G.; Vij, J.; Johari, G. Kinetics of spontaneous change in the localized motions of D-sorbitol glass. *J. Chem. Phys.* **2006**, *124*, 074509.
87. Righetti, M. C.; Tombari, E.; Johari, G. P. Aging kinetics of levoglucosan orientational glass as a rate dispersion process and consequences for the heterogeneous dynamics view. *J. Chem. Phys.* **2016**, *145*.
88. Pathmanathan, K.; Cavaillé, J.-Y.; Johari, G. Dielectric studies of the effects of thermal history on secondary relaxation in bisphenol-a-polycarbonate. *J. Polym. Sci., Part B: Polym. Phys.* **1989**, *27*, 1519–1527.
89. Casalini, R.; Roland, C. M. Aging of the Secondary Relaxation to Probe Structural Relaxation in the Glassy State. *Phys. Rev. Lett.* **2009**, *102*, 035701.
90. Casalini, R.; Roland, C. Aging of a low molecular weight poly (methyl methacrylate). *J. Non-Cryst. Solids* **2011**, *357*, 282–285.
91. Yin, H.; Chapala, P.; Bermeshev, M.; Schönhals, A.; Bohning, M. Molecular mobility and physical aging of a highly permeable glassy polynorbornene as revealed by dielectric spectroscopy. *ACS Macro Lett.* **2017**, *6*, 813–818.
92. Wagner, H.; Richert, R. Equilibrium and non-equilibrium type β-relaxations: d-sorbitol versus o-terphenyl. *J. Phys. Chem. B* **1999**, *103*, 4071–4077.
93. Yardimci, H.; Leheny, R. L. Aging of the Johari-Goldstein relaxation in the glass-forming liquids sorbitol and xylitol. *J. Chem. Phys.* **2006**, *124*, 214503.

94. Kaminski, K.; Adrjanowicz, K.; Kaminska, E.; Paluch, M. Probing of structural relaxation times in the glassy state of sucrose and trehalose based on dynamical properties of two secondary relaxation processes. *Phys. Rev. E* **2011**, *83*, 061502.
95. Kossack, W.; Adrjanowicz, K.; Tarnacka, M.; Kipnusu, W. K.; Dulski, M.; Mapesa, E. U.; Kaminski, K.; Pawlus, S.; Paluch, M.; Kremer, F. Glassy dynamics and physical aging in fucose saccharides as studied by infrared-and broadband dielectric spectroscopy. *Phys. Chem. Chem. Phys.* **2013**, *15*, 20641–20650.
96. Boucher, V. M.; Cangialosi, D.; Alegrìa, A.; Colmenero, J.; González-Irun, J.; Liz-Marzan, L. M. Physical aging in PMMA/silica nanocomposites: enthalpy and dielectric relaxation. *J. Non-Cryst. Solids* **2011**, *357*, 605–609.
97. McGonigle, E.-A.; Daly, J. H.; Jenkins, S. D.; Liggat, J. J.; Pethrick, R. A. Influence of physical aging on the molecular motion and structural relaxation in poly (ethylene terephthalate) and related polyesters. *Macromolecules* **2000**, *33*, 480–489.
98. Cangialosi, D.; Wübbenhorst, M.; Groenewold, J.; Mendes, E.; Schut, H.; van Veen, A.; Picken, S. J. Physical aging of polycarbonate far below the glass transition temperature: Evidence for the diff mechanism. *Phys. Rev. B* **2004**, *70*, 224213.
99. Niss, K.; Dyre, J. C.; Hecksher, T. Long-time structural relaxation of glass-forming liquids: Simple or stretched exponential? *J. Chem. Phys.* **2020**, *152*, 041103.
100. Cangialosi, D. *Handbook of Thermal Analysis and Calorimetry*; Elsevier: 2018; Vol. 6; pp 301–337.
101. Cangialosi, D.; Boucher, V. M.; Alegría, A.; Colmenero, J. Direct Evidence of Two Equilibration Mechanisms in Glassy Polymers. *Phys. Rev. Lett.* **2013**, *111*, 095701.
102. Perez-De Eulate, N. G.; Cangialosi, D. The very long-term physical aging of glassy polymers. *Phys. Chem. Chem. Phys.* **2018**, *20*, 12356–12361.
103. Wang, W. H. Dynamic relaxations and relaxation-property relationships in metallic glasses. *Prog. Mater. Sci.* **2019**, *106*, 100561.
104. Monnier, X.; Cangialosi, D.; Ruta, B.; Busch, R.; Gallino, I. Vitrification decoupling from α-relaxation in a metallic glass. *Sci. Adv.* **2020**, *6*, eaay1454.
105. Yang, Q.; Peng, S.-X.; Wang, Z.; Yu, H.-B. Shadow glass transition as a thermodynamics signature of β relaxation in hyper-quenched metallic glasses. *Nat. Sci. Rev.* **2020**.
106. Boucher, V. M.; Cangialosi, D.; Alegria, A.; Colmenero, J. Complex nonequilibriumdynamics of stacked polystyrene films deep in the glassy state. *J. Chem. Phys.* **2017**, *146*, 203312.
107. Perez-De-Eulate, N. G.; Cangialosi, D. Double Mechanism for Structural Recovery of Polystyrene Nanospheres. *Macromolecules* **2018**, *51*, 3299–3307.
108. Buquet, C. L.; Hamonic, F.; Saiter, A.; Dargent, E.; Langevin, D.; Nguyen, Q. Physical ageing and molecular mobilities of sulfonated polysulfone for proton exchange membranes. *Thermochim. Acta* **2010**, *509*, 18–23.
109. Golovchak, R.; Kozdras, A.; Balitska, V.; Shpotyuk, O. Step-wise kinetics of natural physical ageing in arsenic selenide glasses. *J. Phys.: Condens. Matter* **2012**, *24*, 505106.
110. Golovchak, R.; Kozdras, A.; Shpotyuk, O.; Balitska, V. Crossover between cooperative and fractal relaxation in complex glass-formers. *J. Phys.: Condens. Matter* **2016**, *28*, 355101.

111. Gallino, I.; Cangialosi, D.; Evenson, Z.; Schmitt, L.; Hechler, S.; Stolpe, M.; Ruta, B. Hierarchical aging pathways and reversible fragile-to-strong transition upon annealing of a metallic glass former. *Acta Mater.* **2018**, *144*, 400–410.
112. Miller, R. S.; MacPhail, R. A. Ultraslow nonequilibrium dynamics in supercooled glycerol by stimulated Brillouin gain spectroscopy. *J. Chem. Phys.* **1997**, *106*, 3393–3401.
113. Boucher, V. M.; Cangialosi, D.; Alegría, A.; Colmenero, J. Enthalpy Recovery of Glassy Polymers: Dramatic Deviations from the Extrapolated Liquidlike Behavior. *Macromolecules* **2011**, *44*, 8333–8342.
114. Andreozzi, L.; Faetti, M.; Giordano, M.; Palazzuoli, D. Enthalpy recovery in low molecular weight PMMA. *J. Non-Cryst. Solids* **2003**, *332*, 229–241.
115. McGonigle, E. A.; Cowie, J. M. G.; Arrighi, V.; Pethrick, R. A. Enthalpy relaxationand free volume changes in aged styrene copolymers containing a hydrogen bondingco-monomer. *J. Mater. Sci.* **2005**, *40*, 1869–1881.
116. Berens, A. R.; Hodge, I. M. Effects of annealing and prior history on enthalpy relaxation in glassy polymers. 1. Experimental study on poly(vinyl chloride). *Macromolecules* **1982**, *15*, 756–761.
117. Fan, J.; Cooper, E. I.; Angell, C. A. Glasses with Strong Calorimetric β-Glass Transitions and the Relation to the Protein Glass Transition Problem. *J. Phys. Chem.* **1994**, *98*, 9345–9349.
118. Johari, G. Specific heat relaxation-based critique of isothermal glass transition, zero residual entropy and time-average formalism for ergodicity loss. *Thermochim. Acta* **2011**, *523*, 97–104.
119. Pradipkanti, L.; Chowdhury, M.; Satapathy, D. K. Stratification and two glass-like thermal transitions in aged polymer films. *Phys. Chem. Chem. Phys.* **2017**, *19*, 29263–29270.
120. Androsch, R.; Jariyavidyanont, K.; Schick, C. Enthalpy relaxation of polyamide 11 of different morphology far below the glass transition temperature. *Entropy* **2019**, *21*, 984.
121. Boucher, V. M.; Cangialosi, D.; Alegria, A.; Colmenero, J. Reaching the ideal glass transition by aging polymer films. *Phys. Chem. Chem. Phys.* **2017**, *19*, 961–965.
122. Greiner, R.; Schwarzl, F. R. Thermal contraction and volume relaxation of amorphous polymers. *Rheol. Acta* **1984**, *23*, 378–395.
123. Hecksher, T.; Olsen, N. B.; Dyre, J. C. Fast contribution to the activation energy of a glass-forming liquid. *Proc. Natl. Acad. Sci. U. S. A.* **2019**, *116*, 16736–16741.
124. Cangialosi, D.; Alegria, A.; Colmenero, J. Effect of nanostructure on the thermal glasstransition and physical aging in polymer materials. *Prog. Polym. Sci.* **2016**, *54–55*, 128.
125. Priestley, R. D.; Cangialosi, D.; Napolitano, S. On the equivalence between the thermodynamic and dynamic measurements of the glass transition in confined polymers. *J. Non-Cryst. Solids* **2015**, *407*, 288–295.
126. Ma, M.-C.; Guo, Y.-L. Physical Properties of Polymers Under Soft and Hard Nanoconfinement: A Review. *Chin. J. Polym. Sci.* **2020**, 1–14.
127. Napolitano, S.; Rotella, C.; Wübbenhorst, M. Can Thickness and Interfacial Interactions Univocally Determine the Behavior of Polymers Confined at the Nanoscale? *ACS Macro Lett.* **2012**, *1*, 1189–1193.
128. Napolitano, S.; Glynos, E.; Tito, N. B. Glass transition of polymers in bulk, confined geometries, and near interfaces. *Rep. Prog. Phys.* **2017**, *80*, 036602.

129. Perez-de Eulate, N. G.; Sferrazza, M.; Cangialosi, D.; Napolitano, S. Irreversible adsorption erases the free surface effect on the Tg of supported films of poly(4-tert-butylstyrene). *ACS Macro Lett.* **2017**, *6*, 354–358.
130. Kremer, F.; Tress, M.; Mapesa, E. U. Glassy dynamics and glass transition in nanometric layers and films: A silver lining on the horizon. *J. Non-Cryst. Solids* **2015**, *407*, 277–283.
131. Lupascu, V.; Picken, S. J.; Wübbenhorst, M. Cooperative and non-cooperative dynamics in ultra-thin films of polystyrene studied by dielectric spectroscopy and capacitive dilatometry. *J. Non-Cryst. Solids* **2006**, *352*, 5594–5600.
132. Ramakrishnan, V.; Harsiny, S.; Goossens, J. G. P.; Hoeks, T. L.; Peters, G. W. M. Physical aging in polycarbonate nanocomposites containing grafted nanosilica particles: A comparison between enthalpy and yield stress evolution. *J. Polym. Sci., Part B: Polym. Phys.* **2016**, *54*, 2069–2081.
133. Burroughs, M. J.; Napolitano, S.; Cangialosi, D.; Priestley, R. D. Direct Measurement of Glass Transition Temperature in Exposed and Buried Adsorbed Polymer Nanolayers. *Macromolecules* **2016**, *49*, 4647–4655.
134. Priestley, R. D.; Rittigstein, P.; Broadbelt, L. J.; Fukao, K.; Torkelson, J. M. Evidence for the molecular-scale origin of the suppression of physical ageing in confined polymer: fluorescence and dielectric spectroscopy studies of polymer-silica nanocomposites. *J. Phys.: Condens. Matter* **2007**, *19*, 2996–3005.

Chapter 7

Broadband Dielectric Spectroscopy of Polymer Nanocomposites

Shiwang Cheng*

Department of Chemical Engineering and Materials Science, Michigan State University, East Lansing, Michigan 48824, United States

***Email: chengsh9@msu.edu**

The inclusion of nanoparticles (NPs) to polymer matrices not only introduces mechanical modifiers but also brings hierarchical interactions between the polymer and NPs or among NPs in polymer nanocomposites (PNCs). Consequently, the structure and dynamics of the matrix polymer have been strongly modified, leading to many advanced macroscopic properties of PNCs. In this chapter, recent developments of broadband dielectric spectroscopy (BDS) on quantifying the dynamics of PNCs have been reviewed, spanning from the secondary relaxation to the chain dynamics. Specifically, different methodologies on analyzing the dielectric broadening of the structural relaxation of PNCs have been compared and discussed. Besides, the combination of BDS with other macroscopic measurements has also been emphasized, where BDS serves as an implement capable of selectively quantifying the matrix dynamics of PNCs.

Introduction

Polymer nanocomposites (PNCs) are important lightweight materials with embedded functionalities that have been widely used in the fields of energy, environmental, health, and infrastructure (*1–8*). The advanced properties of PNCs correlate strongly with the structure and dynamics of the matrix polymer that has been profoundly modified by nanoparticles (NPs) (*9–12*). Specifically, the large surface area of NPs induces a large volume of the interfacial layer (IL) with a thickness, l_{int} (Figure 1), that exhibits distinct properties from the bulk polymer (*13–15*). The nanoscale interparticle surface-to-surface distance, d_{IPS} (Figure 1), brings the strong NP-NP colloidal interactions (*16–18*), and imposes a strong nanoconfinement effect (*15, 19, 20*). Additionally, the d_{IPS}, the radius of NP, R_{NP}, and the radius of gyration of the polymer, R_g, are all at nanoscale, which promotes the formation of the polymer-NP complex or the NP-network through polymer bridging (*21–23*). On one hand, these structural characteristics offer an attractive platform to investigate the surface and the nano-confinement effects on the structure and dynamics of polymers, including the sub-glassy and glassy dynamics (*24–41*), the entanglement and the chain dynamics (*10, 42–48*), the polymer desorption/adsorption (*21, 49, 50*), the NP diffusion (*44, 51–63*), and the mechanical reinforcement of PNCs (*22, 23, 64–69*). On the other hand, the complex geometry of PNCs along with the hierarchical interactions between the polymer and NPs or among NPs impose significant challenges to delineate the properties of the matrix and the NPs (*12, 21*). Therefore, new developments of analytical tools on quantification of the structure and dynamics of PNCs are highly demanding, especially for methods capable of selectively quantifying the property of individual component of PNCs.

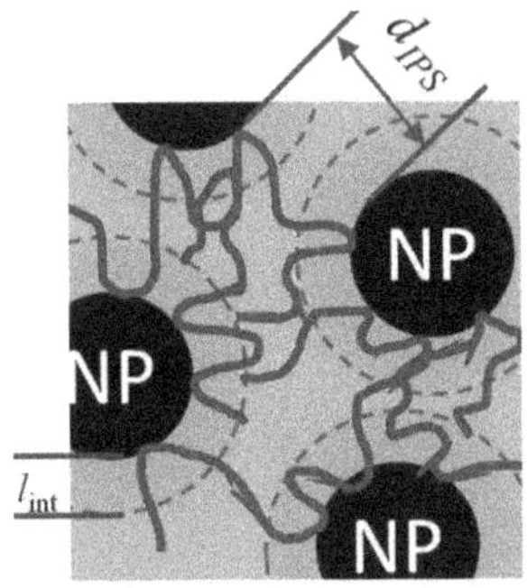

Figure 1. A sketch of the complex geometry of PNCs. The dashed red lines are the virtual boundary between the interfacial layer (green region) and the bulk polymer (grey background). The blue chains represent polymer bridges. The NPs are connected through polymer bridges forming a polymer-NP complex (before percolation) or the NP network (beyond percolation).

In this chapter, broadband dielectric spectroscopy (BDS) as an analytical tool for characterizing the dynamics of the PNCs has been reviewed. The focus is on the recent development of understanding the dielectric spectra of PNCs (Figure 1). For a broader view of the structure-property relationship of PNCs, the readers are referred to several recent review articles (*10–12, 17, 70–73*). The chapter is organized as following: Section II provides a brief introduction to the fundamentals of dielectric relaxation. The basic characteristics of the dielectric spectra of PNCs are presented in Section III. In Section IV, current developments of analyses of dielectric spectra of PNCs are discussed. Section V offers comparisons of results of BDS with H-NMR, neutron scattering, and computer simulations and theories. The combination of dielectric spectroscopy with other macroscopic methods for the structure-property relationship of PNCs is included in Section VI before the summary in Section VII.

Fundamentals of Dielectric Relaxation

In the linear response region, the complex dielectric function, $\varepsilon^*(\omega) = \varepsilon'(\omega) - i\varepsilon''(\omega)$, is associated with the correlation function, $\Phi(t)$, of the polarization fluctuation through (*74–76*)

$$\frac{\varepsilon^*(\omega) - \varepsilon_\infty}{\varepsilon_s - \varepsilon_\infty} = \int_0^\infty \frac{-d\Phi(t)}{dt} \exp(-i\omega t)dt$$

where

$$\Phi(t) = \frac{\langle \Delta P(t)\Delta P(0)\rangle}{\langle (\Delta P(0))^2\rangle}$$

with $\Delta P(t)$ and $\Delta P(0)$ the fluctuation of the polarization P around its equilibrium value at time t and time 0, respectively, ω the angular frequency, i the unit imaginary number, ε_s and ε_∞ the real part of the dielectric permittivity at zero frequency and infinitely high frequency. The correlation function describes the decay of the electric polarization when an external electric field is removed from dielectric materials. Thus, it reflects the characteristic dynamics of the dipolar reorientation of the dielectric materials which is comparable to their structural relaxation or the α relaxation.

For the Debye process of a single relaxation time, τ, the

$$\Phi(t) = \exp(-\frac{t}{\tau})$$

The dielectric dispersion is thus

$$\frac{\varepsilon^*(\omega) - \varepsilon_\infty}{\varepsilon_s - \varepsilon_\infty} = \int_0^\infty \frac{-d\Phi(t)}{dt} \exp(-i\omega t)dt = \frac{1}{1 + i\omega\tau}$$

with the real and imaginary parts are

$$\varepsilon'(\omega) = \varepsilon_\infty + \frac{\Delta\varepsilon}{1 + (\omega\tau)^2}$$

and

$$\varepsilon''(\omega) = \Delta\varepsilon \times \frac{\omega\tau}{1 + (\omega\tau)^2}$$

where

$$\Delta\varepsilon = \varepsilon_s - \varepsilon_\infty \sim g \times \frac{N}{V}$$

is the dielectric amplitude and $\frac{N}{V}$ is the number density of the permanent dipoles and g the Kirkwood-Fröhlich factor. As shown in Figure 2, the Debye process appears as a step in the storage permittivity, $\varepsilon'(\omega)$, and relaxation peak in the loss permittivity, $\varepsilon''(\omega)$, in the frequency domain.

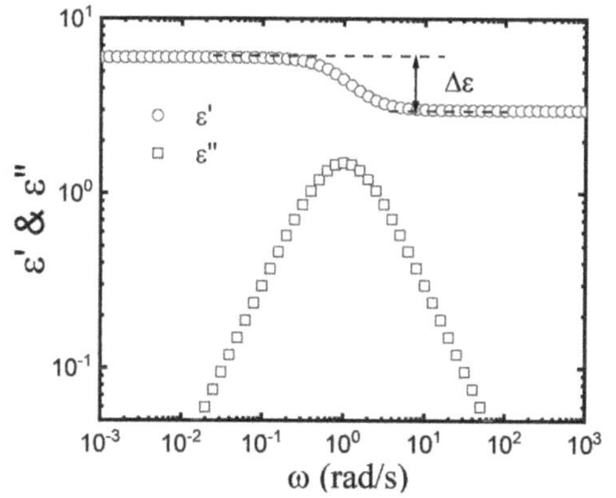

Figure 2. The dielectric spectra of a Debye process with $\varepsilon_\infty = 3.0$, $\tau = 1.0$ s, and $\Delta\varepsilon = 3.0$.

In heterogeneous materials, $\Phi(t)$ is a stretched exponential decay function. The dielectric dispersion is often described by Havriliak-Negami (HN) functions (*75, 77*):

$$\varepsilon^*(\omega) - \varepsilon_\infty = \sum_j \frac{\Delta\varepsilon_j}{(1 + (i\omega\tau_{HN,j})^{\beta_j})^{\gamma_j}}$$

where $\Delta\varepsilon_j$ is the dielectric amplitude of the jth relaxation process, $\tau_{HN,j}$ is the HN relaxation time, and the β_j and the γ_j are the shape parameters of the HN function. To obtain a full fit of the spectra, a dc-conductivity term, $\frac{\sigma_0}{i\omega\varepsilon_0}$, and a Jonscher's law for electrode polarization process, $A\omega^{-n}$, are often included on top of the HN functions with σ_0 the dc-conductivity, ε_0 the vacuum permittivity, A and n are two constants. The HN relaxation time and the relaxation time of the jth process are connected through

$$\tau_j = \tau_{HN,j}\left[sin\frac{\beta_j\pi}{2+2\gamma_j}\right]^{-1/\beta_j}\left[sin\frac{\beta_j\gamma_j\pi}{2+2\gamma_j}\right]^{1/\beta_j}$$

Depending on the time and length scale of the dipolar relaxation, the dielectric measurements can capture both the local dynamics such as the secondary relaxation (also called the sub-glassy dynamics) and the structural relaxation (the glassy dynamics). For polymers with an accumulation of the intrinsic dipolar moment along the main backbone, the dynamics of the whole chain can also be captured as a dielectric normal mode (*78*).

Basic Characteristics of the Dielectric Spectra of Polymer Nanocomposites

PNCs are formed through a physical mixing of the matrix polymer and the NPs. The loading of a PNC is obtained through thermogravimetric analysis. Model PNCs with well-dispersed NPs are often prepared through solution casting followed by a fast solvent evaporation and drying. The details of the model PNC preparation can be found in previous publications (*29, 30, 32*).

The influence of NPs to the dielectric spectra of PNCs has been intensively studied in the past two decades (*15, 21, 27–33, 39, 40, 79–95*), especially for the structural relaxation (the α relaxation). This section presents generic features of the dielectric spectra of PNCs. Figures 3a-3c show a comparison of the dielectric storage permittivity, $\varepsilon'(\omega)$, the derivative spectrum, $\varepsilon'_{der} = -\frac{\pi}{2}\frac{\partial\varepsilon'}{\partial ln\omega}$, and loss permittivity, $\varepsilon''(\omega)$, of neat poly(vinyl acetate) (PVAc) (red) and PVAc/SiO_2 nanocomposites (blue) with $R_{NP} = 12.5\ nm$ and volume fraction $\varphi_{NP} = 23\ \%$. Note the derivative spectra analysis is employed to eliminate the influence of the dc-conductivity to the analysis of the low frequency part of the spectra and it represents an approximation to the $\varepsilon''(\omega)$ (*96*). Several differences between the spectra of the neat PVAc and the spectra of PNCs can be noted: (i) the α relaxation peak is asymmetrically broadened with more changes at the low-frequency wing; (ii) the α relaxation peak position shifts slightly to a lower frequency; (iii) the dielectric amplitude of the α relaxation process of the PNC, $\Delta\varepsilon$, is smaller than the neat polymer; (iv) an additional relaxation process associated with the interfacial polarization shows up, i.e. the Maxwell-Wagner-Sillar's (MWS) polarization; and (v) a shift in σ_0. In addition, the spectra shape and position change in the β relaxation (one type of the secondary relaxation) are observed (Figure 3d). The inset of the

Figure 3d presents the frequency domain of the β relaxation of the neat PVAc and the PVAc/SiO_2 nanocomposites. For PNCs with *type-A* polymer matrices, a significant broadening of the normal mode has also been observed as shown in Figure 3e. Note that a weak MWS process could emerge due to the trapped air inside the sample (97), especially for highly entangled polymers. The weak peak of the neat PVAc at 1 – 100 rad/s in the derivative spectra of Figure 3b might be due to the trapped air bubbles in the PVAc.

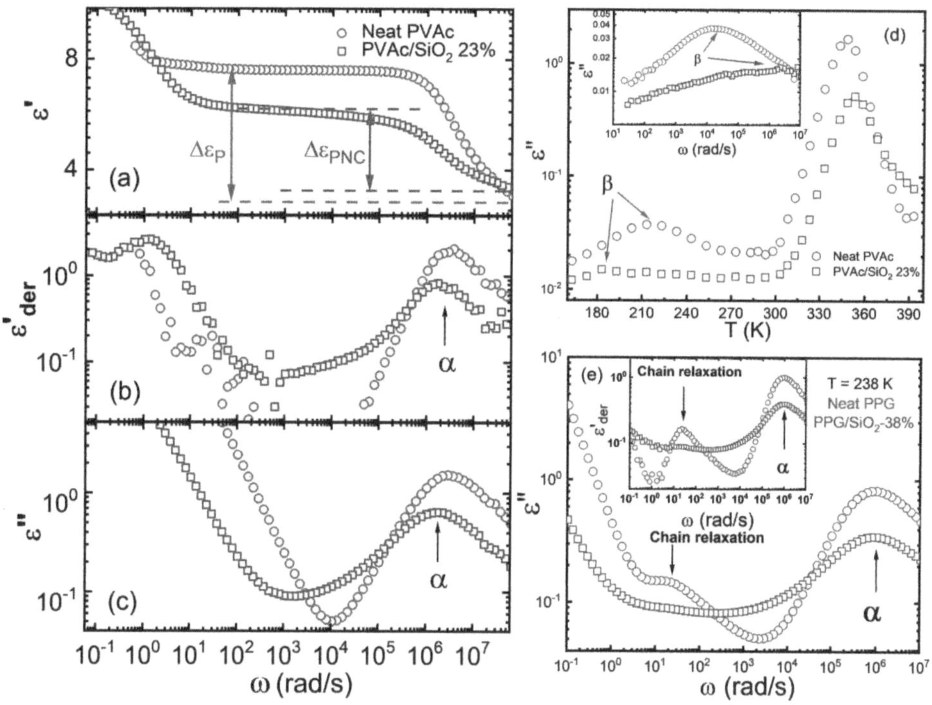

Figure 3. The (a) storage, ε′, (b) derivative, ε'_{der}, and (c) loss, ε″, spectra of the neat PVAc and PVAc/SiO_2-23% (R_{NP} = 12.5 nm). A significant broadening of the α process and the emergence of a MWS process can be observed. (d) The comparison of the ε″ of the neat PVAc and the PVAc/SiO_2-23% at T = 163-393 K and ω = 6.28 × 10^5 rad/s. The inset shows the β relaxation of the neat PVAc and the PVAc/SiO_2-23% in the frequency domain at T = 203 K. (e) The ε″ (mainframe) and the ε'_{der} (inset) of the α and the chain relaxation spectra of the PPG and PPG/SiO_2-38% (R_{NP} = 12.5 nm). Dielectric broadening shows up in the chain relaxation mode as well. Panels (a)-(d) are reproduced with permission from reference (30). Copyright 2017 American Chemical Society. Panel (e) is reproduced with permission from reference (21). Copyright 2020 American Chemical Society.

Analysis of the Dielectric Spectra of the Polymer Nanocomposites

Although the basic features of the dielectric spectra of PNCs have been well-established, their detailed interpretation remains challenging. In this section, the current development of analyzing the dielectric spectra of PNCs will be reviewed and summarized.

The Dielectric Broadening of the Structural Relaxation

The broadening of the structural relaxation of PNCs is often attributed to the presence of the IL between the matrix polymer and the surface of NPs (*32, 33, 80, 87, 89, 91*). For PNCs with attractive polymer-NP interactions, the IL slows down the α relaxation that leads to broadening at the low-frequency side of the spectra. The characteristics of the structural relaxation of PNCs have received intensive attention in the past decade due to their close relationship with the glass transition, the polymer dynamics at the interface, the nanoconfinement effect, and the mechanical reinforcement of the PNCs. Three types of approaches had been proposed to extract the dynamics and the thickness of the IL: the one-HN function approach (*89*), the two-HN function approach (*39, 80, 87*), and the heterogeneous model analysis (HMA) (*29, 30, 32, 33*).

The One-HN Function Approach

In one-HN function fit for the structural relaxation (*89*), the complex permittivity function of PNCs is fit through

$$\varepsilon^*(\omega) = \varepsilon_\infty + \frac{\Delta\varepsilon_\alpha}{\left[1 + \left(i\omega\tau_{HN,\alpha}\right)^\beta\right]^\gamma}$$

where ε_0 is the vacuum permittivity and $\Delta\varepsilon_\alpha$ is the dielectric amplitude of the α process. The two shape parameters are assumed to connect through $\gamma = 1 - 0.812 \times (1 - \beta)^{0.387}$ and $\delta = (\beta\gamma)^{1/1.23}$ with δ being the KWW stretching exponent of the polarization function, $\Phi(t)$ (*89*). Thus, there are three free parameters in the one-HN function fit: the HN relaxation time, $\tau_{HN,\alpha}$, the dielectric amplitude of the relaxation process, $\Delta\varepsilon_\alpha$, and the stretching exponent, δ.

The one-HN approach assumes the IL is dielectric inactive in the frequency window and the broadening of the structural relaxation is purely from the increased dynamic heterogeneity of the matrix polymer. Since the dielectric amplitude is directly proportional to the number density of the permanent dipoles of PNCs, the ratio of the $\Delta\varepsilon_\alpha/\Delta\varepsilon_P$ represents the volume fraction of the bulk polymer, $\varphi_b = \Delta\varepsilon_\alpha/\Delta\varepsilon_P$, with $\Delta\varepsilon_P$ being the dielectric amplitude of the neat matrix. As a result, the volume fraction of the IL is $\varphi_{int} = 1 - \varphi_b - \varphi_{NP}$ and the interfacial layer thickness

$$l_{int} = \left(\left(\frac{\varphi_{int} + \varphi_{NP}}{\varphi_{NP}}\right)^{1/3} - 1\right) R_{NP}$$

The advantage of this approach is a simple analysis and a small number of free parameters. The one-HN function fit assumes no interfacial layer contribution to the dielectric spectra of PNCs even for temperatures well above the glass transition temperature of the PNC, T_g. Thus, it does not provide explicit dynamics of the IL and often over-estimates the φ_b. Moreover, the one-HN function can only fit well at a limited frequency range. Figures 4a-4c show a detailed one-HN function fit of the dielectric spectra of the PVAc and PVAc/SiO_2-23%. A clear underestimate of the dielectric amplitude of the PNC has been observed (Figure 4a). When σ_0 was strongly suppressed, the dielectric broadening at the intermediate frequency cannot be fully described (Figures 4b-4c).

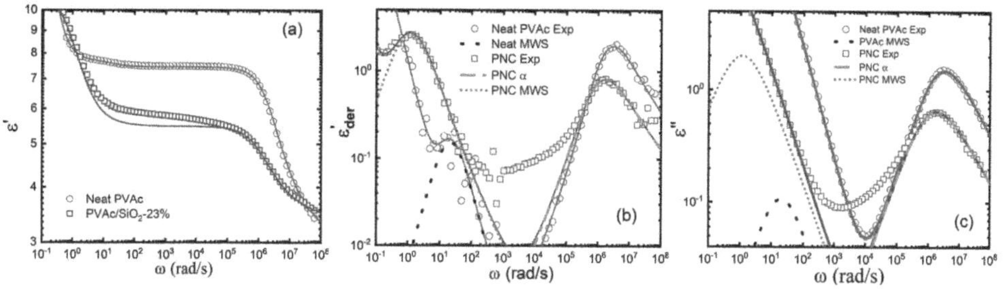

Figure 4. One-HN function fit of (a) the ε', (b) the ε'_{der}, and (c) the ε" of the neat PVAc and the PVAc/SiO_2-23% nanocomposites. The solid lines are the sum of all relaxation processes. The MWS process and the conductivity contribution are included to fit the spectra of the whole frequency range. The orange dashed-lines in panels (b) and (c) are the one-HN fit of the α relaxation process. Reproduced with permission from reference (30). Copyright 2017 American Chemical Society.

The Two-HN Function Approach

The two-HN function approach receives the most attention and has been widely employed for the analysis of the dielectric spectra of PNCs (*37–40, 80, 87, 90, 91, 93, 98, 99*). In this approach, both the IL and the bulk polymer contribute to the dielectric dispersion. More specifically, the IL has distinct dynamics. The structural relaxation of the IL broadens the dielectric dispersion of the PNCs. The complex dielectric function over the whole frequency window is

$$\varepsilon^*(\omega) = \varepsilon_\infty + \frac{\Delta\varepsilon_{int}}{\left[1+\left(i\omega\tau_{HN}^{int}\right)^{\beta_{int}}\right]^{\gamma_{int}}} + \frac{\Delta\varepsilon_b}{\left[1+\left(i\omega\tau_{HN}^{b}\right)^{\beta_b}\right]^{\gamma_b}} + \frac{\Delta\varepsilon_{MWS}}{[1+(i\omega\tau_{HN}^{MWS})^{\beta_{MWS}}]^{\gamma_{MWS}}} + \frac{\sigma_0}{i\varepsilon_0\omega}$$

where the $\Delta\varepsilon_x$ and the τ_{HN}^x $(x = int, b, and\ MWS)$ are the dielectric amplitudes and the HN relaxation time of the IL, the bulk polymer, and the MWS process. We denote the α relaxation time of the bulk polymer and the interfacial layer as τ_α and τ_{int}. In the two-HN function fit, the shape parameters of the bulk polymer relaxation are set to be identical to the neat polymer. Thus, the volume fraction of the interfacial layer is

$$\varphi_{int} = (1-\varphi_{NP}) \times \frac{\Delta\varepsilon_{int}}{\Delta\varepsilon_{int} + \Delta\varepsilon_b}$$

with $\Delta\varepsilon_{int} + \Delta\varepsilon_b = (1-\varphi_{NP})\Delta\varepsilon_P$ (*80*). The latter indicates all the active permanent dipoles of the matrix contribute to the dielectric relaxation of PNCs and there is no "dead" layer in PNCs. The two-HN function can provide a much better fit of the dielectric spectra of PNCs with two important outcomes: the characteristic time, τ_{HN}^{int}, and the volume fraction, φ_{int}. Figures 5a-5c present the two-HN function fit of the PVAc and PVAc/SiO_2-23%. The orange, purple, and green dashed-lines represent the dielectric dispersion of the bulk polymer, the IL, and the MWS polarization, respectively. The solid lines are the sum of them.

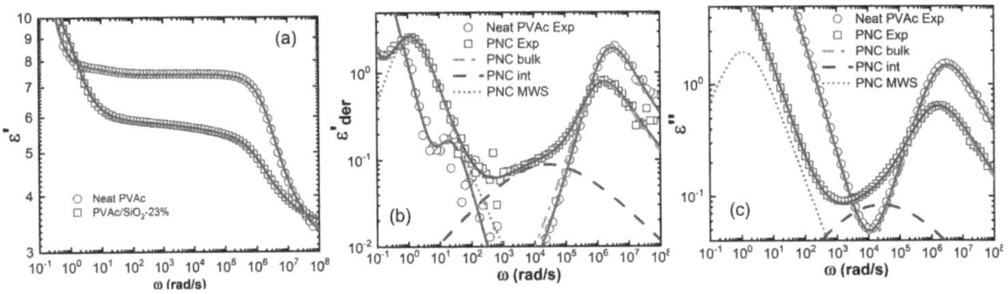

Figure 5. The two-HN function fit of (a) the ε', (b) the ε'_{der}, and (c) the ε" of the neat PVAc and the PVAc/SiO_2-23%. The solid lines are the sum of the contributions of the individual relaxation process. The MWS process and the conductivity contribution are included to fit the spectra of the whole frequency range. The orange dashed-lines and the purple lines in panels (b) and (c) represent the relaxation of the bulk polymer and the interfacial layer. Reproduced with permission from reference (30). Copyright 2017 American Chemical Society.

The two-HN function approach underlies the following assumptions: (i) the NP does not contribute to the dielectric function of the PNCs, (ii) the dielectric response of the IL and the bulk polymer contribute independently to the dielectric spectra, and (iii) the IL and the bulk polymer share an identical Kirkwood-Fröhlich factor, g, which was proven to be not valid recently (27). As shown in following sections, the first two assumptions are also found not appropriate.

The Heterogeneous Model Analysis (HMA)

The heterogeneous model analysis (HMA) computes the effective dielectric functions (*30, 33, 100*) of the PNCs by taking into account the contributions of the NP and the matrix polymer, based on either a two-phase model (TPM, Figure 6a) or an interfacial layer model (ILM, Figure 6b).

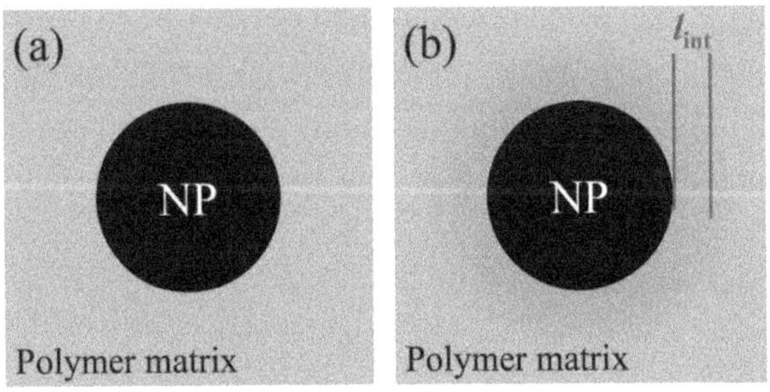

Figure 6. The geometry of the (a) two-phase model (TPM), and the (b) interfacial layer model (ILM).

In the TPM model,

$$\begin{aligned}\varepsilon^*_{PNC}(\omega) &= \varepsilon^*_b(\omega) \times \frac{(1+2\varphi_{NP})\varepsilon^*_{NP}(\omega) + (2-2\varphi_{NP})\varepsilon^*_b(\omega)}{(1-\varphi_{NP})\varepsilon^*_{NP}(\omega) + (2+\varphi_{NP})\varepsilon^*_b(\omega)} \\ &= (1-\varphi_{NP})\varepsilon^*_b(\omega) + \varphi_{NP}\varepsilon^*_{NP}(\omega) - (1-\varphi_{NP})\varphi_{NP} \times \frac{\left(\varepsilon^*_b(\omega) - \varepsilon^*_{NP}(\omega)\right)^2}{(1-\varphi_{NP})\varepsilon^*_{NP}(\omega) + (2+\varphi_{NP})\varepsilon^*_b(\omega)} \\ &= (1-\varphi_{NP})\varepsilon^*_b(\omega) + \varphi_{NP}\varepsilon^*_{NP}(\omega) - (1-\varphi_{NP})\varphi_{NP}\varepsilon^*_{cross}(\omega)\end{aligned} \quad (1)$$

Here, $\varepsilon^*_{NP}(\omega)$ is the dielectric function of the NP, and

$$\varepsilon^*_{cross}(\omega) \equiv \frac{\left(\varepsilon^*_b(\omega) - \varepsilon^*_{NP}(\omega)\right)^2}{(1-\varphi_{NP})\varepsilon^*_{NP}(\omega) + (2+\varphi_{NP})\varepsilon^*_b(\omega)}$$

Note that the first two terms of eq.1 represent the additive contributions of the matrix and the NP and the third term signifies the interference term between them. Since the dielectric amplitude of $\varepsilon^*_{cross}(\omega)$ is non-negative, eq.1 indicates the dielectric amplitude of the PNCs is smaller than the sum of the individual component, in sharp contrast to the basic assumption of the two-HN function. $\varepsilon^*_b(\omega)$, $\varepsilon^*_{NP}(\omega)$, and φ_{NP} are parameters that can be determined in advance. Thus, there are no free parameters of the TPM model. As shown in following sections, the simple TPM model reproduces many features of the dielectric response of PNCs and offers a benchmark for further discussion of the dielectric properties of PNCs.

To take into account the IL, the ILM model predicts (*100*):

$$\varepsilon^*_{PNC}(\omega) = \frac{\varepsilon^*_{NP}(\omega)\varphi_{NP} + \varepsilon^*_{int}(\omega)\varphi_{int}R^* + \varepsilon^*_b(\omega)\varphi_b S^*}{\varphi_{NP} + \varphi_{int}R^* + \varphi_b S^*} \tag{2}$$

with

$$R^* = \frac{2\varepsilon^*_{int}(\omega) + \varepsilon^*_{NP}(\omega)}{3\varepsilon^*_{int}(\omega)}$$

$$\varphi_{NP} + \varphi_{int} + \varphi_b = 1.0$$

and

$$S^* = \frac{\left(\varepsilon^*_{int}(\omega) + 2\varepsilon^*_b(\omega)\right)\left(\varepsilon^*_{NP}(\omega) + 2\varepsilon^*_{int}(\omega)\right) + 2d(\varepsilon^*_{int}(\omega) - \varepsilon^*_b(\omega))(\varepsilon^*_{NP}(\omega) - \varepsilon^*_{int}(\omega))}{9\varepsilon^*_{int}(\omega)\varepsilon^*_b(\omega)}$$

where $d = \varphi_{NP}/(\varphi_b + \varphi_{NP})$. The free parameters of the ILM are the $\varepsilon^*_{int}(\omega)$ and the φ_{int}.

Figure 7 shows the TPM fit of the glycerol/SiO_2 ($\varphi_{NP} = 23.6\%$) (Figure 7a) and PVAc/SiO_2 ($\varphi_{NP} = 23\%$) (Figures 7b-7d). Although the TPM model cannot fully describe the dielectric dispersion of nanocomposites, several important features are worth noting: (i) compared with the neat matrix, an increase in the dielectric broadening of the PNCs has been observed for both the glycerol/SiO_2 and the PVAc/SiO_2. The broadening is especially notable in glycerol/SiO_2 due to the large dielectric amplitude of the neat glycerol. (ii) The TPM predicts a reduction in $\Delta\varepsilon$. Thus, despite the naïve assumption of the identical dynamics of the matrix with the neat substance, the TPM qualitatively captures the leading features of the dielectric broadening and dielectric amplitude reduction of PNCs.

The IL is introduced to account for the extra broadening of the structural relaxation of the PNCs. As shown in Figure 7, the ILM offers a much better fit for the dielectric functions of the PNC over the entire frequency range. In the ILM, the $\varepsilon^*_{int}(\omega)$ is described by one-HN function with structural relaxation, τ_{int}. In the ILM analysis, the structural relaxation time of the bulk matrix is assigned as τ_α. The fit of the ILM to the dielectric spectra of the PNC is very sensitive to the explicit value of φ_{int} (*30*).

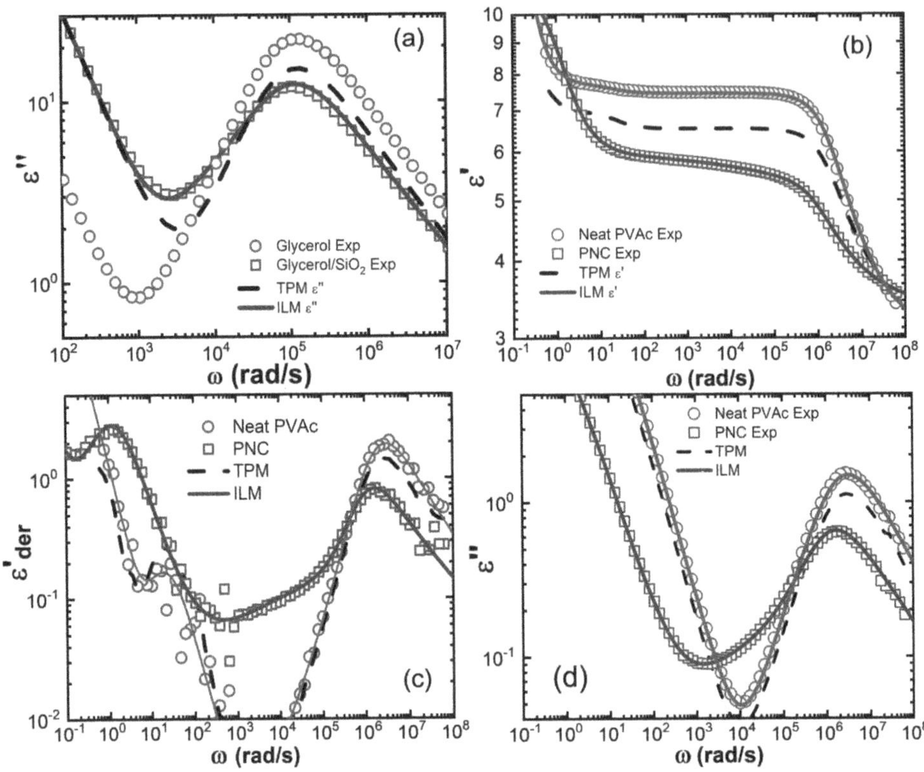

Figure 7. (a) The heterogeneous model analysis (including both the TPM and the ILM) of the dielectric loss spectra, ε", of glycerol/SiO_2-23.6% at T = 233 K. (b-d) The heterogeneous model analysis of the ε' (b), the ε'_{der} (c), and the ε" (d) of the neat PVAc and PVAc/SiO_2-23% at T = 373 K. The solid red line in Panels (b-d) are the HN function fit to the spectra of the neat PVAc. Panel (a) is reproduced with permission from reference (33). Copyright 2015 American Institute of Physics. Panels (b-d) are reproduced with permission from reference (30). Copyright 2017 American Chemical Society.

Additionally, the IL as an additional relaxation process can also be justified by the HMA analysis. Note that there are no free parameters in the TPM model. Taken the experimental measurements of the $\varepsilon^*_{PNC}(\omega)$ as an input in eq.1, the effective dielectric response of the matrix, $\varepsilon^*_{eff}(\omega)$, can be obtained by solving the reverse TPM model. Conversion of the, $\varepsilon^*_{eff}(\omega)$, in the frequency domain into the time domain enables better resolving overlapping relaxation processes (*101*). For example, the dielectric function can be described by a superposition of Debye functions, $\varepsilon^*_{eff}(\omega) = \varepsilon_\infty + \Delta\varepsilon \int \frac{g(ln\tau)}{1+i\omega\tau} dln\tau + \frac{\sigma}{i\varepsilon_0\omega}$ with $\int g(ln\tau)dln\tau = 1$, where $g(ln\tau)$ is the relaxation time spectra. Note that the relaxation time analysis does not mean the structural relaxation of the matrix is a simple superposition of multiple Debye relaxation processes. It is applied to qualitatively differentiate the relaxation processes of $\varepsilon^*_{eff}(\omega)$. Figure 8 shows the relaxation time spectra of the PVAc/SiO_2 through a generalized regularization method (*102*). The emergence of a shoulder peak at the long τ tail supports the presence of the IL in PNCs.

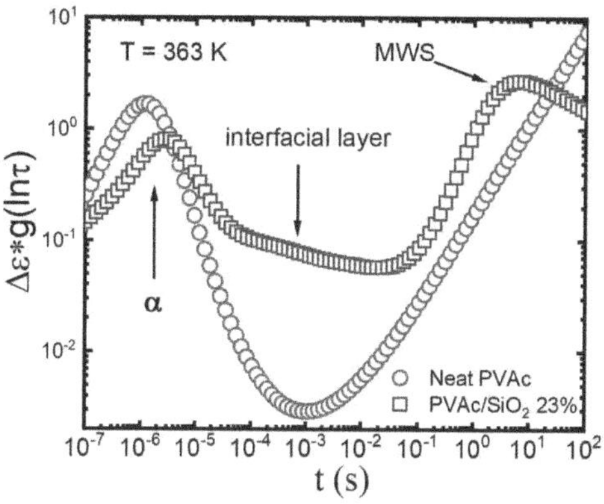

Figure 8. Relaxation time distribution analysis of the neat PVAc (red circles) and the PVAc/SiO_2-23% nanocomposites (blue squares) at T = 363 K. A shoulder peak shows up in the relaxation time distribution spectra of the PNC that is attributed to the structural relaxation of the interfacial layer. The MWS process can also be identified from the relaxation time distribution analysis.

The Comparison of the Three Methods

The aforementioned three methods focus on the dynamics and thickness of the IL. The one-HN function approach assumes a dielectric inactive layer of the IL and does not offer the relaxation time of the IL. The two-HN function approach assumes additive contributions of the bulk polymer and the IL. The HMA models adopt the effective medium approach with the dynamics and the volume fraction of the IL as two independent free parameters.

Figures 9a-9b show the comparisons of the $l_{int}(T)$, the $\tau_{int}(T)$, and the $\tau_\alpha(T)$ from the three analyses for the same PVAc/SiO_2. The two-HN function shows the smallest values of the $l_{int}(T)$ that remains almost unchanged with temperature. Both the ILM and the one-HN function exhibit an increase in the $l_{int}(T)$ upon cooling, and much stronger temperature dependence than the two-HN function. The two HN-function and the ILM share an almost identical temperature dependence of the $\tau_{int}(T)$ and the τ_α. The two-HN function has a slightly higher $\tau_{int}(T)$ and $\tau_{int}(T)/\tau_\alpha$ that the ILM.

To further compare the different analyses of the dynamics of the IL, one can extract the non-adsorbed polymers from the PNC and make direct measurements of the PNCs of NPs with adsorbed polymers. The dielectric function of the NPs with adsorbed polymers should be a close approximation of that of the IL. Figure 10 shows such a measurement of PVAc/SiO_2 before (blue squares) and after (green diamonds) taking out the non-adsorbed polymers. The red circles in Figure 10 represent the dielectric function of the neat polymer. Compared with the dielectric function of the IL of the two-HN function fit, the IL of the ILM offers better agreement with the experiments of the polymer adsorbed NPs (Ad-NPs). Therefore, the ILM provides the most accurate analysis of the dielectric spectra of PNCs and will be employed in the discussions in a following section.

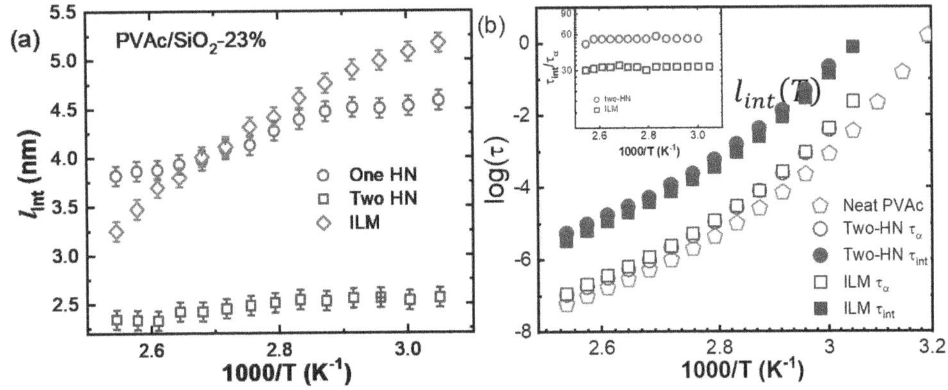

Figure 9. (a) The comparison of the thickness of the interfacial layer, $l_{int}(T)$, from the three different methods on the same set of dielectric spectra of the neat PVAc and the PVAc/SiO_2-23%. (b) The temperature dependence of the structural relaxation time of the interfacial layer, $\tau_{int}(T)$, and the bulk polymer, $\tau_{\alpha}(T)$, of the neat PVAc and PVAc/SiO_2-32%. The inset shows the $\tau_{int}(T)/\tau_{\alpha}(T)$ of the two-HN fit and the ILM analysis. Reproduced with permission from reference (30). Copyright 2017 American Chemical Society.

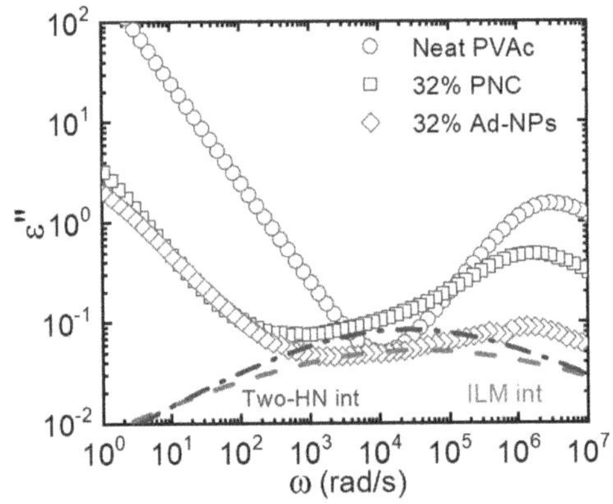

Figure 10. The dielectric loss of the neat PVAc (red circle), the PVAc/SiO_2-32% (blue square), and the polymer adsorbed NPs (Ad-NPs) after extracting all the non-adsorbed polymers from the PVAc/SiO_2-32%. Reproduced with permission from reference (30). Copyright 2017 American Chemical Society.

The Chain Dynamics

The dielectric normal mode offers the chain dynamics of the matrix. However, both the adsorbed polymers and the non-adsorbed polymers contribute to the dielectric normal modes (*103*), rendering it difficult to obtain a detailed interpretation of the chain dynamics of PNCs. These complications include but not limited to: (i) the chain dynamics of the adsorbed polymers depend strongly on the structural details of the adsorbed polymers, especially the length and population of the dangling chain ends (*103, 104*); (ii) The structure and dielectric dispersion of the adsorbed polymers (one or two ends anchored on to the surface of the NPs) can be very different from the neat polymer (*103, 105*); (iii) The frictional coefficient among segments at the IL can be much larger than the bulk polymer (*106*); (iv) The chain dynamics of the non-adsorbed polymers are strongly mediated by the nano-confinement effect (*20*). Therefore, the current analysis of the chain dynamics of PNCs with type-A polymer matrix mostly focuses on a qualitative analysis of the average relaxation time of the polymer as well as their temperature dependence (*21, 95*).

The Secondary Dynamics

Since the secondary relaxation only involves local dynamics, it has been assumed to remain unchanged in PNCs previously (*80*). The amplitude of the secondary relaxation peak has been employed as an internal indicator for accurate quantification of the sample thickness of the PNC in the dielectric measurements. However, recent studies showed the NPs modify strongly the secondary relaxation of the PNCs (*30*). As shown in Figure 3d and Figure 11, the shape, the relaxation time, and the activation energy of the β relaxation, E_β, of the PVAc/SiO_2 change with φ_{NP}. Interestingly, the peak value of E_β occurs at

$$d_{IPS} = \left(\left(\frac{16}{\pi\varphi_{NP}}\right)^{1/3} - 2\right) R_{NP} \approx 2R_g$$

where the frustrated chain packing starts to emerge (*32*). At $d_{IPS} > 2R_g$, the attractive interfacial interactions increase the activation energy of the secondary relaxation. At $d_{IPS} < 2R_g$, frustrated polymer packing can speed up the secondary relaxation and reduce the activation energy (*30–32, 41*). Thus, the non-monotonic change of the E_β with φ_{NP} suggests strong influences of NPs on the sub-glassy dynamics of the matrix polymer.

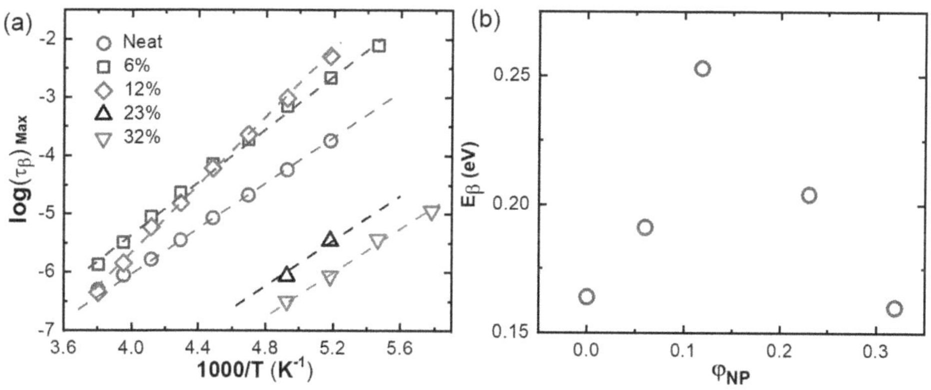

Figure 11. (a) The temperature dependence of the β relaxation time, τ_β, of PNCs of different NP loadings, φ_{NP}. The activation energy of the β relaxation, E_β, can be calculated from the slope of the log(τ_β) vs 1000/T. (b) The activation energy of PNCs of different φ_{NP}, where a non-monotonic NP loading dependence of the E_β has been observed. Reproduced with permission from reference (30). Copyright 2017 American Chemical Society.

The Maxwell-Wagner-Sillar's (MWS) Interfacial Polarization

The MWS interfacial polarization is another unique feature of the dielectric response of heterogeneous materials that impacts the entire spectra (*100*). A leading feature of the MWS polarization is the emergence of the MWS process from the dielectric or conductivity mismatch between the matrix and NPs (*77, 100*). The surface sensitivity of the MWS process enables a quantification of the surface and the microstructure of the PNCs (*82*, *107*, *108*). The NP aggregation and the formation of the NP physical network correlate strongly with the dielectric amplitude of the MWS process (*108*). Thus, the percolation threshold of the NPs can be identified through tracking the dielectric amplitude of the MWS that was found comparable with the results from rheological measurements (*108*).

The Comparison of Dielectric Measurements with Other Techniques.

The Interfacial Layer Thickness

The IL has been found in many other measurements, including H-NMR measurement (*24*, *26*, *66*), differential scanning calorimetry (DSC) (*29*, *30*, *33*, *80*), small-angle x-ray scattering (SAXS) (*29*) or small-angle neutron scattering (SANS) (*109*, *110*), and atomic force microscopy (AFM) (*64*). Figure 12a shows a summary of the thickness of the IL analyzed with the HMA, including glycerol/SiO_2, poly(propylene glycol) (PPG)/SiO_2, poly(methyl methacrylate) (PMMA)/SiO_2, poly(vinyl acetate)(PVAc)/SiO_2, and poly(2-vinyl pyridine) (P2VP)/SiO_2. The l_{int} are presented in terms of the characteristic ratio of the matrix, C_∞. For glycerol, $C_\infty = 1.0$. Interestingly, the results of DSC, SAXS, and AFM measurements are all comparable with each other and slightly smaller than the BDS measurements. A similar IL thickness identified from BDS and DSC has also been found in simple liquids (*111*). However, different from the two-step glass transition of simple liquids, the glass transition of PNCs typically appears as a very broad step (*29*, *30*). The systematic differences of the results among different techniques have been attributed to the sensitivity of the different techniques (*29*).

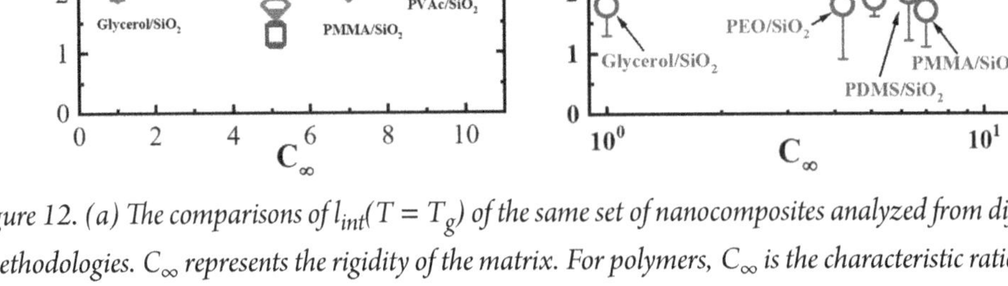

Figure 12. (a) The comparisons of $l_{int}(T = T_g)$ of the same set of nanocomposites analyzed from different methodologies. C_∞ represents the rigidity of the matrix. For polymers, C_∞ is the characteristic ratio. (b) Comparison of the $l_{int}(T)$ at $T = 1.20 - 1.30T_g$ from BDS analysis with literature results of other types of polymers from H-NMR or neutron scattering measurements. The dashed line is the result of molecular dynamics simulations. PDMS: polydimethylsiloxane; PEO: poly(ethylene oxide). Panel (a) contains data from reference (29). Panel (b) reproduced with permission from reference (29). Copyright 2017 American Chemical Society.

Figure 12b summarizes the thickness of the IL from almost all available results up to now. Due to the different temperature range and different dynamic window of different types of measurements, the comparison is made at $T = 1.20 - 1.30\ T_g$. The symbols are experimental data with error bars covering the results from different research groups, and the dashed lines are the results of recent molecular dynamics simulations (*112*). The results of the BDS are also consistent with NMR, SANS, and computer simulations. These results also reveal the chain rigidity as an important parameter controlling the thickness of the IL (*15, 29, 112*).

The Temperature Dependence of l_{int} and τ_{int}

Analyzing the dielectric spectra of PNCs at different temperatures allows the quantification of the temperature dependence of $l_{int}(T)$ and $\tau_{int}(T)$. As shown in Figure 13, the $l_{int}(T)$ decreases with heating up for a number of nanocomposites. The temperature dependence of $l_{int}(T)$ is quite generic according to experiments from different groups and different methods (*91, 113*). Note that in analyses of the H-NMR spectra, the "dead" layer or "glassy" layer has often been invoked. However, recent studies found no solid evidence of the "dead" or the "glassy" layer. Instead, the slowing down in polymer dynamics is a generic feature of IL. The discrepancies between H-NMR and dielectric measurements may be from the capability to access slow dynamics up to 100 s in dielectric measurements, while H-NMR typically measures the dynamics faster than 10^{-6} s.

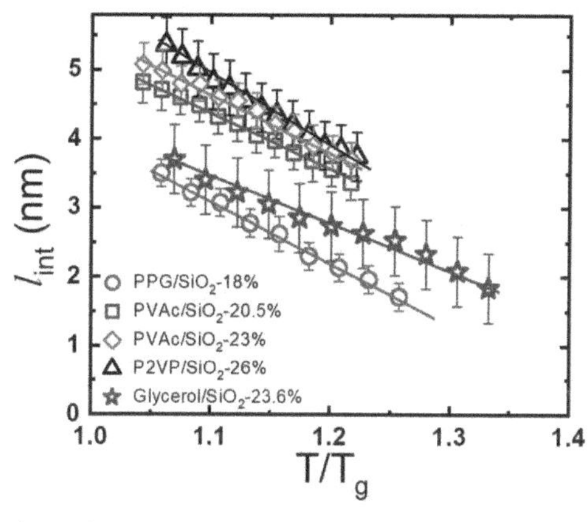

Figure 13. The temperature dependence of $l_{int}(T)$ of different PNCs from the dielectric measurements with the ILM analysis. A linear relationship $l_{int}(T) = -BT/T_g + C$ can be visualized for all PNCs, which has been reproduced by recent theoretical developments. Reproduced with permission from reference (29). Copyright 2017 American Chemical Society.

The characteristics of $l_{int}(T)$ and τ_{int} have also been a focus of computer simulations (*114, 115*) and theories (*14, 34, 35, 116–120*). For instance, a close look at the $l_{int}(T)$ reveals a linear relationship of $l_{int}(T) = -BT/T_g + C$ where *B* and *C* are constants (Figure 13) (*29*). A similar linear relationship of the thickness of the interfacial layer and the temperature has also been found in recent computer simulations (*114*). Recent theoretical development enabled predictions of the linear temperature dependence of the $l_{int}(T)$ and the almost constant ratio of τ_{int}/τ_α of PNC (*117*), consistent with the dielectric measurements. These results point to the molecular origin of the formation of the IL in PNCs (*14*).

The Chain Dynamics

The chain dynamics of PNCs are very important for the rheology and processing of PNCs, yet it is very challenging to quantify. Figure 3e shows the dielectric measurements of PPG/SiO_2 with a clear normal mode. Interestingly, the characteristic time of the normal mode of the PPG/SiO_2 is found to be identical with the neat polymer (Figure 14), regardless of the loading and the molecular weight of the PNC (*21*, *95*). However, chain diffusion measurements from Rutherford backscattering show significant slowing down in the chain diffusion of PNCs, which rely strongly on the nano-confinement effect in PNCs (*20*, *121*). It is not clear what leads to the discrepancy on the outcomes of the different measurements. Future development of a detailed analysis of the dielectric normal mode of PNCs should offer new insights.

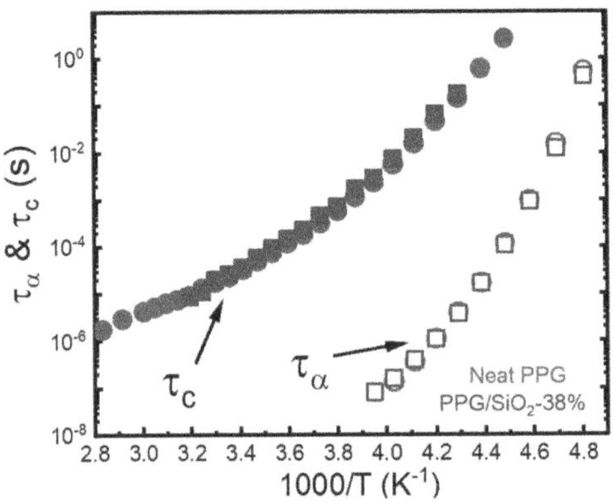

Figure 14. Comparison of the segmental, τ_α, and the chain relaxation time, τ_C, of the neat PPG and the PPG/SiO_2-38% nanocomposite. The characteristic time of the dielectric normal mode in PNC is almost identical to the neat polymer. Reproduced with permission from reference (21). Copyright 2020 American Chemical Soceity.

The Combination of Dielectric and Other Techniques

In this section, the implementation of BDS to other macroscopic measurements is presented. Three examples will be presented here: (i) decoupling the polymer dynamics and the NP network dynamics; (ii) unraveling the characteristics of the desorption dynamics; and (iii) delineating the mechanical reinforcement effect of glassy PNCs.

Decoupling the Polymer Dynamics and NP-Network Dynamics

PNCs exhibit very rich slow dynamics, including the Rouse dynamics of the adsorbed polymer, the reptation dynamics of the non-adsorbed polymer, and the multi-time scale relaxation of the polymer-NP complex or the NP-network. All of them are strongly coupled, and separation their contribution to dielectric spectra is a challenging task (*21*). Dielectric measurements that can selectively probe the dynamics of the matrix polymer offer a unique way of decoupling the polymer dynamics from the dynamics of the PNCs. Despite the challenge of detailed analyses of the dielectric spectra of PNCs, the temperature dependence of the segmental and the chain relaxation can be obtained due to the negligible temperature dependence of the shapes of the relaxation processes (*21*). Figure 15 shows the shift factors, a_T^R and a_T^B, of PNCs from the linear viscoelastic measurements (symbols) and the dielectric measurements (lines). The a_T^R are from the time-temperature superposition and the a_T^B are from direct measurements.

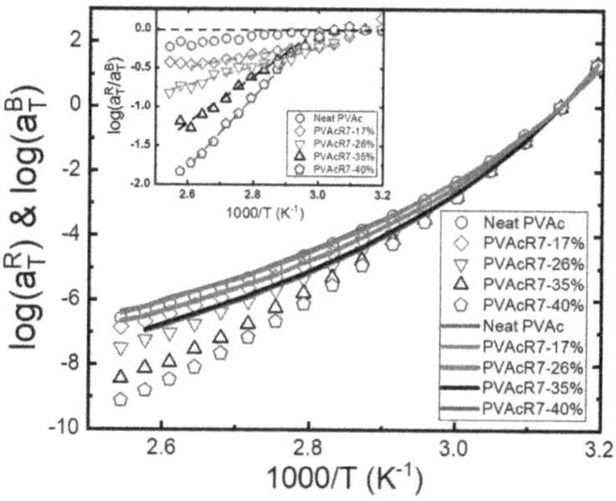

Figure 15. The shift factors from rheological measurement (a_T^R, empty symbols) and the BDS (a_T^B, lines) of PNCs of different φ_{NP}. The inset shows the linear relationship between the $\log(a_T^R/a_T^B)$ and the 1000/T above a certain temperature for PNCs. Reproduced with permission from reference (21). Copyright 2020 American Chemical Society.

The comparison between a_T^R and a_T^B is revealing: (i) a_T^R varies much more with loading than a_T^B. (ii) a_T^R and a_T^B follow very different temperature dependencies. Since a_T^B reflects the dynamics of the polymer matrix, the different temperature dependence of a_T^R indicates the rheological measurements captures the contributions of both the polymer and the NPs. Applying the pre-deformation to destruct the NP network with polymer bridges can restore the shift factors, a_T^R, to that of the neat polymer (*21*). Thus, the shift factors obtained from rheological measurements contain the contribution of both the matrix polymer and the NP-network. The characteristics of the dynamics of the matrix polymer can be decoupled from that of the PNCs through a combination of the rheology and BDS.

Unraveling the Characteristics of the Desorption Dynamics

For PNCs with strong NP-network, the dynamics of the PNCs at high temperatures (or low frequencies) are primarily controlled by the desorption of the polymer bridges, which is an activation process. A detailed analysis of the comparison of the a_T^R and the a_T^B can offer a decoupling of the dynamics of the matrix polymer and allow a quantification of the dynamics of polymer desorption. $\tau_{ad}(T) = \tau_\alpha(T)\exp\left(\frac{\Delta E}{RT}\right)$ with ΔE the average desorption energy of the polymer bridge, $\tau_{ad}(T)$ the desorption time, $\tau_\alpha(T)$ the α time of the matrix, and R the gas constant. The a_T^R at high temperatures reflect the temperature dependence of the $\tau_{ad}(T)$:

$$a_T^R = \frac{\tau_{ad}(T)}{\tau_{ad}(T_{ref})} = \frac{\tau_\alpha(T)\exp\left(\frac{\Delta E}{RT}\right)}{\tau_\alpha(T_{ref})\exp\left(\frac{\Delta E}{RT_{ref}}\right)} = a_T^B \exp\left(\frac{\Delta E}{R}\left(\frac{1}{T} - \frac{1}{T_{ref}}\right)\right)$$

In other words,

$$\ln\left(\frac{a_T^R}{a_T^B}\right) = \frac{\Delta E}{R}\left(\frac{1}{T} - \frac{1}{T_{ref}}\right)$$

and the slope of the plot of $\left(\frac{a_T^R}{a_T^B}\right)$ vs $1000/T$ offers a quantification of the activation energy, ΔE. Figure 16 presents a detailed analysis with $\frac{\Delta E}{E_0}\sim 1-6$ with $E_0 = 20\ kJ/mol$ being the energy of one hydrogen bond. Moreover, the absolute values of the $\Delta E/E_0$ scale with the average number of polymer bridges between two adjacent NPs, $\left(d_{IPS}/R_g\right)^{-2}$, as shown in the inset of Figure 16, further confirming the polymer desorption dominated dynamics in a_T^R. Thus, the combination of BDS and rheology enables a detailed quantification of desorption dynamics (*21*) that otherwise cannot be easily obtained.

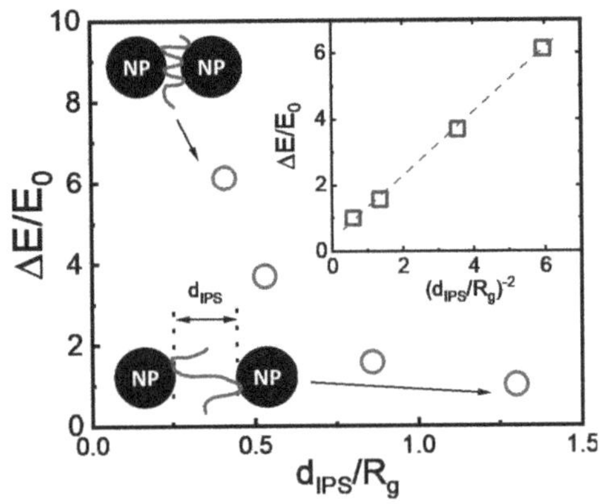

Figure 16. Activation energy estimated from the slope of the (a_T^R/a_T^B) vs 1000/T line shown in the inset of Figure 15. The inset shows a linear relationship between the $\Delta E/E_0$ and the $(d_{ips}/R_g)^{-2}$, implying polymer bridge desorption as the leading contribution to the activation energy. Data are from reference (21).

Delineating the Mechanical Reinforcement of Glassy PNCs

The dynamic-mechanical coupling indicates a much higher modulus at the IL than the bulk polymer away from the NPs that is known to play a significant role in the high mechanical strength of PNCs (*64*). The thickness of the IL is thus important to quantify the mechanical reinforcement of PNCs.

Figure 17a shows the shear (mainframe) and bulk (inset) modulus of PVAc/SiO_2 at different loadings from Brillouin light scattering (*64*). The black and the blue lines are the predictions from the simple mixing rule (the Wood's law) and the continuum calculations based on the geometry of the TPM, both of which predict significantly smaller outcomes than experiments. In glassy PNCs with well-dispersed NPs, the higher modulus from experiments is attributed to the high modulus of the IL. However, to understand the mechanical properties of the IL and their roles to the mechanical reinforcement, one has to know the l_{int} first. Adopting the thickness of the l_{int} from the BDS measurements, the shear, bulk, and Young's modulus of the IL can be identified analytically from the continuum calculations based on the same geometry of the ILM model. The blue line and the dashed orange line are the results from the ILM model and the finite element analysis (FEA), yielding similar results. The thickness of the IL from BDS is found consistent with that of the AFM measurement at the nanoscale (Figure 17b) from mechanical properties. Moreover, taking l_{int} from

BDS measurements as an input parameter, the Young's modulus of the IL from continnum calculation agrees excellent with the AFM measurements through indentation. Thus, BDS can also help delineate the mechanics of the IL and the mechanical reinforcement of PNCs in the glassy state.

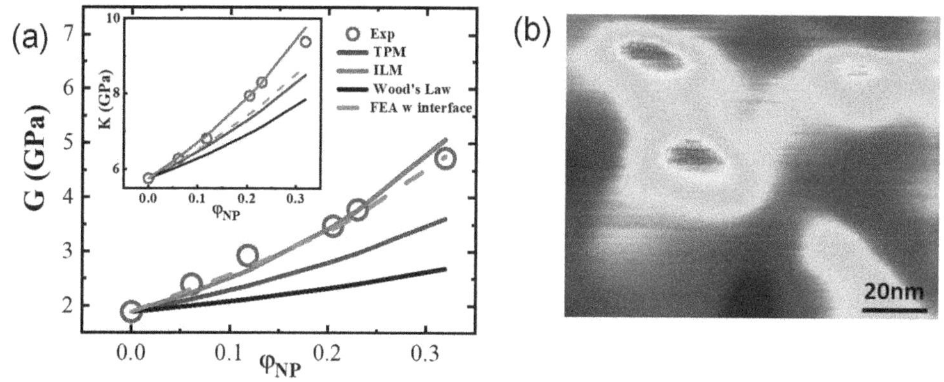

Figure 17. (a) The shear modulus, G (mainframe), and the bulk modulus, K (inset), of PVAc/SiO_2 nanocomposites of different nanoparticle loadings, φ_{NP}. The red circles are the modulus from Brillouin light scattering. The black lines are the results of the simple mixing rule. The blue lines are continuum calculations based on the geometry of the TPM. The green lines are continuum calculations based on the geometry of the ILM with the l_{int} = 3 nm. The dashed orange lines are the outcome of the finite element analysis calculation with l_{int} = 3 nm. (b) A map of the heterogeneous mechanical properties of PNCs from high-resolution atomic force microscopy. The red and the yellow regions are the NPs. The green and cyan regions represent the interfacial layer. The blue background is the matrix polymer. Reproduced with permission from reference (64). Copyright 2016 American Chemical Society.

Summary and Outlook

This chapter describes the current development of BDS in analyzing the dynamics of PNCs. The wide frequency and temperature range of BDS measurements offers a versatile platform for selectively quantifying the matrix dynamics. Although the basic characteristics of the influences of the NPs on the dielectric dispersion of PNCs have been well established, detailed analysis of the polymer dynamics, especially the slow dynamics, is generally a challenge. Recent work focused on the dielectric broadening of the structural relaxation with the thickness and the average dynamics of the IL. The HMA is found to be the best among available methodologies and agrees with results from measurements such as H-NMR, SAXS, SANS, and AFM. Besides, intriguing changes in the dielectric functions of the secondary relaxation have also been revealed, suggesting strong perturbation of the NPs to the local chain packing. Future development on quantifying the dielectric dispersion of the secondary relaxation might help understand the physical aging behavior of PNCs (*122*). The understanding of the dielectric normal mode of PNCs remains another challenge due to the strong coupling of the dynamics of the adsorbed and non-adsorbed polymer. Further development of an in-depth analysis of the dielectric normal mode of PNCs should offer valuable insight into the slow dynamics. Lastly, the combination of BDS with other macroscopic measurements has been demonstrated to be very effective to answer important scientific questions that cannot be easily tackled otherwise. Thus, developments in integrating BDS with other macroscopic measurements are also anticipated in the future.

Acknowledgments

This work was supported, in part, by the Michigan State University.

References

1. Yang, C.; Wei, H.; Guan, L.; Guo, J.; Wang, Y.; Yan, X.; Zhang, X.; Wei, S.; Guo, Z. Polymer nanocomposites for energy storage, energy saving, and anticorrosion. *J. Mater. Chem. A* **2015**, *3* (29), 14929–14941.
2. Majka, T. M.; Leszczyńska, A.; Pielichowski, K. Thermal Stability and Degradation of Polymer Nanocomposites. In *Polymer Nanocomposites: Electrical and Thermal Properties*; Huang, X., Zhi, C., Eds.; Springer International Publishing: Cham, 2016; pp 167−190.
3. Hule, R. A.; Pochan, D. J. Polymer Nanocomposites for Biomedical Applications. *MRS Bull.* **2007**, *32* (4), 354–358.
4. Huang, X.; Zhi, C. *Polymer Nanocomposites*; 1st ed.; Springer International Publishing: 2016.
5. Qi, W.; Zhang, X.; Wang, H. Self-assembled polymer nanocomposites for biomedical application. *Curr. Opin. Colloid Interface Sci.* **2018**, *35*, 36–41.
6. Kessler, M. R.; Goertzen, W. K. Polymer Nanocomposites for Infrastructure Rehabilitation. *Nanotechnology in Construction* 3; Bittnar, Z., Bartos, P. J. M., Němeček, J., Šmilauer, V., Zeman, J. , Eds.; Springer: Berlin, Heidelberg, 2009; pp 241−250.
7. Silvestre, J.; Silvestre, N.; de Brito, J. Polymer nanocomposites for structural applications: Recent trends and new perspectives. *Mech. Adv. Mater. Struct.* **2016**, *23* (11), 1263–1277.
8. Mittal, V.; Kim, J. K.; Pal, K. *Recent Adances in Elastomeric Nanocomposites*; Springer: 2011.
9. Vilgis, T. A.; Heinrich, G.; Klüppel, M. *Reinforcement of Polymer Nano-Composites: Theory, Experiments and Applications*; Cambridge University Press: Cambridge, 2009.
10. Bailey, E. J.; Winey, K. I. Dynamics of Polymer Segments, Polymer Chains, and Nanoparticles in Polymer Nanocomposite Melts: A Review. *Prog. Polym. Sci.* **2020**, 101242.
11. Kumar, S. K.; Ganesan, V.; Riggleman, R. A. Perspective: Outstanding theoretical questions in polymer-nanoparticle hybrids. *J. Chem. Phys.* **2017**, *147* (2), 020901.
12. Jancar, J.; Douglas, J. F.; Starr, F. W.; Kumar, S. K.; Cassagnau, P.; Lesser, A. J.; Sternstein, S. S.; Buehler, M. J. Current issues in research on structure-property relationships in polymer nanocomposites. *Polymer* **2010**, *51* (15), 3321–3343.
13. Schadler, L. S.; Kumar, S. K.; Benicewicz, B. C.; Lewis, S. L.; Harton, S. E. Designed Interfaces in Polymer Nanocomposites: A Fundamental Viewpoint. *MRS Bull.* **2007**, *32* (04), 335–340.
14. Schweizer, K. S.; Simmons, D. S. Progress towards a phenomenological picture and theoretical understanding of glassy dynamics and vitrification near interfaces and under nanoconfinement. *J. Chem. Phys.* **2019**, *151* (24), 240901.
15. Cheng, S.; Carroll, B.; Bocharova, V.; Carrillo, J.-M.; Sumpter, B. G.; Sokolov, A. P. Focus: Structure and dynamics of the interfacial layer in polymer nanocomposites with attractive interactions. *J. Chem. Phys.* **2017**, *146* (20), 203201.
16. Okamoto, M. Rheology and Processing of Polymer/Layered Silicate Nanocomposites. In *Rheology and Processing of Polymer Nanocomposites*; Thomas, S., Muller, R., Abraham, J., Eds.; Wiley: 2016; Chapter 5, pp 187−234.

17. Song, Y.; Zheng, Q. Concepts and Conflicts in Nanoparticles Reinforcement to Polymers beyond Hydrodynamics. *Prog. Mater Sci.* **2016**, *84*, 1–58.
18. Sun, R.; Melton, M.; Zuo, X.; Cheng, S. Nonmonotonic Strain Rate Dependence on the Strain Hardening of Polymer Nanocomposites. *ACS Macro Lett.* **2020**, *9* (9), 1224–1229.
19. Bansal, A.; Yang, H. C.; Li, C. Z.; Cho, K. W.; Benicewicz, B. C.; Kumar, S. K.; Schadler, L. S. Quantitative equivalence between polymer nanocomposites and thin polymer films. *Nat. Mater.* **2005**, *4* (9), 693–698.
20. Choi, J.; Hore, M. J. A.; Meth, J. S.; Clarke, N.; Winey, K. I.; Composto, R. J. Universal Scaling of Polymer Diffusion in Nanocomposites. *ACS Macro Lett.* **2013**, *2* (6), 485–490.
21. Yang, J.; Melton, M.; Sun, R.; Yang, W.; Cheng, S. Decoupling the Polymer Dynamics and the Nanoparticle Network Dynamics of Polymer Nanocomposites through Dielectric Spectroscopy and Rheology. *Macromolecules* **2020**, *53* (1), 302–311.
22. Chen, Q.; Gong, S.; Moll, J.; Zhao, D.; Kumar, S. K.; Colby, R. H. Mechanical Reinforcement of Polymer Nanocomposites from Percolation of a Nanoparticle Network. *ACS Macro Lett.* **2015**, *4* (4), 398–402.
23. Baeza, G. P.; Dessi, C.; Costanzo, S.; Zhao, D.; Gong, S.; Alegria, A.; Colby, R. H.; Rubinstein, M.; Vlassopoulos, D.; Kumar, S. K. Network dynamics in nanofilled polymers. *Nat. Commun.* **2016**, *7*, 11368.
24. Papon, A.; Montes, H.; Hanafi, M.; Lequeux, F.; Guy, L.; Saalwächter, K. Glass-Transition Temperature Gradient in Nanocomposites: Evidence from Nuclear Magnetic Resonance and Differential Scanning Calorimetry. *Phys. Rev. Lett.* **2012**, *108* (6), 065702.
25. Kim, S. Y.; Meyer, H. W.; Saalwächter, K.; Zukoski, C. F. Polymer Dynamics in PEG-Silica Nanocomposites: Effects of Polymer Molecular Weight, Temperature and Solvent Dilution. *Macromolecules* **2012**, *45* (10), 4225–4237.
26. Papon, A.; Saalwächter, K.; Schäler, K.; Guy, L.; Lequeux, F.; Montes, H. Low-Field NMR Investigations of Nanocomposites: Polymer Dynamics and Network Effects. *Macromolecules* **2011**, *44* (4), 913–922.
27. Popov, I.; Carroll, B.; Bocharova, V.; Genix, A.-C.; Cheng, S.; Khamzin, A.; Kisliuk, A.; Sokolov, A. P. Strong Reduction in Amplitude of the Interfacial Segmental Dynamics in Polymer Nanocomposites. *Macromolecules* **2020**, *53* (10), 4126–4135.
28. Cheng, S.; Xie, S.-J.; Carrillo, J.-M. Y.; Carroll, B.; Martin, H.; Cao, P.-F.; Dadmun, M. D.; Sumpter, B. G.; Novikov, V. N.; Schweizer, K. S.; Sokolov, A. P. Big Effect of Small Nanoparticles: A Shift in Paradigm for Polymer Nanocomposites. *ACS Nano* **2017**, *11* (1), 752–759.
29. Cheng, S.; Carroll, B.; Lu, W.; Fan, F.; Carrillo, J.-M. Y.; Martin, H.; Holt, A. P.; Kang, N.-G.; Bocharova, V.; Mays, J. W.; Sumpter, B. G.; Dadmun, M.; Sokolov, A. P. Interfacial Properties of Polymer Nanocomposites: Role of Chain Rigidity and Dynamic Heterogeneity Length Scale. *Macromolecules* **2017**, *50* (6), 2397–2406.
30. Carroll, B.; Cheng, S.; Sokolov, A. P. Analyzing the Interfacial Layer Properties in Polymer Nanocomposites by Broadband Dielectric Spectroscopy. *Macromolecules* **2017**, *50* (16), 6149–6163.
31. Holt, A. P.; Bocharova, V.; Cheng, S.; Kisliuk, A. M.; White, B. T.; Saito, T.; Uhrig, D.; Mahalik, J. P.; Kumar, R.; Imel, A. E.; Etampawala, T.; Martin, H.; Sikes, N.; Sumpter, B. G.;

Dadmun, M. D.; Sokolov, A. P. Controlling Interfacial Dynamics: Covalent Bonding versus Physical Adsorption in Polymer Nanocomposites. *ACS Nano* **2016,** *10* (7), 6843–6852.

32. Cheng, S.; Holt, A. P.; Wang, H.; Fan, F.; Bocharova, V.; Martin, H.; Etampawala, T.; White, B. T.; Saito, T.; Kang, N.-G.; Dadmun, M. D.; Mays, J. W.; Sokolov, A. P. Unexpected Molecular Weight Effect in Polymer Nanocomposites. *Phys. Rev. Lett.* **2016,** *116* (3), 038302.
33. Cheng, S.; Mirigian, S.; Carrillo, J.-M. Y.; Bocharova, V.; Sumpter, B. G.; Schweizer, K. S.; Sokolov, A. P. Revealing spatially heterogeneous relaxation in a model nanocomposite. *J. Chem. Phys.* **2015,** *143* (19), 194704.
34. Pazmino Betancourt, B. A.; Douglas, J. F.; Starr, F. W. Fragility and cooperative motion in a glass-forming polymer-nanoparticle composite. *Soft Matter* **2013,** *9* (1), 241–254.
35. Starr, F. W.; Douglas, J. F. Modifying Fragility and Collective Motion in Polymer Melts with Nanoparticles. *Phys. Rev. Lett.* **2011,** *106* (11), 115702.
36. Starr, F. W.; Douglas, J. F.; Meng, D.; Kumar, S. K. Bound Layers "Cloak" Nanoparticles in Strongly Interacting Polymer Nanocomposites. *ACS Nano* **2016,** *10* (12), 10960–10965.
37. Klonos, P.; Kyritsis, A.; Pissis, P. Interfacial and confined dynamics of PDMS adsorbed at the interfaces and in the pores of silica–gel: Effects of surface modification and thermal annealing. *Polymer* **2016,** *84,* 38–51.
38. Klonos, P.; Kyritsis, A.; Pissis, P. Interfacial dynamics of polydimethylsiloxane adsorbed on fumed metal oxide particles of a wide range of specific surface area. *Polymer* **2015,** *77,* 10–13.
39. Klonos, P.; Panagopoulou, A.; Kyritsis, A.; Bokobza, L.; Pissis, P. Dielectric studies of segmental dynamics in poly(dimethylsiloxane)/titania nanocomposites. *J. Non-Cryst. Solids* **2011,** *357* (2), 610–614.
40. Fragiadakis, D.; Pissis, P. Glass transition and segmental dynamics in poly(dimethylsiloxane)/silica nanocomposites studied by various techniques. *J. Non-Cryst. Solids* **2007,** *353* (47–51), 4344–4352.
41. Holt, A. P.; Bocharova, V.; Cheng, S.; Kisliuk, A. M.; Ehlers, G.; Mamontov, E.; Novikov, V. N.; Sokolov, A. P. Interplay between local dynamics and mechanical reinforcement in glassy polymer nanocomposites. *Phys. Rev. Mater.* **2017,** *1* (6), 062601.
42. Kalathi, J. T.; Kumar, S. K.; Rubinstein, M.; Grest, G. S. Rouse mode analysis of chain relaxation in polymer nanocomposites. *Soft Matter* **2015,** *11* (20), 4123–4132.
43. Kalathi, J. T.; Grest, G. S.; Kumar, S. K. Universal Viscosity Behavior of Polymer Nanocomposites. *Phys. Rev. Lett.* **2012,** *109* (19)
44. Karatrantos, A.; Composto, R. J.; Winey, K. I.; Clarke, N. Polymer and spherical nanoparticle diffusion in nanocomposites. *J. Chem. Phys.* **2017,** *146* (20), 203331.
45. Choi, J.; Clarke, N.; Winey, K. I.; Composto, R. J. Polymer Diffusion from Attractive and Athermal Substrates. *Macromolecules* **2017,** *50* (7), 3038–3042.
46. Karatrantos, A.; Clarke, N.; Composto, R. J.; Winey, K. I. Entanglements in polymer nanocomposites containing spherical nanoparticles. *Soft Matter* **2016,** *12,* 2567–2574.
47. Tung, W.-S.; Composto, R. J.; Riggleman, R. A.; Winey, K. I. Local Polymer Dynamics and Diffusion in Cylindrical Nanoconfinement. *Macromolecules* **2015,** *48* (7), 2324–2332.
48. Schneider, G. J.; Nusser, K.; Willner, L.; Falus, P.; Richter, D. Dynamics of Entangled Chains in Polymer Nanocomposites. *Macromolecules* **2011,** *44* (15), 5857–5860.

49. Sarvestani, A. S.; Picu, C. R. Network model for the viscoelastic behavior of polymer nanocomposites. *Polymer* **2004**, *45* (22), 7779–7790.
50. Voylov, D. N.; Holt, A. P.; Doughty, B.; Bocharova, V.; Meyer, H. M.; Cheng, S.; Martin, H.; Dadmun, M.; Kisliuk, A.; Sokolov, A. P. Unraveling the Molecular Weight Dependence of Interfacial Interactions in Poly(2-vinylpyridine)/Silica Nanocomposites. *ACS Macro Lett.* **2017**, 68–72.
51. Ge, T.; Kalathi, J. T.; Halverson, J. D.; Grest, G. S.; Rubinstein, M. Nanoparticle Motion in Entangled Melts of Linear and Nonconcatenated Ring Polymers. *Macromolecules* **2017**, *50* (4), 1749–1754.
52. Kalathi, J. T.; Yamamoto, U.; Schweizer, K. S.; Grest, G. S.; Kumar, S. K. Nanoparticle Diffusion in Polymer Nanocomposites. *Phys. Rev. Lett.* **2014**, *112* (10), 108301.
53. Ge, T.; Grest, G. S.; Rubinstein, M. Nanorheology of Entangled Polymer Melts. *Phys. Rev. Lett.* **2018**, *120* (5), 057801.
54. Ge, T.; Rubinstein, M.; Grest, G. S. Effects of Tethered Polymers on Dynamics of Nanoparticles in Unentangled Polymer Melts. *Macromolecules* **2020**.
55. Cai, L.-H.; Panyukov, S.; Rubinstein, M. Hopping Diffusion of Nanoparticles in Polymer Matrices. *Macromolecules* **2015**, *48* (3), 847–862.
56. Cai, L.-H.; Panyukov, S.; Rubinstein, M. Mobility of Nonsticky Nanoparticles in Polymer Liquids. *Macromolecules* **2011**, *44* (19), 7853–7863.
57. Yamamoto, U.; Carrillo, J.-M. Y.; Bocharova, V.; Sokolov, A. P.; Sumpter, B. G.; Schweizer, K. S. Theory and Simulation of Attractive Nanoparticle Transport in Polymer Melts. *Macromolecules* **2018**, *51* (6), 2258–2267.
58. Carroll, B.; Bocharova, V.; Carrillo, J.-M. Y.; Kisliuk, A.; Cheng, S.; Yamamoto, U.; Schweizer, K. S.; Sumpter, B. G.; Sokolov, A. P. Diffusion of Sticky Nanoparticles in a Polymer Melt: Crossover from Suppressed to Enhanced Transport. *Macromolecules* **2018**, *51* (6), 2268–2275.
59. Yamamoto, U.; Schweizer, K. S. Microscopic Theory of the Long-Time Diffusivity and Intermediate-Time Anomalous Transport of a Nanoparticle in Polymer Melts. *Macromolecules* **2015**, *48* (1), 152–163.
60. Bailey, E. J.; Griffin, P. J.; Composto, R. J.; Winey, K. I. Multiscale Dynamics of Small, Attractive Nanoparticles and Entangled Polymers in Polymer Nanocomposites. *Macromolecules* **2019**, *52* (5), 2181–2188.
61. Griffin, P. J.; Bocharova, V.; Middleton, L. R.; Composto, R. J.; Clarke, N.; Schweizer, K. S.; Winey, K. I. Influence of the Bound Polymer Layer on Nanoparticle Diffusion in Polymer Melts. *ACS Macro Lett.* **2016**, *5* (10), 1141–1145.
62. Choi, J.; Cargnello, M.; Murray, C. B.; Clarke, N.; Winey, K. I.; Composto, R. J. Fast Nanorod Diffusion through Entangled Polymer Melts. *ACS Macro Lett.* **2015**, 952–956.
63. Park, J.; Bailey, E. J.; Composto, R. J.; Winey, K. I. Single-Particle Tracking of Nonsticky and Sticky Nanoparticles in Polymer Melts. *Macromolecules* **2020**, *53* (10), 3933–3939.
64. Cheng, S.; Bocharova, V.; Belianinov, A.; Xiong, S.; Kisliuk, A.; Somnath, S.; Holt, A. P.; Ovchinnikova, O. S.; Jesse, S.; Martin, H.; Etampawala, T.; Dadmun, M.; Sokolov, A. P. Unraveling the Mechanism of Nanoscale Mechanical Reinforcement in Glassy Polymer Nanocomposites. *Nano Lett.* **2016**, *16* (6), 3630–3637.

65. Berriot, J.; Montes, H.; Martin, F.; Mauger, M.; Pyckhout-Hintzen, W.; Meier, G.; Frielinghaus, H. Reinforcement of model filled elastomers: synthesis and characterization of the dispersion state by SANS measurements. *Polymer* **2003**, *44* (17), 4909–4919.
66. Papon, A.; Montes, H.; Lequeux, F.; Oberdisse, J.; Saalwachter, K.; Guy, L. Solid particles in an elastomer matrix: impact of colloid dispersion and polymer mobility modification on the mechanical properties. *Soft Matter* **2012**, *8* (15), 4090–4096.
67. Mujtaba, A.; Keller, M.; Ilisch, S.; Radusch, H. J.; Beiner, M.; Thurn-Albrecht, T.; Saalwächter, K. Detection of Surface-Immobilized Components and Their Role in Viscoelastic Reinforcement of Rubber–Silica Nanocomposites. *ACS Macro Lett.* **2014**, *3* (5), 481–485.
68. Zhu, Z.; Thompson, T.; Wang, S.-Q.; von Meerwall, E. D.; Halasa, A. Investigating Linear and Nonlinear Viscoelastic Behavior Using Model Silica-Particle-Filled Polybutadiene. *Macromolecules* **2005**, *38* (21), 8816–8824.
69. Sternstein, S. S.; Zhu, A.-J. Reinforcement Mechanism of Nanofilled Polymer Melts As Elucidated by Nonlinear Viscoelastic Behavior. *Macromolecules* **2002**, 35 (19), 7262–7273.
70. Song, Y.; Zheng, Q. A Guide for Hydrodynamic Reinforcement Effect in Nanoparticle-filled Polymers. *Crit. Rev. Solid State Mater. Sci.* **2016**, *41* (4), 318–346.
71. Zeng, Q. H.; Yu, A. B.; Lu, G. Q. Multiscale modeling and simulation of polymer nanocomposites. *Prog. Polym. Sci.* **2008**, *33* (2), 191–269.
72. Vogiatzis, G. G.; Theodorou, D. N. Multiscale Molecular Simulations of Polymer-Matrix Nanocomposites. *Arch. Comput. Meth. Eng.* **2018**, *25* (3), 591–645.
73. Rueda, M. M.; Auscher, M.-C.; Fulchiron, R.; Périé, T.; Martin, G.; Sonntag, P.; Cassagnau, P. Rheology and applications of highly filled polymers: A review of current understanding. *Prog. Polym. Sci.* **2017**, *66*, 22–53.
74. Frohlich, H. *Theory of dielectrics: dielectric constant and dielectric loss*; At the Clarendon Press: Oxford, 1987.
75. Kremer, F.; Schohals, A. *Broadband Dielectric Spectroscopy*; Springer-Verlag: Berlin: 2002.
76. Bottcher, C. J. F.; Bordewijk, P. *Theory of Electric Polarization: Dielectrics in time-dependent fields*; Elsevier: 1978; Vol. II.
77. Asami, K. Characterization of heterogeneous systems by dielectric spectroscopy. *Prog. Polym. Sci.* **2002**, *27* (8), 1617–1659.
78. Adachi, K.; Kotaka, T. Dielectric normal mode relaxation. *Prog. Polym. Sci.* **1993**, *18* (3), 585–622.
79. Holt, A. P.; Sangoro, J. R.; Wang, Y.; Agapov, A. L.; Sokolov, A. P. Chain and Segmental Dynamics of Poly(2-vinylpyridine) Nanocomposites. *Macromolecules* **2013**, *46* (10), 4168–4173.
80. Holt, A. P.; Griffin, P. J.; Bocharova, V.; Agapov, A. L.; Imel, A. E.; Dadmun, M. D.; Sangoro, J. R.; Sokolov, A. P. Dynamics at the Polymer/Nanoparticle Interface in Poly(2-vinylpyridine)/Silica Nanocomposites. *Macromolecules* **2014**, *47* (5), 1837–1843.
81. Cheng, S.; Sokolov, A. P. Correlation between the temperature evolution of the interfacial region and the growing dynamic cooperativity length scale. *J. Chem. Phys.* **2020**, *152* (9), 094904.

82. Böhning, M.; Goering, H.; Fritz, A.; Brzezinka, K.-W.; Turky, G.; Schönhals, A.; Schartel, B. Dielectric Study of Molecular Mobility in Poly(propylene-graft-maleic anhydride)/Clay Nanocomposites. *Macromolecules* **2005**, *38* (7), 2764–2774.
83. Hao, N.; Böhning, M.; Goering, H.; Schönhals, A. Nanocomposites of Polyhedral Oligomeric Phenethylsilsesquioxanes and Poly(bisphenol A carbonate) as Investigated by Dielectric Spectroscopy. *Macromolecules* **2007**, *40* (8), 2955–2964.
84. Hao, N.; Böhning, M.; Schönhals, A. Dielectric Properties of Nanocomposites Based on Polystyrene and Polyhedral Oligomeric Phenethyl-Silsesquioxanes. *Macromolecules* **2007**, *40* (26), 9672–9679.
85. Schönhals, A.; Schick, C.; Huth, H.; Frick, B.; Mayorova, M.; Zorn, R. Molecular dynamics in glass-forming poly(phenyl methyl siloxane) as investigated by broadband thermal, dielectric and neutron spectroscopy. *J. Non-Cryst. Solids* **2007**, *353* (41–43), 3853–3861.
86. Schönhals, A.; Goering, H.; Costa, F. R.; Wagenknecht, U.; Heinrich, G. Dielectric Properties of Nanocomposites Based on Polyethylene and Layered Double Hydroxide. *Macromolecules* **2009**, *42* (12), 4165–4174.
87. Füllbrandt, M.; Purohit, P. J.; Schönhals, A. Combined FTIR and Dielectric Investigation of Poly(vinyl acetate) Adsorbed on Silica Particles. *Macromolecules* **2013**, *46* (11), 4626–4632.
88. Konnertz, N.; Böhning, M.; Schönhals, A. Dielectric investigations of nanocomposites based on Matrimid and polyhedral oligomeric phenethyl-silsesquioxanes (POSS). *Polymer* **2016**, *90*, 89–101.
89. Gong, S.; Chen, Q.; Moll, J. F.; Kumar, S. K.; Colby, R. H. Segmental Dynamics of Polymer Melts with Spherical Nanoparticles. *ACS Macro Lett.* **2014**, *3* (8), 773–777.
90. Fragiadakis, D.; Pissis, P.; Bokobza, L. Glass transition and molecular dynamics in poly(dimethylsiloxane)/silica nanocomposites. *Polymer* **2005**, *46* (16), 6001–6008.
91. Fragiadakis, D.; Pissis, P.; Bokobza, L. Modified chain dynamics in poly(dimethylsiloxane)/silica nanocomposites. *J. Non-Cryst. Solids* **2006**, *352* (42–49), 4969–4972.
92. Pissis, P.; Fragiadakis, D.; Kanapitsas, A.; Delides, K. Broadband Dielectric Relaxation Spectroscopy in Polymer Nanocomposites. *Macromol. Symp.* **2008**, *265* (1), 12–20.
93. Klonos, P.; Pandis, C.; Kripotou, S.; Kyritsis, A.; Pissis, P. Interfacial effects in polymer nanocomposites studied by dielectric and thermal techniques. *IEEE Trans. Dielectr. Electr. Insul.* **2012**, *19* (4), 1283–1290.
94. Bogoslovov, R. B.; Roland, C. M.; Ellis, A. R.; Randall, A. M.; Robertson, C. G. Effect of Silica Nanoparticles on the Local Segmental Dynamics in Poly(vinyl acetate). *Macromolecules* **2008**, *41* (4), 1289–1296.
95. Casalini, R.; Roland, C. M. Local and Global Dynamics in Polypropylene Glycol/Silica Composites. *Macromolecules* **2016**, *49* (10), 3919–3924.
96. Wübbenhorst, M.; van Turnhout, J. Analysis of complex dielectric spectra. I. One-dimensional derivative techniques and three-dimensional modelling. *J. Non-Cryst. Solids* **2002**, *305* (1), 40–49.
97. Richert, R.; Agapov, A.; Sokolov, A. P. Appearance of a Debye process at the conductivity relaxation frequency of a viscous liquid. *J. Chem. Phys.* **2011**, *134* (10), 104508.

98. Klonos, P.; Sulym, I. Y.; Kyriakos, K.; Vangelidis, I.; Zidropoulos, S.; Sternik, D.; Borysenko, M. V.; Kyritsis, A.; Deryło-Marczewska, A.; Gun'ko, V. M.; Pissis, P. Interfacial phenomena in core–shell nanocomposites of PDMS adsorbed onto low specific surface area fumed silica nanooxides: Effects of surface modification. *Polymer* **2015**, *68*, 158–167.
99. Klonos, P.; Kulyk, K.; Borysenko, M. V.; Gun'ko, V. M.; Kyritsis, A.; Pissis, P. Effects of Molecular Weight below the Entanglement Threshold on Interfacial Nanoparticles/Polymer Dynamics. *Macromolecules* **2016**, *49* (24), 9457–9473.
100. Steeman, P. A. M.; Maurer, F. H. J. An interlayer model for the complex dielectric constant of composites. *Colloid Polym. Sci.* **1990**, *268* (4), 315–325.
101. Schäfer, H.; Sternin, E.; Stannarius, R.; Arndt, M.; Kremer, F. Novel Approach to the Analysis of Broadband Dielectric Spectra. *Phys. Rev. Lett.* **1996**, *76* (12), 2177–2180.
102. Roths, T.; Marth, M.; Weese, J.; Honerkamp, J. A generalized regularization method for nonlinear ill-posed problems enhanced for nonlinear regularization terms. *Comput. Phys. Commun.* **2001**, *139* (3), 279–296.
103. Koch, M.; Sommer, J.-U.; Blumen, A. Polymer chains tethered to impenetrable interfaces: Broadening of relaxation spectra. *J. Chem. Phys.* **1997**, *106* (3), 1248–1256.
104. Scheutjens, J. M. H. M.; Fleer, G. J. Statistical theory of the adsorption of interacting chain molecules. 2. Train, loop, and tail size distribution. *J. Phys. Chem.* **1980**, *84* (2), 178–190.
105. Guiselin, O. Irreversible Adsorption of a Concentrated Polymer Solution. *Europhys. Lett.* **1992**, *17* (3), 225.
106. Zheng, X.; Sauer, B. B.; Van Alsten, J. G.; Schwarz, S. A.; Rafailovich, M. H.; Sokolov, J.; Rubinstein, M. Reptation Dynamics of a Polymer Melt near an Attractive Solid Interface. *Phys. Rev. Lett.* **1995**, *74* (3), 407–410.
107. Baeza, G. P.; Oberdisse, J.; Alegria, A.; Couty, M.; Genix, A.-C. A high-temperature dielectric process as a probe of large-scale silica filler structure in simplified industrial nanocomposites. *Phys. Chem. Chem. Phys.* **2015**, *17* (3), 1660–1666.
108. Baeza, G. P.; Oberdisse, J.; Alegria, A.; Saalwächter, K.; Couty, M.; Genix, A.-C. Depercolation of aggregates upon polymer grafting in simplified industrial nanocomposites studied with dielectric spectroscopy. *Polymer* **2015**, 73, 131–138.
109. Kim, S. Y.; Schweizer, K. S.; Zukoski, C. F. Multiscale Structure, Interfacial Cohesion, Adsorbed Layers, and Thermodynamics in Dense Polymer-Nanoparticle Mixtures. *Phys. Rev. Lett.* **2011**, *107* (22), 225504.
110. Krutyeva, M.; Wischnewski, A.; Monkenbusch, M.; Willner, L.; Maiz, J.; Mijangos, C.; Arbe, A.; Colmenero, J.; Radulescu, A.; Holderer, O.; Ohl, M.; Richter, D. Effect of Nanoconfinement on Polymer Dynamics: Surface Layers and Interphases. *Phys. Rev. Lett.* **2013**, *110* (10), 108303.
111. Talik, A.; Tarnacka, M.; Geppert-Rybczyńska, M.; Hachuła, B.; Bernat, R.; Chrzanowska, A.; Kaminski, K.; Paluch, M. Are hydrogen supramolecular structures being suppressed upon nanoscale confinement? The case of monohydroxy alcohols. *J. Colloid Interface Sci.* **2020**, *576*, 217–229.
112. Carrillo, J.-M. Y.; Cheng, S.; Kumar, R.; Goswami, M.; Sokolov, A. P.; Sumpter, B. G. Untangling the Effects of Chain Rigidity on the Structure and Dynamics of Strongly Adsorbed Polymer Melts. *Macromolecules* **2015**, *48* (12), 4207–4219.

113. Berriot, J.; Lequeux, F.; Monnerie, L.; Montes, H.; Long, D.; Sotta, P. Filler–elastomer interaction in model filled rubbers, a 1H NMR study. *J. Non-Cryst. Solids* **2002**, *307–310*, 719–724.

114. Lang, R. J.; Simmons, D. S. Interfacial Dynamic Length Scales in the Glass Transition of a Model Freestanding Polymer Film and Their Connection to Cooperative Motion. *Macromolecules* **2013**, *46* (24), 9818–9825.

115. Diaz-Vela, D.; Hung, J.-H.; Simmons, D. S. Temperature-Independent Rescaling of the Local Activation Barrier Drives Free Surface Nanoconfinement Effects on Segmental-Scale Translational Dynamics near Tg. *ACS Macro Lett.* **2018**, 7 (11), 1295–1301.

116. Phan, A. D.; Schweizer, K. S. Theory of Spatial Gradients of Relaxation, Vitrification Temperature and Fragility of Glass-Forming Polymer Liquids Near Solid Substrates. *ACS Macro Lett.* **2020**, 448–453.

117. Phan, A. D.; Schweizer, K. S. Influence of Longer Range Transfer of Vapor Interface Modified Caging Constraints on the Spatially Heterogeneous Dynamics of Glass-Forming Liquids. *Macromolecules* **2019**, *52* (14), 5192–5206.

118. Phan, A. D.; Schweizer, K. S. Theory of the spatial transfer of interface-nucleated changes of dynamical constraints and its consequences in glass-forming films. *J. Chem. Phys.* **2019**, *150* (4), 044508.

119. Phan, A. D.; Schweizer, K. S. Dynamic Gradients, Mobile Layers, Tg Shifts, Role of Vitrification Criterion, and Inhomogeneous Decoupling in Free-Standing Polymer Films. *Macromolecules* **2018**, *51* (15), 6063–6075.

120. Hanakata, P. Z.; Douglas, J. F.; Starr, F. W. Interfacial mobility scale determines the scale of collective motion and relaxation rate in polymer films. *Nat. Commun.* **2014**, 5, 4163.

121. Lin, C.-C.; Gam, S.; Meth, J. S.; Clarke, N.; Winey, K. I.; Composto, R. J. Do Attractive Polymer–Nanoparticle Interactions Retard Polymer Diffusion in Nanocomposites? *Macromolecules* **2013**, *46* (11), 4502–4509.

122. Cangialosi, D.; Boucher, V. M.; Alegria, A.; Colmenero, J. Physical aging in polymers and polymer nanocomposites: recent results and open questions. *Soft Matter* **2013**, *9* (36), 8619–8630.

Chapter 8

Glassy Dynamics in Nanometric Confinement of Various Topologies, a Comparison for the Case of Poly(2-vinylpyridine)

Wycliffe K. Kipnusu,* Martin Tress, and Friedrich Kremer*

Peter Debye Institute for Soft Matter Physics, University of Leipzig, Linnéstr. 5, 04103 Leipzig, Germany

***Email: kipnusu@physik.uni-leipzig.de; friedrich.kremer@physik.uni-leipzig.de**

The impact of various topologies of confinement on the glassy dynamics of polymers is investigated for the example of poly(2-vinylpyridine) (P2VP). A spin-coated nanometric film (>3 nm) and a grafted layer are studied by means of Broadband Dielectric Spectroscopy in a wide temperature (300–450K) and frequency (10 Hz–1 MHz) ranges. This is enabled by novel nanostructured electrodes with insulating spacers (35 nm height) which avoid evaporation of metal counter electrodes. Furthermore P2VP deposited into cylindrical pores having a diameter of 10 nm and a length of 50 µm is examined. The preparations are exercised with special care in order to remove remaining solvent, to exclude the adsorption of water, and to anneal the polymers. In all three types of confinement the mean relaxation rate of the glassy dynamics remains bulk-like in the entire investigated dynamic range; an analysis of the relaxation time distributions reveals the existence of additional slower and faster relaxation modes caused by attractive interfacial interactions and less efficient packing (reduced density) in confined space, respectively. For P2VP in nano-pores both effects are present but in brush geometry, only effects of packing frustration are detected in addition to dominating bulk-like dynamics. These results are in accord with the fact, that glassy dynamics takes place on the length scale of a few repeat units and is determined by density.

Introduction

Nanometric geometrical confinement (*1*) of polymers can be realized in a variety of different topologies, in thin polymer layers and self-supporting films (*2–35*), in grafted polymers (*36, 37*), in nanodroplets containing just a single separated polymer chain (*38*), in cylindrical pores having nanometric diameter (*33, 39–42*), but as well in the three-dimensional constraints of polymer/nanoparticle composites (*43*). Glassy dynamics (in other words the dynamic glass transition) is in all cases the dominating fluctuation and hence understanding glassy dynamics is intimately linked with the exploration of the glass transition (*39*) under these conditions. It is evident that in such studies, many experimental approaches that are sensitive to quite different properties of glassy systems are utilized. They can be separated into dynamic methods, such as (alternating current) calorimetry (ACC) (*14, 15, 18*), broadband dielectric spectroscopy (BDS) (*1, 39, 44–49*), nuclear magnetic resonance (NMR) (*50*), photon correlation spectroscopy (PCS) (*51*), neutron scattering (*52, 53*), mechanical spectroscopy (*20*) and spectroscopies using specific dyes or labels (*16, 17, 25, 26*) and thermodynamic techniques, such as ellipsometry (*3, 4*), measurements of thermal expansion (*54*) or the lifetime of specific fluorescent labels (*25, 26*). Naturally the former measure (primarily) in the liquid state and enable one to trace glassy dynamics often over 8 decades or more (ACC, BDS, NMR, PCS) while the latter are usually restricted to the cross-over from liquid to glassy regime in determining typically a kink in a temperature dependence of ellipsometric constants, or the expansion coefficient from which a glass transition temperature is deduced. Broadband Dielectric Spectroscopy in combination with nanostructured electrode arrangements (*1, 38*) has turned out to be especially versatile due to its extraordinary sensitivity which enables one to determine even the dynamics of isolated polymer coils in a wide frequency- and temperature range. Furthermore, it offers the possibility to extract the relaxation time distribution function (*39*) as well of polymers having two different interfaces: one to a solid substrate and one to a free interface.

Since the first reports claiming confinement induced shifts of the glass transition temperature[2] numerous studies were published delivering often strongly diverging results (*55*). It is not the intention to evaluate and discuss all these works; instead, the authors will focus on the glassy dynamics of poly(2-vinylpyridine) (P2VP) as model system under conditions of nanometric confinement having widely varying topologies. It will turn out that the dynamics is predominately bulk-like with small modifications due to interfacial effects.

Experimental

Preparation of Polymer Thin Films

The thin polymer film was prepared by spin-coating P2VP (250,000 g/mol) from a 0.03 wt% chloroform solution onto a piece of highly doped silicon serving as substrate and bottom electrode. The conductive silicon purchased from Si-Mat has a specific resistance $\rho < 0.01\ \Omega$ cm; in order to achieve a proper electrical contact, the backside of the silicon electrodes is coated with a 200 nm-thick aluminum layer. The as-received silicon wafers were covered with a 700 nm thick protective layer of photoresist (microresist technology GmbH) before they were cut with a diamond pen along the crystallographic axes into small rectangular pieces of 4×10 mm^2. After cutting, the photoresist layer was removed in an ultrasonic bath of pure acetone and then dried with compressed nitrogen flow. As a second step, the cleaned substrates were inserted into a plasma cleaner (Plasma cleaner Femto from Diener GmbH) for 5 min then placed on top of a hot plate (at 150 °C) followed by cleaning with supercritical CO_2 (Applied surface technologies). The polymer was spin-coated on cleaned silicon substrates at 3000 rpm for 20 s.

Preparation of Polymer Brushes

For the preparation of P2VP brushes, wafer dice of 10×20 mm^2 size were sonicated in ethanol and then treated by an oxygen plasma. As first step of a two-step procedure (*56–59*) a thin layer ($d \approx 2.5$ nm) of poly(glycidyl-methacrylate) (PGMA) ($M_n = 17{,}500$ g/mol, $M_w = 29{,}700$ g/

mol, obtained from Polymer Source Inc.) was deposited from a 0.02 wt% chloroform solution and annealed at 100 °C for 20 min. The PGMA creates an anchoring layer where carboxyl groups can chemically bind. In the second step, carboxyl terminated poly(2-vinyl-pyridine) (P2VP–COOH, $M_n = 40{,}600$ g/mol, $M_w = 43{,}800$ g/mol, obtained from Polymer Source, Inc.) was spincast from a 1 wt% THF solution. By applying annealing times from 1 min to 16 h, the grafting process of the P2VP–COOH chains was allowed to form brushes of different thickness. Then, the brushes were rinsed in THF to remove excess polymer chains which were not grafted. An identical electrode as in the thin-film preparation was used to finalize the sample.

Sample (Thin Films and Brushes) Arrangement for BDS Measurement

The prototype sample setup with novel nanostructured counter electrodes used in this work (Figure 1) was first developed by Serghei et al. (*60*). The idea emerged out of the need to measure ultrathin polymer films without evaporating counter metal electrodes on top of the soft polymer film. To achieve this, a regular matrix of insulating silica nanostructures (acting as spacers) were fabricated onto a highly conductive and polished silicon wafer to serve as the top electrode. The quadratic nano–spacers (Figure 1 b, c), having lateral dimensions of 5 μm and heights of 35 nm (Figure 1 d) are produced by thermal oxidation and standard optical lithography on the surface of a conductive silicon wafer having resistivity $\rho < 0.01$ Ω cm (Si Mat silicon materials) similar to the bottom electrodes. By keeping the sample thicknesses much smaller than the spacer height, a free upper interface (air) is realized (Figure 1 b). This arrangement circumvents the problem of electrical shorts and the possible artifacts that can result from evaporated metal electrodes. This arrangement enables BDS measurements of ultra-thin polymer films (*34*), brushes (*36*) or even isolated polymer coils (*38*).

Before the dielectric measurement, each sample was annealed at 150°C (50 K above T_g) for 24 hours at 10^{-5} mbar. To ensure homogeneity of the polymer brushes and confirm the film thickness as well as the absence of de-wetting, Atomic Force Microscopy (AFM) is carried out employing a Vecco Dimension 3000 AFM in tapping mode. For the determination of the film thickness, the film surface is scratched with a stainless steel syringe needle and the step height at the edge of the scratch is measured at several spots in the active sample area. On a bare silicon dice it has been verified that the damage to the silicon surface by such a scratch is negligible. The thickness of the brushes was determined by ellipsometry (Sentech SE-402 scanning microfocus ellipsometer).

Preparation of Nanoporous Silica

The straightforward geometry to investigate the effect of 2D constraints on molecular dynamics by BDS requires uni-directional nanopores. This was achieved by preparing nanoporous silica ($pSiO_2$) membranes in which porous silicon (Si) was first obtained by electrochemical partial dissolution in the presence of hydrofluoric (HF) acid. Anisotropic dissolution of silicon is aided by the presence of hole charge carriers which polarizes the surface Si-H bonds. When immersed in HF electrolyte, fluoride ions displace hydrogen atoms leading to the formation of Si-F as shown in Figure 2 (a) (step 1 and 2). Further attack by fluoride ions (step 3) results in the release of hydrogen gas and the dissolution of silicon in the form of $H_2SiF_6^{-2}$ occurring in steps 4 and 5, respectively (Figure 2 a).

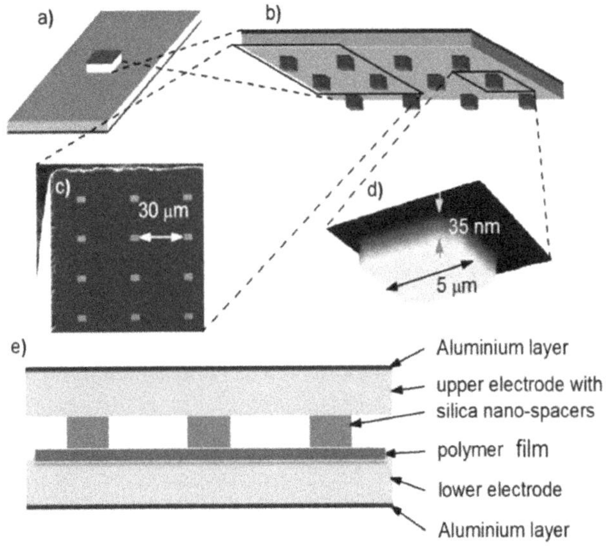

Figure 1. Sample arrangement: the capacitor (a) is composed of two highly doped silicon electrodes (ρ < 0.01 Ω cm, bottom electrode: 10 × 20 mm^2, top electrode: 1 × 1 mm^2; (b) cross section of the sample capacitor). The top electrode is covered by an array of insulating silica nanostructures, serving as spacers (c). The spacers have a height of 35 nm, as confirmed by AFM scans (d), and enable one for dielectric measurements without evaporating a metal counter electrode. Adapted with permission from reference (1). Copyright 2014 Springer Nature.

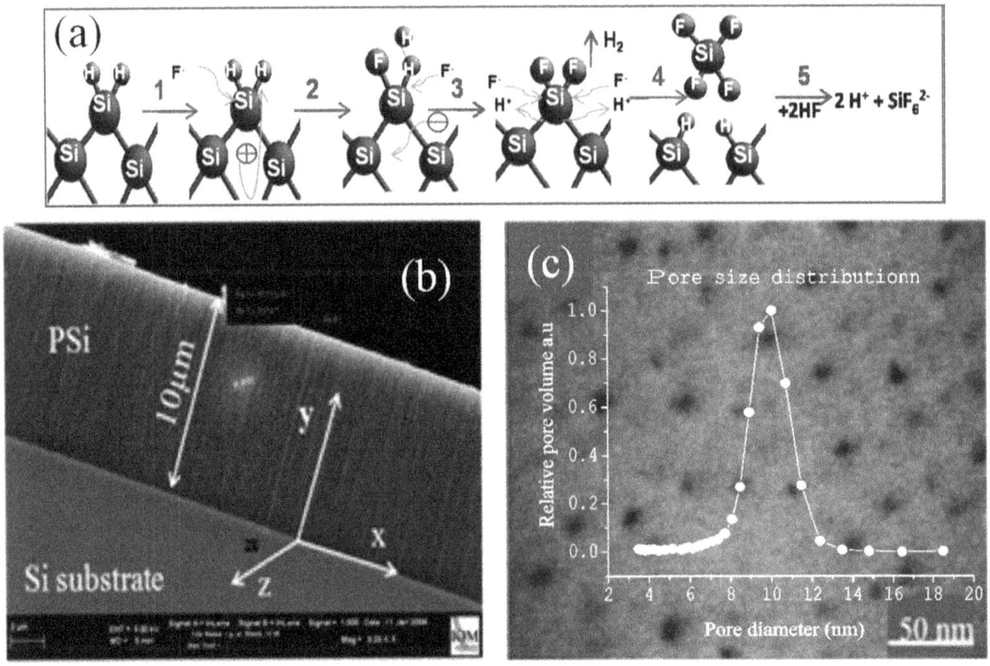

Figure 2. (a) Anodic dissolution mechanism of Silicon in the presence of HF electrolyte solution. Different steps in the reaction scheme are explained in the text. (b) Scanning electron micrographs (SEM) of the cross-section of etched silicon showing uni-directional nanopores (c) SEM image of the surface of formed porous silica ($pSiO_2$) membrane with embedded pore size distribution obtained by nuclear magnetic resonance (NMR) cryoporometry revealing average pore size of 10 nm.

The $pSiO_2$ samples employed in this work were prepared by the electrochemical etching of highly doped (0.005 Ω cm) p–type <100> oriented mono-crystalline silicon wafers in a home built anodization cell. 48% HF and Ethanol (C_2H_5OH-99%) purchased from sigma Aldrich were mixed in the ratio of 1:1 to form the electrolyte. A current density (j) of 120 $mAcm^{-2}$ was applied to obtain porous silicon (pSi) membranes having average pore diameter of 10 nm with porosity of 23 %. The etching time of about 120 min achieved thickness of the pSi of 50 μm. A final electro–polishing step (j = 700 $mAcm^{-2}$) was applied for 3–4 s to lift the pSi from the Si substrate. The obtained pSi membranes were subsequently oxidized thermally in an oven (Vulcan 3–550, Neytech) at 1100 K (heating rate 3 K/min) for 6 h to form completely transparent and insulating nanoporous silica ($pSiO_2$) with unidirectional pores. SEM images of the cross–section of pSi (Figure 2 b) reveals non–intersecting pores. The SEM image of the top surface of $pSiO_2$ is shown in Figure 2 (c) where the Pore size distribution as determined by nuclear magnetic resonance (NMR) cryoporometry is also depicted.

Infiltration of Polymer Sample into Nanoporous Silica

A Fractionated infiltration procedure was employed to fill the $pSiO_2$ with P2VP (Mw= 23,000 g/mol obtained from Polymer source). 0.5 % wt solution of P2VP in chloroform was used in this procedure. Before filling the $pSiO_2$ membranes, they were first annealed in a high-vacuum chamber (10^{-6} mbar) at 573 K for 24 h to remove water and other volatile impurities. This was done in a home–built vacuum chamber (Figure 3 a) which is evacuated in two steps. The first step involves a membrane pump that brings the pressure in the chamber down to 10^{-3} mbar before a turbo-molecular system pump is engaged in the second step to achieve a high vacuum of 10^{-6} mbar. In the process, the $pSiO_2$ samples contained in fused silica glass tube was heated to 573 K using an oven equipped with a temperature controller. After 24 h, the temperature was decreased to 300 K before the polymer solution was injected into the closed vacuum where the pores were filled by capillary wetting. The natural rubber (or diaphragm) ensures that the vacuum chamber remains air–tight even after the injection of the sample. This was followed by a solvent removal step done at 374 K after 24 h of infiltration (Figure 3 b). This procedure of infiltration and solvent removal (here referred to as fractionated filling) was repeated over a period of 3 weeks. The surface of the filled $pSiO_2$ membrane was then thoroughly cleaned using tissue paper soaked in chloroform and then annealed in vacuum (10^{-3} mbar) at 424 K for 24 h to remove remaining solvent. This procedure achieved about 70 % pore filling fraction as determined gravimetrically. The filling fraction was determined as follows: After annealing in high vacuum (10^{-6} mbar) the empty porous matrix was weighed with a microbalance and its mass (m_1) was recorded. This was repeated for the porous silica infiltrated with the polymer sample (m_2). During this procedure, the sample was handled in dry nitrogen atmosphere. The mass of infiltrated material given by (m_2 - m_1) is then obtained. Volume (V_m) of infiltrated material is given by $V_m = (m_2 - m_1)$/density of polymer. Next the total volume (V_p) of the voids/pores in the sample is determined: V_p = (Surface area × thickness of the porous matrix used) × porosity. Filling fraction, Φ (about 70 % ± 5 in this case) was finally obtained by $\Phi = (V_m/V_p) \times 100$. The remaining approximately 30 %, which is mainly nitrogen gas, has a negligible influence on the mean dielectric relaxation peak of the confined polymer, especially given that the nanopores are unidirectional implying parallel capacitance representation. To improve electrical contacts for the

BDS measurement, the sample was finally sandwiched by thin aluminum foil having a thickness of 800 nm.

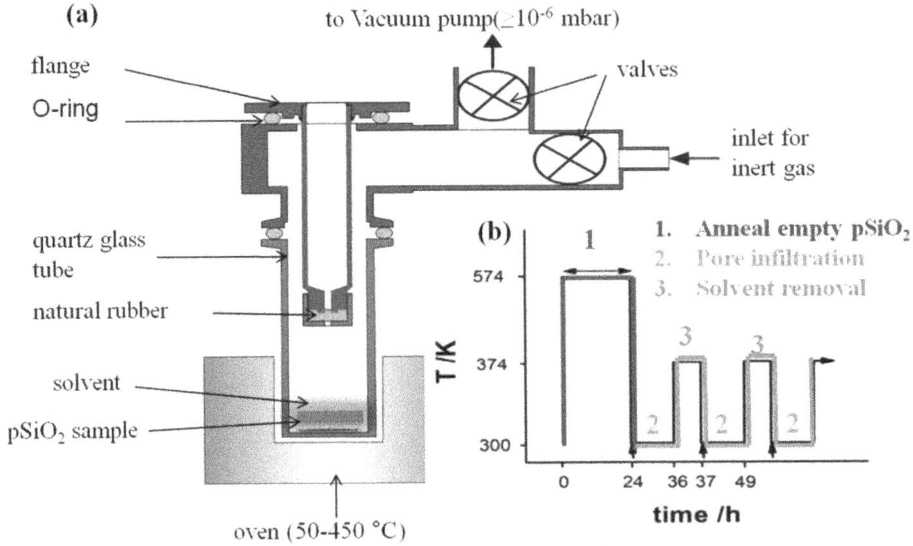

Figure 3. (a) A schematic representation vacuum annealing (10^{-6} mbar) and infiltration of P2VP polymer into the $pSiO_2$ nanopores. (b) Fractionated pore filling procedure employed in this work which involves first step (1) of annealing empty $pSiO_2$ membrane followed by cycles of pore infiltration and solvent removal steps (2) and (3), respectively. Adapted with permission from reference (47). Copyright 2014 Taylor & Francis.

Results and Discussion

The isothermal dielectric loss data ε″(ω) (the imaginary part of the complex function $\varepsilon^*(\omega)=\varepsilon'(\omega)-i\varepsilon''(\omega)$) of all samples were fitted by the superposition of Havriliak–Negami (HN) functions and conductivity contribution given by:

$$\epsilon''(\omega) = -\sum_{j=1}^{n} \mathrm{Im}\left[\frac{\Delta\epsilon_j}{\left[1+\left(i\omega\tau_{HN}\right)^{\alpha_j}\right]^{\gamma_j}}\right] + \frac{\sigma_0}{\omega\epsilon_0} \tag{1}$$

where $\Delta\varepsilon$ is the dielectric relaxation strength, σ_0 is the dc conductivity, ε_0 is the permittivity in free space, ω is the angular frequency (ω = 2p*f*) of the external applied electric field, a and g are the shape parameters describing symmetric and asymmetric broadening of the complex dielectric function, and τ_{HN} is the characteristic relaxation time which is related to the position at maximum loss by

$$\omega_{max} = \frac{1}{\tau_{HN}}\left[\sin\frac{\pi\alpha}{2+2\gamma}\right]^{1/\alpha}\left[\sin\frac{\pi\alpha\gamma}{2+2\gamma}\right]^{-1/\alpha} \tag{2}$$

The dielectric loss spectra of the thin film contain a strong conductivity contribution at low frequencies as well as a signature of the silicon electrodes at high frequencies which have been considered in a fit function together with a Havriliak Negami function to describe the segmental relaxation process of P2VP in-between. (Figure 4). Coinciding with many other studies on thin polymer films employing the same setup, the position of the segmental relaxation peak differs only slightly from the maximum in the bulk (*23*, *55*, *61*, *62*).

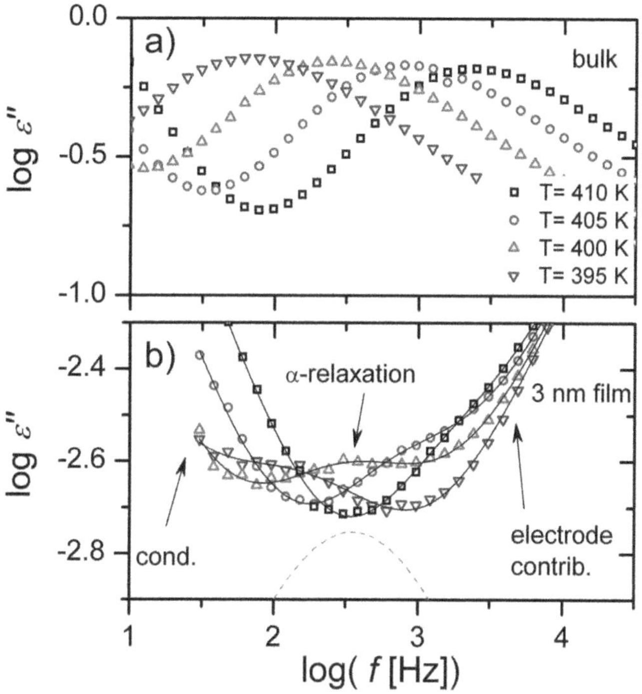

Figure 4. Loss spectra of P2VP (250,000 g/mol) in bulk (200-nm layer between evaporated aluminum electrodes), in a 3-nm film (using nanostructured electrodes) at different temperatures as indicated. The solid lines are fits taking into consideration different contributions as labeled; the dashed line represents the fit of the α -relaxation peak of P2VP at 400 K.

In the dielectric loss spectra of the brushes (Figure 5), a conductivity contribution similar to that in the thin film data exists. In addition to the segmental relaxation process of the P2VP, there is also a relaxation assigned to the PGMA layer at high frequencies (*36*). All these contributions were accounted for in a fit function. From these loss spectra it is evident that also for the brushes the maxima of the segmental relaxation peak (i.e. the mean relaxation rate) is basically unchanged from the bulk position (Figure 5).

The analysis of the data for P2VP confined in nanopores follows a similar approach as discussed above. In this case the low frequency wing is dominated by dc conductivity contribution while the high frequency shows a signal of an additional (b) relaxation process assigned to the local fluctuation of the pyridine moiety (*63*) (Figure 6). This relaxation process is more pronounced at lower temperatures. The peak position of the main (segmental) relaxation is identical for both the bulk and confined samples with the exception that it is more broadened for the latter.

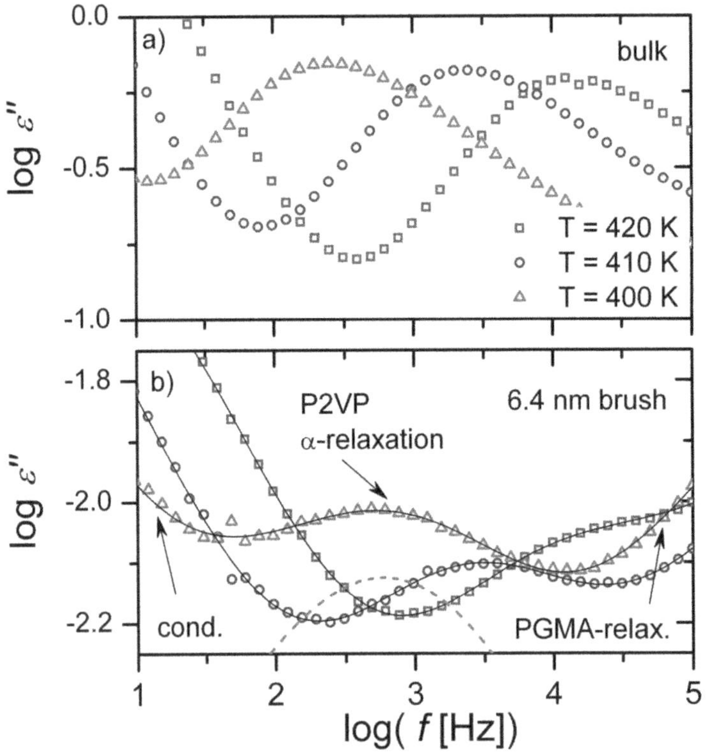

Figure 5. Loss spectra of P2VP bulk (200-nm layer between evaporated aluminum electrodes, 250,000 g/mol) and the corrected loss spectra of carboxylic acid terminated P2VP (43,800 g/mol) grafted to a silica surface (with a PGMA layer as adhesive) of different thicknesses and temperatures as indicated. The solid lines are fits taking into consideration different contributions as labeled; the dashed line represents the fit of the alpha-relaxation peak of P2VP at 400 K.

This broadening effect is more evident when the data is displayed in terms of relaxation time distribution (RTD) (Figure 7). The RTD is computed from the HN fit parameters using the following expression (*64, 65*):

$$\mathbf{g}_{HN}\,(\log\tau) = \frac{\Delta\epsilon}{\pi}\frac{\left(\frac{\tau}{\tau_{HN}}\right)^{\alpha\gamma}\sin(\gamma\Phi)}{\left[1+2\left(\frac{\tau}{\tau_{HN}}\right)^{\alpha}\cos(\pi\alpha)+\left(\frac{\tau}{\tau_{HN}}\right)^{2\alpha}\right]^{\gamma/2}} \tag{3}$$

where

$$\Phi = \arctan\left[\frac{\sin(\pi\alpha)}{\left(\frac{\tau}{\tau_{HN}}\right)^{\alpha}+\cos(\pi\alpha)}\right], \Phi \in [0,\pi]$$

The RTD data for P2VP in the bulk state and under confinement in brushes and in nanopores are compared for the selected temperature of 410 K. This is the temperature at which the segmental relaxation process is more visible for all data within the experimental frequency window. Notably, the RTD depict differences that are associated with the local density defects and interfacial interactions with the confining substrates. The sample in brush geometry depicts a broader RTD on the short time scales due to the fact that the chains at the polymer-air interface do not achieve preferred conformation owing to their chemical anchoring on the support. This also limits the formation of dense adsorbed layers at the substrate. In nanopores, the RTD is broader than bulk on both wings. This is consistent with 2D confinement which often exhibits strong contributions of both finite-size and surface effects (*33*, *39*, *42*, *46*). Whereas the attractive substrate interfacial interactions slow the dynamics down, in this consideration, the geometry also creates regions of low density hence increasing number of segments with faster relaxation times. This effect is persistent for molecular liquids and low molecular weight polymers (*42*, *46*, *66*, *67*) which corroborate the role played by packing density in determining glassy dynamics (*68*, *69*).

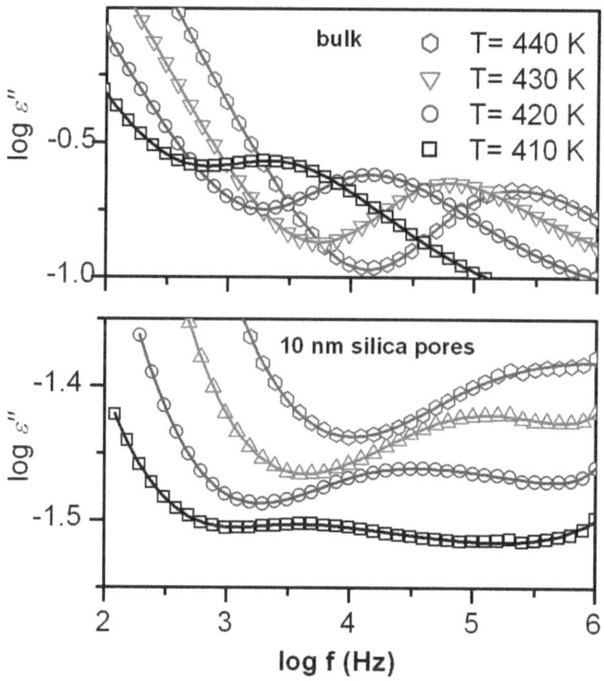

Figure 6. Dielectric loss spectra (at the selected temperatures) of P2VP (Mw 21000 g/mole) in the bulk state and when confined in silica nanopores having average pore diameters of 10 nm. The lines are the HN fit functions to the data.

However, the peak position of the dielectric relaxation times (τ) for all data are identical within experimental accuracy (Figure 7). In order to follow this τ over a wide temperature range, we analyzed the data measured at different temperatures as discussed above and plotted the relaxation rates ($1/\tau_{max}$) versus inverse temperature in a semi logarithmic scale as shown in Figure 8. In this activation plot, all data for P2VP in bulk, thin film, brush and in nanopores collapse onto the single curve (within experimental uncertainty) over a wide temperature range. This dependence is described by the Vogel-Fulcher-Tamman (VFT) equation given by

$$\omega_{max} = \frac{1}{\tau_{max}} = \omega_{\infty} \exp\left[\frac{DT_0}{T - T_0}\right] \quad (4)$$

where ω_{∞} is the high temperature limit of the relaxation rate, D is the fragility parameter, and T_0 is the Vogel temperature. The fit of the VFT function to the data is represented by the red line in Figure 8. This demonstrates that the mean relaxations times of the dynamic glass transition of P2VP remain bulk-like even under extreme geometrical confinement and in different geometries. This is in accord with many other published results in the literature (*1, 35, 41, 43, 55, 61, 71*) and conforms to the fact that the glass transition takes place at a length scale of about 2-3 repeat units (<~ 2 nm) which is still smaller than the finite-size of the external confinement realized in the different topologies.

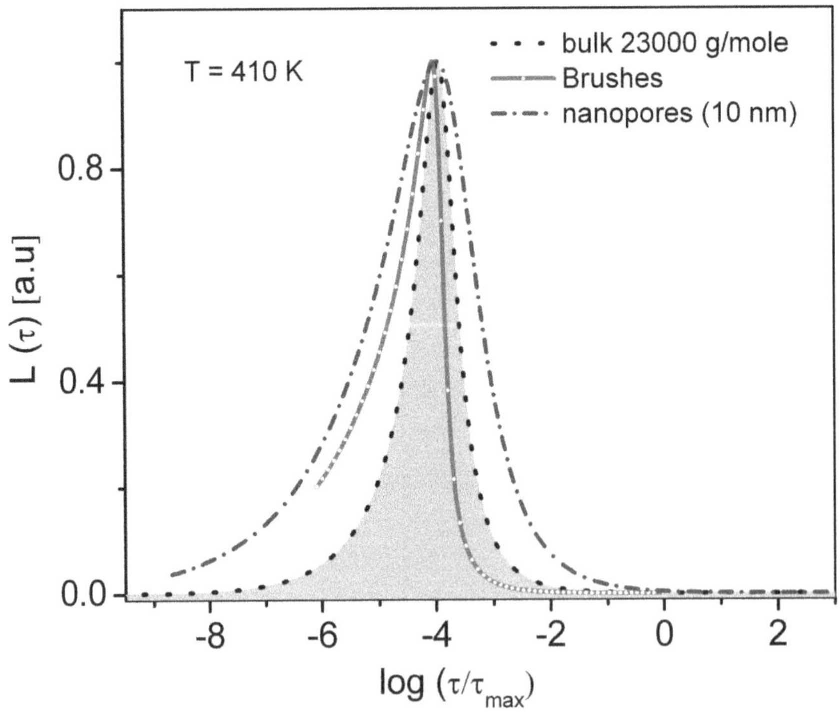

Figure 7. Normalized relaxation time distribution (RTD) of the segmental relaxation as calculated from Havriliak-Negami fit parameters for P2VP in the bulk state and when confined in brushes and in silica nanopores (average pore diameter 10 nm). The gray shade represents the area under RTD for bulk the sample.

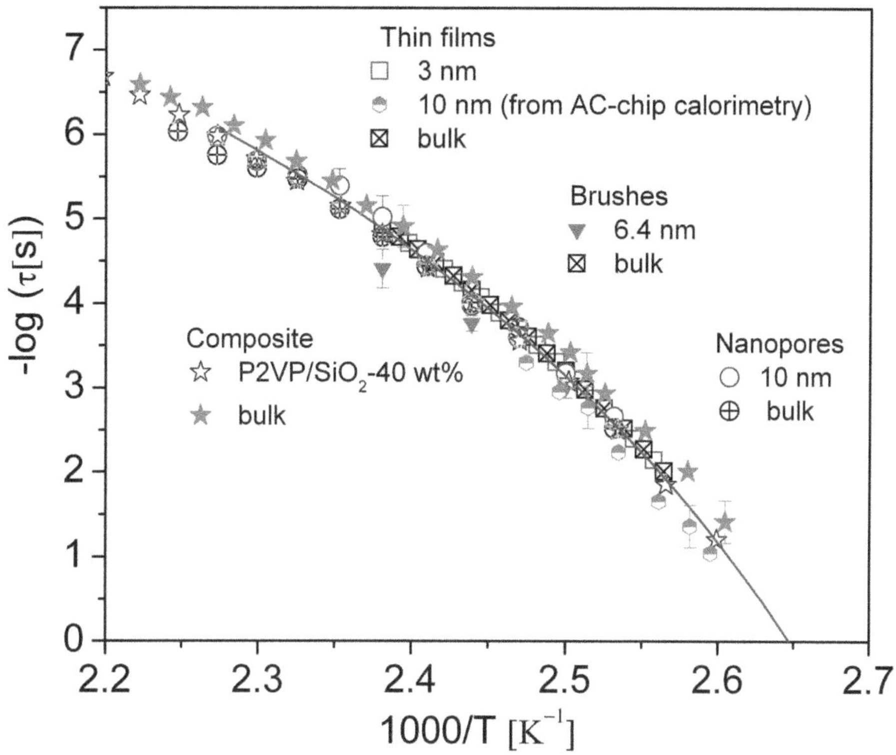

Figure 8. Activation plot compared for the alpha-relaxation rates for P2VP bulk data and similar data for P2VP when confined in thin film (thickness 3 nm), brushes (thicknesses 6.4 nm) and in nanopores (pores diameters 10 nm). The data for P2VP confined in silica nanocomposite was obtained from reference (43). The AC-chip calorimetry data was obtained from reference (70). Representative error bars are shown.

When there is increased impact of interfaces, for instance in nanocomposites consisting of silica nanoparticles dispersed in a P2VP matrix (43), an interfacial layer of a few nanometers thickness was found. This layer thickness is in the order of the persistence length of the polymer (72) which confirms the conclusion that interfacial effects propagate only to the length comparable to the correlation length of the polymer.

Conclusions

In this study, broadband dielectric spectroscopy (BDS) was employed to investigate the impact of different geometries of spatial confinement on glassy dynamics of a model polymer Poly(2vinylpyridne). For the studied cases, [a thin film (~3 nm thickness), a surface-grafted polymer brush and confined to uni-directional nanopores (~10 nm diameter)], the mean relaxation time remains bulk-like in a wide temperature range. A more detailed analysis of the data reveals differences in the relaxation time distribution from the bulk. This indicates the existence of fractions of polymer segments with slower dynamics due to attractive interaction with the interface as well as faster modes due to frustrated packing, i.e. lower density in confined spaces. Nevertheless, the bulk-like fraction by far dominates the glassy dynamics which demonstrates the limited effects induced by interfaces and finite-sizes of confinements in diverse geometries.

This study also exemplifies the following advantages of BDS in the investigation of not only confined systems but general material characterization related to dipolar fluctuations: (i) due to the fact that the capacitance (i.e. the measured signal) scales inversely to the sample thickness, its sensitivity *increases* with *decreasing* amount of sample material (as long as an electrical field is applied while avoiding electrical shorts). State of the art in this case involves spacers as small as 35 nm as realized in the above described nanostructured electrode arrangements; (ii) BDS measurements can be carried out in a broad frequency range and at widely varying temperatures; (iii) BDS measurements are carried out in thermodynamic equilibrium; (iv) BDS enables detailed insights through the relaxation-time distribution function; (v) BDS measures directly the dynamics of all polar moieties in a molecular system without the need of markers or labels (which may aggregate or decouple dynamically); (vi) BDS can be readily adapted to determine glassy dynamics in confinement of quite different topologies such as nanometric supported layers, in nanopores, in powders or composites.

References

1. *Dynamics in Geometrical Confinement*; Kremer, F., Ed.; Advances in Dielectrics; Springer International Publishing: Cham, 2014. https://doi.org/10.1007/978-3-319-06100-9.
2. Keddie, J. L.; Jones, R. A. L.; Cory, R. A. Size-Dependent Depression of the Glass Transition Temperature in Polymer Films. *Europhys. Lett. EPL* **1994**, *27* (1), 59–64. https://doi.org/10.1209/0295-5075/27/1/011.
3. Forrest, J. A.; Dalnoki-Veress, K.; Stevens, J. R.; Dutcher, J. R. Effect of Free Surfaces on the Glass Transition Temperature of Thin Polymer Films. *Phys. Rev. Lett.* **1996**, *77* (19), 4108–4108. https://doi.org/10.1103/PhysRevLett.77.4108.
4. Forrest, J. A.; Dalnoki-Veress, K.; Dutcher, J. R. Interface and Chain Confinement Effects on the Glass Transition Temperature of Thin Polymer Films. *Phys. Rev. E* **1997**, *56* (5), 5705–5716. https://doi.org/10.1103/PhysRevE.56.5705.
5. Fukao, K.; Miyamoto, Y. Glass Transition Temperature and Dynamics of α-Process in Thin Polymer Films. *Europhys. Lett.* **1999**, *46* (5), 649–654. https://doi.org/10.1209/epl/i1999-00309-6.
6. Fryer, D. S.; Nealey, P. F.; de Pablo, J. J. Thermal Probe Measurements of the Glass Transition Temperature for Ultrathin Polymer Films as a Function of Thickness. *Macromolecules* **2000**, *33* (17), 6439–6447. https://doi.org/10.1021/ma0003349.
7. Fukao, K.; Miyamoto, Y. Slow Dynamics near Glass Transitions in Thin Polymer Films. *Phys. Rev. E* **2001**, *64* (1), 011803. https://doi.org/10.1103/PhysRevE.64.011803.
8. Roth, C. B.; Pound, A.; Kamp, S. W.; Murray, C. A.; Dutcher, J. R. Molecular-Weight Dependence of the Glass Transition Temperature of Freely-Standing Poly(Methyl Methacrylate) Films. *Eur. Phys. J. E* **2006**, *20* (4), 441–448. https://doi.org/10.1140/epje/i2006-10034-0.
9. Ellison, C. J.; Kim, S. D.; Hall, D. B.; Torkelson, J. M. Confinement and Processing Effects on Glass Transition Temperature and Physical Aging in Ultrathin Polymer Films: Novel Fluorescence Measurements. *Eur. Phys. J. E* **2002**, *8* (2), 155–166. https://doi.org/10.1140/epje/i2001-10057-y.
10. Efremov, M. Yu.; Olson, E. A.; Zhang, M.; Zhang, Z.; Allen, L. H. Glass Transition in Ultrathin Polymer Films: Calorimetric Study. *Phys. Rev. Lett.* **2003**, *91* (8), 085703. https://doi.org/10.1103/PhysRevLett.91.085703.

11. Efremov, M. Yu.; Olson, E. A.; Zhang, M.; Zhang, Z.; Allen, L. H. Probing Glass Transition of Ultrathin Polymer Films at a Time Scale of Seconds Using Fast Differential Scanning Calorimetry. *Macromolecules* **2004**, *37* (12), 4607–4616. https://doi.org/10.1021/ma035909r.
12. Fakhraai, Z.; Sharp, J. S.; Forrest, J. A. Effect of Sample Preparation on the Glass-Transition of Thin Polystyrene Films. *J. Polym. Sci., Part B: Polym. Phys.* **2004**, *42* (24), 4503–4507. https://doi.org/10.1002/polb.20281.
13. Lupaşcu, V.; Huth, H.; Schick, C.; Wübbenhorst, M. Specific Heat and Dielectric Relaxations in Ultra-Thin Polystyrene Layers. *Thermochim. Acta* **2005**, *432* (2), 222–228. https://doi.org/10.1016/j.tca.2005.04.022.
14. Koh, Y. P.; McKenna, G. B.; Simon, S. L. Calorimetric Glass Transition Temperature and Absolute Heat Capacity of Polystyrene Ultrathin Films. *J. Polym. Sci., Part B: Polym. Phys.* **2006**, *44* (24), 3518–3527. https://doi.org/10.1002/polb.21021.
15. Huth, H.; Minakov, A. A.; Schick, C. Differential AC-Chip Calorimeter for Glass Transition Measurements in Ultrathin Films. *J. Polym. Sci., Part B: Polym. Phys.* **2006**, *44* (20), 2996–3005. https://doi.org/10.1002/polb.20921.
16. Mundra, M. K.; Ellison, C. J.; Behling, R. E.; Torkelson, J. M. Confinement, Composition, and Spin-Coating Effects on the Glass Transition and Stress Relaxation of Thin Films of Polystyrene and Styrene-Containing Random Copolymers: Sensing by Intrinsic Fluorescence. *Polymer* **2006**, *47* (22), 7747–7759. https://doi.org/10.1016/j.polymer.2006.08.064.
17. Mundra, M. K.; Ellison, C. J.; Rittigstein, P.; Torkelson, J. M. Fluorescence Studies of Confinement in Polymer Films and Nanocomposites: Glass Transition Temperature, Plasticizer Effects, and Sensitivity to Stress Relaxation and Local Polarity. *Eur. Phys. J.: Spec. Top.* **2007**, *141* (1), 143. https://doi.org/10.1140/epjst/e2007-00032-0.
18. Huth, H.; Minakov, A. A.; Serghei, A.; Kremer, F.; Schick, C. Differential AC-Chip Calorimeter for Glass Transition Measurements in Ultra Thin Polymeric Films. *Eur. Phys. J.: Spec. Top.* **2007**, *141* (1), 153–160. https://doi.org/10.1140/epjst/e2007-00033-y.
19. Napolitano, S.; Wübbenhorst, M. Dielectric Signature of a Dead Layer in Ultrathin Films of a Nonpolar Polymer. *J. Phys. Chem. B* **2007**, *111* (31), 9197–9199. https://doi.org/10.1021/jp072868i.
20. Koh, Y. P.; Simon, S. L. Structural Relaxation of Stacked Ultrathin Polystyrene Films. *J. Polym. Sci., Part B: Polym. Phys.* **2008**, *46* (24), 2741–2753. https://doi.org/10.1002/polb.21598.
21. Serghei, A.; Huth, H.; Schick, C.; Kremer, F. Glassy Dynamics in Thin Polymer Layers Having a Free Upper Interface. *Macromolecules* **2008**, *41* (10), 3636–3639. https://doi.org/10.1021/ma702381t.
22. Inoue, R.; Kanaya, T.; Nishida, K.; Tsukushi, I.; Telling, M. T. F.; Gabrys, B. J.; Tyagi, M.; Soles, C.; Wu, W.-l. Glass Transition and Molecular Mobility in Polymer Thin Films. *Phys. Rev. E* **2009**, *80* (3), 031802. https://doi.org/10.1103/PhysRevE.80.031802.
23. Mapesa, E. U.; Erber, M.; Tress, M.; Eichhorn, K.-J.; Serghei, A.; Voit, B.; Kremer, F. Glassy Dynamics in Nanometer Thin Layers of Polystyrene. *Eur. Phys. J.: Spec. Top.* **2010**, *189* (1), 173–180. https://doi.org/10.1140/epjst/e2010-01320-2.
24. Tress, M.; Erber, M.; Mapesa, E. U.; Huth, H.; Müller, J.; Serghei, A.; Schick, C.; Eichhorn, K.-J.; Voit, B.; Kremer, F. Glassy Dynamics and Glass Transition in Nanometric Thin Layers

of Polystyrene. *Macromolecules* **2010**, *43* (23), 9937–9944. https://doi.org/10.1021/ma102031k.

25. Paeng, K.; Swallen, S. F.; Ediger, M. D. Direct Measurement of Molecular Motion in Freestanding Polystyrene Thin Films. *J. Am. Chem. Soc.* **2011**, *133* (22), 8444–8447. https://doi.org/10.1021/ja2022834.
26. Paeng, K.; Ediger, M. D. Molecular Motion in Free-Standing Thin Films of Poly(Methyl Methacrylate), Poly(4-Tert-Butylstyrene), Poly(α-Methylstyrene), and Poly(2-Vinylpyridine). *Macromolecules* **2011**, *44* (17), 7034–7042. https://doi.org/10.1021/ma201266r.
27. Boucher, V. M.; Cangialosi, D.; Yin, H.; Schönhals, A.; Alegría, A.; Colmenero, J. T_g Depression and Invariant Segmental Dynamics in Polystyrene Thin Films. *Soft Matter* **2012**, *8* (19), 5119–5122. https://doi.org/10.1039/C2SM25419K.
28. Efremov, M. Yu.; Kiyanova, A. V.; Last, J.; Soofi, S. S.; Thode, C.; Nealey, P. F. Glass Transition in Thin Supported Polystyrene Films Probed by Temperature-Modulated Ellipsometry in Vacuum. *Phys. Rev. E* **2012**, *86* (2), 021501. https://doi.org/10.1103/PhysRevE.86.021501.
29. Wang, J.; McKenna, G. B. Viscoelastic and Glass Transition Properties of Ultrathin Polystyrene Films by Dewetting from Liquid Glycerol. *Macromolecules* **2013**, *46* (6), 2485–2495. https://doi.org/10.1021/ma400040j.
30. Yin, H.; Cangialosi, D.; Schönhals, A. Glass Transition and Segmental Dynamics in Thin Supported Polystyrene Films: The Role of Molecular Weight and Annealing. *Thermochim. Acta* **2013**, *566*, 186–192. https://doi.org/10.1016/j.tca.2013.05.034.
31. Mapesa, E. U.; Tress, M.; Schulz, G.; Huth, H.; Schick, C.; Reiche, M.; Kremer, F. Segmental and Chain Dynamics in Nanometric Layers of Poly(Cis-1,4-Isoprene) as Studied by Broadband Dielectric Spectroscopy and Temperature-Modulated Calorimetry. *Soft Matter* **2013**, *9* (44), 10592–10598. https://doi.org/10.1039/C3SM51311D.
32. Boucher, V. M.; Cangialosi, D.; Alegría, A.; Colmenero, J. Enthalpy Recovery in Nanometer to Micrometer Thick Polystyrene Films. *Macromolecules* **2012**, *45* (12), 5296–5306. https://doi.org/10.1021/ma300622k.
33. Kipnusu, W. K.; Elsayed, M.; Krause–Rehberg, R.; Kremer, F. Glassy Dynamics of Polymethylphenylsiloxane in One- and Two-Dimensional Nanometric Confinement—A Comparison. *J. Chem. Phys.* **2017**, *146* (20), 203302. https://doi.org/10.1063/1.4974767.
34. Tress, M.; Erber, M.; Mapesa, E. U.; Huth, H.; Müller, J.; Serghei, A.; Schick, C.; Eichhorn, K.-J.; Voit, B.; Kremer, F. Glassy Dynamics and Glass Transition in Nanometric Thin Layers of Polystyrene. *Macromolecules* **2010**, *43* (23), 9937–9944. https://doi.org/10.1021/ma102031k.
35. Kipnusu, W. K.; Elmahdy, M. M.; Tress, M.; Fuchs, M.; Mapesa, E. U.; Smilgies, D.-M.; Zhang, J.; Papadakis, C. M.; Kremer, F. Molecular Order and Dynamics of Nanometric Thin Layers of Poly(Styrene-b-1,4-Isoprene) Diblock Copolymers. *Macromolecules* **2013**, *46* (24), 9729–9737. https://doi.org/10.1021/ma4019334.
36. Neubauer, N.; Winkler, R.; Tress, M.; Uhlmann, P.; Reiche, M.; Kiprop Kipnusu, W.; Kremer, F. Glassy Dynamics of Poly(2-Vinyl-Pyridine) Brushes with Varying Grafting Density. *Soft Matter* **2015**, *11* (15), 3062–3066. https://doi.org/10.1039/C5SM00213C.

37. Neubauer, N.; Treß, M.; Winkler, R.; Mapesa, E. U.; Kipnusu, W. K.; Uhlmann, P.; Kremer, F. Molecular Dynamics of Swollen Poly(2-Vinylpyridine) Brushes. *Macromolecules* **2016**, *49* (16), 6101–6105. https://doi.org/10.1021/acs.macromol.6b00363.
38. Tress, M.; Mapesa, E. U.; Kossack, W.; Kipnusu, W. K.; Reiche, M.; Kremer, F. Glassy Dynamics in Condensed Isolated Polymer Chains. *Science* **2013**, *341* (6152), 1371–1374. https://doi.org/10.1126/science.1238950.
39. Kremer, F.; Schönhals, A. *Broadband Dielectric Spectroscopy*; Springer Science & Business Media: 2002.
40. Schwartz, G. A.; Bergman, R.; Swenson, J. Relaxation Dynamics of a Polymer in a 2D Confinement. *J. Chem. Phys.* **2004**, *120* (12), 5736–5744. https://doi.org/10.1063/1.1650289.
41. Kipnusu, W. K.; Elmahdy, M. M.; Mapesa, E. U.; Zhang, J.; Böhlmann, W.; Smilgies, D.-M.; Papadakis, C. M.; Kremer, F. Structure and Dynamics of Asymmetric Poly(Styrene-b-1,4-Isoprene) Diblock Copolymer under 1D and 2D Nanoconfinement. *ACS Appl. Mater. Interfaces* **2015**, *7* (23), 12328–12338. https://doi.org/10.1021/am506848s.
42. Kipnusu, W. K.; Kossack, W.; Iacob, C.; Jasiurkowska, M.; Rume Sangoro, J.; Kremer, F. Molecular Order and Dynamics of Tris(2-Ethylhexyl)Phosphate Confined in Uni-Directional Nanopores. *Z. Für Phys. Chem.* **2012**, *226* (7–8), 797–805. https://doi.org/10.1524/zpch.2012.0287.
43. Holt, A. P.; Griffin, P. J.; Bocharova, V.; Agapov, A. L.; Imel, A. E.; Dadmun, M. D.; Sangoro, J. R.; Sokolov, A. P. Dynamics at the Polymer/Nanoparticle Interface in Poly(2-Vinylpyridine)/Silica Nanocomposites. *Macromolecules* **2014**, *47* (5), 1837–1843. https://doi.org/10.1021/ma5000317.
44. Serghei, A.; Kremer, F. Broadband Dielectric Spectroscopy on Ultrathin Organic Layers Having One Free (Upper) Interface. *Rev. Sci. Instrum.* **2006**, *77* (11), 116108. https://doi.org/10.1063/1.2372740.
45. Serghei, A.; Kremer, F. Broadband Dielectric Studies on the Interfacial Dynamics Enabled by Use of Nanostructured Electrodes. *Rev. Sci. Instrum.* **2008**, *79* (2), 026101. https://doi.org/10.1063/1.2839021.
46. Kipnusu, W. K.; Elmahdy, M. M.; Elsayed, M.; Krause-Rehberg, R.; Kremer, F. Counterbalance between Surface and Confinement Effects As Studied for Amino-Terminated Poly(Propylene Glycol) Constraint in Silica Nanopores. *Macromolecules* **2019**, *52* (4), 1864–1873. https://doi.org/10.1021/acs.macromol.8b02687.
47. Mapesa, E. U.; Popp, L.; Kipnusu, W. K.; Tress, M.; Kremer, F. Molecular Dynamics in 1- and 2-D Confinement as Studied for the Case of Poly(Cis-1,4-Isoprene). *Soft Mater.* **2014**, *12* (sup1), S22–S30. https://doi.org/10.1080/1539445X.2014.928320.
48. Serghei, A.; Hartmann, L.; Kremer, F. Molecular Dynamics in Thin Films of Isotactic Poly(Methylmethacrylate) – Revisited. *J. Non-Cryst. Solids* **2007**, *353* (47), 4330–4333. https://doi.org/10.1016/j.jnoncrysol.2007.01.080.
49. Serghei, A.; Mikhailova, Y.; Huth, H.; Schick, C.; Eichhorn, K.-J.; Voit, B.; Kremer, F. Molecular Dynamics of Hyperbranched Polyesters in the Confinement of Thin Films. *Eur. Phys. J. E* **2005**, *17* (2), 199–202. https://doi.org/10.1140/epje/i2005-10009-7.
50. Papon, A.; Montes, H.; Hanafi, M.; Lequeux, F.; Guy, L.; Saalwächter, K. Glass-Transition Temperature Gradient in Nanocomposites: Evidence from Nuclear Magnetic Resonance and

Differential Scanning Calorimetry. *Phys. Rev. Lett.* **2012**, *108* (6), 065702. https://doi.org/10.1103/PhysRevLett.108.065702.

51. Dutcher, J. R.; Dalnoki-Veress, K.; Forrest, J. A. Optical Probes of the Glass Transition in Thin Polymer Films. In *Supramolecular Structure in Confined Geometries*; ACS Symposium Series; American Chemical Society: 1999; Vol. 736, pp 127–139. https://doi.org/10.1021/bk-1999-0736.ch008.
52. Inoue, R.; Kawashima, K.; Matsui, K.; Kanaya, T.; Nishida, K.; Matsuba, G.; Hino, M. Distributions of Glass-Transition Temperature and Thermal Expansivity in Multilayered Polystyrene Thin Films Studied by Neutron Reflectivity. *Phys. Rev. E* **2011**, *83* (2), 021801. https://doi.org/10.1103/PhysRevE.83.021801.
53. Krutyeva, M.; Martin, J.; Arbe, A.; Colmenero, J.; Mijangos, C.; Schneider, G. J.; Unruh, T.; Su, Y.; Richter, D. Neutron Scattering Study of the Dynamics of a Polymer Melt under Nanoscopic Confinement. *J. Chem. Phys.* **2009**, *131* (17), 174901. https://doi.org/10.1063/1.3258329.
54. Bhattacharya, M.; Sanyal, M. K.; Geue, Th.; Pietsch, U. Glass Transition in Ultrathin Polymer Films: A Thermal Expansion Study. *Phys. Rev. E* **2005**, *71* (4), 041801. https://doi.org/10.1103/PhysRevE.71.041801.
55. Kremer, F.; Tress, M.; Mapesa, E. U. Glassy Dynamics and Glass Transition in Nanometric Layers and Films: A Silver Lining on the Horizon. *J. Non-Cryst. Solids* **2015**, *407*, 277–283. https://doi.org/10.1016/j.jnoncrysol.2014.08.016.
56. Ionov, L.; Sidorenko, A.; Stamm, M.; Minko, S.; Zdyrko, B.; Klep, V.; Luzinov, I. Gradient Mixed Brushes: "Grafting To" Approach. *Macromolecules* **2004**, *37* (19), 7421–7423. https://doi.org/10.1021/ma049147r.
57. Iyer, K. S.; Zdyrko, B.; Malz, H.; Pionteck, J.; Luzinov, I. Polystyrene Layers Grafted to Macromolecular Anchoring Layer. *Macromolecules* **2003**, *36* (17), 6519–6526. https://doi.org/10.1021/ma034460z.
58. Damiron, D.; Mazzolini, J.; Cousin, F.; Boisson, C.; D'Agosto, F.; Drockenmuller, E. Poly(Ethylene) Brushes Grafted to Silicon Substrates. *Polym. Chem.* **2012**, *3* (7), 1838–1845. https://doi.org/10.1039/C1PY00459J.
59. Zdyrko, B.; Klep, V.; Luzinov, I. Synthesis and Surface Morphology of High-Density Poly(Ethylene Glycol) Grafted Layers. *Langmuir* **2003**, *19* (24), 10179–10187. https://doi.org/10.1021/la034974r.
60. Serghei, A.; Kremer, F. Broadband Dielectric Studies on the Interfacial Dynamics Enabled by Use of Nanostructured Electrodes. *Rev. Sci. Instrum.* **2008**, *79* (2), 026101. https://doi.org/10.1063/1.2839021.
61. Erber, M.; Tress, M.; Mapesa, E. U.; Serghei, A.; Eichhorn, K.-J.; Voit, B.; Kremer, F. Glassy Dynamics and Glass Transition in Thin Polymer Layers of PMMA Deposited on Different Substrates. *Macromolecules* **2010**, *43* (18), 7729–7733. https://doi.org/10.1021/ma100912r.
62. Serghei, A.; Tress, M.; Kremer, F. The Glass Transition of Thin Polymer Films in Relation to the Interfacial Dynamics. *J. Chem. Phys.* **2009**, *131* (15), 154904. https://doi.org/10.1063/1.3248368.
63. Papadopoulos, P.; Peristeraki, D.; Floudas, G.; Hadjichristidis, N. Origin of Glass Transition of Poly(2-vinylpyridine). A Temperature- and Pressure-Dependent Dielectric Spectroscopy

Study. *Macromolecules* **2004**, *37*, 8116–8122. https://pubs.acs.org/doi/abs/10.1021/ma048555s

64. Richert, R. Supercooled Liquids and Glasses by Dielectric Relaxation Spectroscopy. In *Advances in Chemical Physics*; John Wiley & Sons, Ltd.: 2014; pp 101−195. https://doi.org/10.1002/9781118949702.ch4.
65. Zorn, R. Applicability of Distribution Functions for the Havriliak–Negami Spectral Function. *J. Polym. Sci., Part B: Polym. Phys.* **1999**, 37 (10), 1043–1044. https://doi.org/10.1002/(SICI)1099-0488(19990515)37:10<1043::AID-POLB9>3.0.CO;2-H.
66. Kipnusu, W. K.; Elsayed, M.; Krause–Rehberg, R.; Kremer, F. Glassy Dynamics of Polymethylphenylsiloxane in One- and Two-Dimensional Nanometric Confinement—A Comparison. *J. Chem. Phys.* **2017**, *146* (20), 203302. https://doi.org/10.1063/1.4974767.
67. Kipnusu, W. K.; Elsayed, M.; Kossack, W.; Pawlus, S.; Adrjanowicz, K.; Tress, M.; Mapesa, E. U.; Krause-Rehberg, R.; Kaminski, K.; Kremer, F. Confinement for More Space: A Larger Free Volume and Enhanced Glassy Dynamics of 2-Ethyl-1-Hexanol in Nanopores. *J. Phys. Chem. Lett.* **2015**, *6* (18), 3708–3712. https://doi.org/10.1021/acs.jpclett.5b01533.
68. Kipnusu, W. K.; Elsayed, M.; Iacob, C.; Pawlus, S.; Krause-Rehberg, R.; Paluch, M. Glassy Dynamics Predicted by Mutual Role of Free and Activation Volumes. *Soft Matter* **2019**, *15* (23), 4656–4661. https://doi.org/10.1039/C9SM00363K.
69. White, R. P.; Lipson, J. E. G. To Understand Film Dynamics Look to the Bulk. *Phys. Rev. Lett.* **2020**, *125* (5), 058002. https://doi.org/10.1103/PhysRevLett.125.058002.
70. Madkour, S.; Yin, H.; Füllbrandt, M.; Schönhals, A. Calorimetric Evidence for a Mobile Surface Layer in Ultrathin Polymeric Films: Poly(2-Vinyl Pyridine). *Soft Matter* **2015**, *11* (40), 7942–7952. https://doi.org/10.1039/C5SM01558H.
71. Mapesa, E. U.; Erber, M.; Tress, M.; Eichhorn, K.-J.; Serghei, A.; Voit, B.; Kremer, F. Glassy Dynamics in Nanometer Thin Layers of Polystyrene. *Eur. Phys. J.: Spec. Top.* **2010**, *189* (1), 173–180. https://doi.org/10.1140/epjst/e2010-01320-2.
72. Cheng, S.; Carroll, B.; Lu, W.; Fan, F.; Carrillo, J.-M. Y.; Martin, H.; Holt, A. P.; Kang, N.-G.; Bocharova, V.; Mays, J. W.; Sumpter, B. G.; Dadmun, M.; Sokolov, A. P. Interfacial Properties of Polymer Nanocomposites: Role of Chain Rigidity and Dynamic Heterogeneity Length Scale. *Macromolecules* **2017**, *50* (6), 2397–2406. https://doi.org/10.1021/acs.macromol.6b02816.

Chapter 9

Polymers under 2-D Confinement: Flow of Polymer Melts at the Nanoscale

Chien-Hua Tu,[1] Martin Steinhart,[2] Hans-Juergen Butt,[1] and George Floudas[*,1,3]

[1]Max Planck Institute for Polymer Research, D-55128 Mainz, Germany

[2]Institut für Chemie neuer Materialien, Universität Osnabrück, D-49069 Osnabrück, Germany

[3]Department of Physics, University of Ioannina, 45110 Ioannina, Greece

[*]Email: gfloudas@uoi.gr.

With the recently developed nanofluidic method one is able to follow *in situ* the imbibition of polymers into nanotubes. Therefore, pieces of nanoporous anodic aluminum oxide (AAO) are coated with gold electrodes on the top and bottom face. The dielectric response is recorded while the polymer imbibes into the cylindrical nanopores of ≈ 100 μm length and 12–200 nm radius. The method provides simultaneous access to the imbibition kinetics and the associated polymer dynamics (at the segmental and chain length scales) during polymer flow. Two models- a parallel and a series model- with different interactions between the polymer/AAO are investigated with respect to the evolution of the dielectric strength and of the characteristic relaxation times. The model predictions are compared and tested against existing results from two polymers: a polar polymer (poly(*n*-butyl methacrylate), PnBMA) and a non-polar polymer (*cis*-1,4-polyisoprene, PI). In the case of PI, where the longest normal mode was probed, it was shown that the dielectric strength during imbibition deviates from the usual dependence. The results are discussed in terms of increasing adsorption sites with time. We further discuss the potential of the method in current polymer physics problems of interest.

Introduction

Polymer properties may deviate from bulk when the polymer is confined into sizes close to or even below the radius of gyration. Most of these deviations originate from the interactions between the polymer and the confining medium at the vicinity of the interface (*1*). Discerning these effects can bring forth certain advantages in the design and eventually fabrication of nanostructured devices. Several confining geometries (1D, 2D, and 3D) have been employed to address the issue of confinement (*1*). A relatively recent development in this direction has been the construction of model 2D confining tubes based on self-ordered nanoporous anodic aluminum oxide, better known as AAO (*2, 3*). AAO templates contain arrays of parallel cylindrical nanopores uniform in length and diameter where polymers can be infiltrated by capillary action (*2, 3*).

Our concern here is the exact process of polymer imbibition in AAO. For Newtonian liquids, nearly a century ago, Lucas and Washburn proposed an equation (known as LWE) to describe the capillary imbibition (*4, 5*):

$$h(t) = \sqrt{\frac{\gamma R \cos\theta}{2\eta}}\sqrt{t} \tag{1}$$

Here, $h(t)$ is the imbibition length (up to ~ 100 μm in the present system), R is the capillary radius (200 nm and 20 nm in the present study), γ is the surface tension, θ is the advancing contact angle, η is the viscosity and t is the imbibition time. Although successfully capturing the imbibition of Newtonian liquids, eq. 1 fails to describe the capillary imbibition of complex liquids like polymers. Recently, a unified theory constructed by Doi (*6*) proposed that the imbibition of polymer melts inside nanopores can be described via two mechanisms: one is the *dead-layer* effect due to the strong adsorption of polymer chains. This effect creates a smaller effective pore radius ($R_{effective}<R_{AAO}$) and gives rise to a higher effective viscosity ($\eta_{effective}>\eta_{bulk}$) during flow. The other mechanism relies on the *reptation* model under a pressure gradient. As pores reduce to a size comparable to the thickness of the dead zone (extreme confinement), polymer transport proceeds by the *reptation* of free polymer chains in a network. It is driven by the pressure gradient originating from the capillary pressure at the front. This mechanism leads to a lower effective viscosity ($\eta_{effective}< \eta_{bulk}$). The competition of the two mechanisms gives rise to a non-monotonic behavior for the dependence of the effective viscosity on inverse pore size. Compared with the experimental results (*6, 7*), the reversal in imbibition of melts of poly(ethylene oxide) was qualitatively captured by the theory.

Despite several studies of polymer dynamics under 2D confinement (including shifts in the glass temperature and the global chain motion) little is known on how polymers flow within narrow pores. Comprehensive understanding of the imbibition process requires not only knowledge of the effective viscosity but also the relevant polymer dynamics *during* flow. The only problem is that in such experiments the polymer chains are away from equilibrium. However, most processes in our daily life are away from equilibrium and certainly the design of nanomaterials and nanostructures requires a deep understanding of how polymers flow next to surfaces and in the presence of strong boundary interactions. One possible method of following polymer imbibition *in situ* was proposed recently (*8*). The method is based on an interferometric technique by detecting changes of the refractive index of the polymer/AAO (*8*). However, despite its simplicity, it requires several calibration steps with liquids of known viscosity.

An important step in this direction has been the development of a nanofluidic method, based on dielectric spectroscopy (DS) (*9–15*). The principle of the method is based on measuring the capacitance. The top and bottom side of a piece of AAO are coated with a metal and form the electrodes of a plate capacitor, C ($C=\varepsilon\varepsilon_0 A/d$; where ε is the dielectric permittivity of the polymer, ε_0 the vacuum permittivity, A the area and d is the polymer thickness) during flow. As more polymer is dragged into the pores, the capacitance is increased by virtue of the dielectric permittivity. The method provides access to the imbibition length through the measured dielectric strength. The latter can be compared with independent ex situ measurements with e.g. reflection optical microscopy. At the same time, DS is capable of measuring the polymer dynamics at the segmental or chain lengths during flow. Despite the obvious importance of the method, there exist only a few studies on polymer imbibition: mainly on poly(2-vinyl pyridine) (*9*), and poly(vinylidene fluoride-*co*-trifluoroethylene) (PVDF-TrFE) (*12*). In these studies the imbibition kinetics were found to scale as $t^{1/2}$, as predicted by the LWE. However, the effective viscosity was not extracted, which precluded a stringent test of the LWE.

In the present investigation, we employ different models that take into account the interactions of the polymer with the AAO surface and discuss the evolution of the dielectric strength and of the characteristic times during flow. The model predictions are compared and tested against existing results from two polymers. For this purpose we employ a polar and a non-polar polymer based, respectively, on poly(*n*-butyl methacrylate) (PnBMA) and *cis*-1,4-polyisoprene (PI) and discuss the different possibilities and limitations of the technique. At the end we provide an outlook of the potential of the method to address some important polymer physics questions.

The Materials

A poly(*n*-butyl methacrylate) (PnBMA) homopolymer with M_n=4.9 kg·mol^{-1}, M_w=5.4 kg·mol^{-1}, T_g=285 K and narrow dispersity $Đ$ (M_w/M_n)=1.11 was purchased from PSS Polymer Standards Service GmbH and used without further purification. A *cis*-1,4-polyisoprene (PI) homopolymer with M_n=42 kg·mol^{-1}, M_w=43 kg·mol^{-1}, and narrow dispersity $Đ$ (M_w/M_n)=1.04, T_g=203 K was synthesized in house. Two types of AAO templates were used (Figure 1) as these are used from various research groups: (i) self-ordered AAO (pore diameters of 25 , 35, 65 and 400 nm; pore depth about 100 μm) with one closed end (the other was attached to an Al substrate) prepared following previously reported procedures (*2*, *3*). (ii) AAO templates with both-ends open purchased from InRedox (Longmont, USA). Pore diameters and corresponding porosities were 100 nm (24±3%), 40 nm (12±2%) and 20 nm (11±2%), respectively. Prior to infiltration, all AAO templates were placed in an oven under vacuum at a temperature of 443 K for 8-10 h. This procedure is essential as it produces reproducible surface properties and removes the majority of OH groups from the AAO surface.

The Method–Nanodielectric Spectroscopy

Dielectric spectroscopy (DS) is employed to determine the dynamics of bulk and confined PnBMA and PI. The instrument is equipped with a Novocontrol Alpha frequency analyzer composed of a broadband dielectric converter and an active sample head. For bulk samples, polymers are embedded between two stainless steel electrodes with a diameter of 20 mm accompanied with a Teflon spacer of 50 μm to maintain the thickness. The sample preparation for *in situ* nanodieletric spectroscopy is more complicated (Figure 2). First, the AAO templates are sputtered with a gold layer with a thickness of 35 nm on the top side (or even rear side for templates from InRedox templates). The aim of the sputtered Au layers is to serve as electrodes, which can build a complete circuit across the template. Sputtering was made under high vacuum (better than 2×10^{-5} Pa) by Bal-tec MED 020 with a current density of 40 mA. The high vacuum environment is to ensure the uniformity of the sputtered Au layer. The current density is optimized to produce Au grains with appropriate size (too big will block the AAO pores; too small will hinder the current transport). Following sputtering, scanning electron microscopy (SEM) is used to determine the distribution of Au grains on the inner surface of the AAO templates (Figure 2). In addition to a uniform film deposited on top of AAO templates, Au particles are inevitably found to fall within pores (depth of about 380 nm). However, given the high aspect ratio ($> 10^3$) of nanopores, any inhomogeneities in the field lines can be neglected. Subsequently, a thin film or a drop of polymer is deposited on the top of the template. The frequency range also needs to be selected. A broad frequency range from 10^{-2}–10^7 Hz is normally used in bulk DS measurements. In contrast, a narrower frequency range, from 10–10^6 Hz, is applied to probe the faster imbibition kinetics since a shorter time is needed. Typically, the applied voltage is below 1 Volt. The dielectric function depends on frequency, temperature and time. At a given time and temperature, it can be fitted by the empirical equation of Havriliak and Negami (HN):

$$\varepsilon_{HN}^*(\omega,T,t) = \varepsilon_\infty(T,t) + \sum_{k=1}^{2} \frac{\Delta\varepsilon_k(T,t)}{\left[1+\left(i\omega\tau_{HN,k}(T,t)\right)^{\alpha(t)}\right]^{\gamma(t)}} + \frac{\sigma_0(T,t)}{i\varepsilon_0\omega} \tag{2}$$

Here, k indicates the process under investigation, $\Delta\varepsilon_k(T,t)$ is the relaxation strength, $\tau_{HN,k}$ is the Havriliak-Negami relaxation time, α, γ ($0 < \alpha, \alpha\gamma \leq 1$) are shape parameters describing the symmetric and asymmetric broadening of dielectric function (with the high frequency slope in a log-log representation being $\alpha\gamma$), and ε_∞ is the dielectric permittivity at the limit of high frequencies. At lower frequencies, ε'' rises due to the conductivity ($\varepsilon'' = \sigma_0/(\omega\varepsilon_0)$, where σ_0 is the dc conductivity and ε_0 the vacuum permittivity).

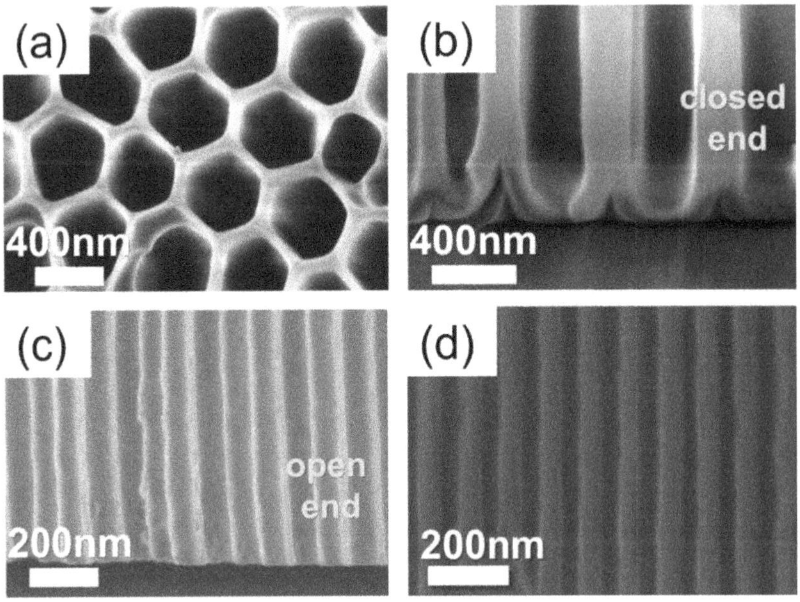

Figure 1. SEM figures showing the (a) top surface of AAO 400 nm pores and cross sections of AAO pores with diameter of (b) 400 nm, (c) 65 nm, and (d) 40 nm. (d) represents the template from InRedox.

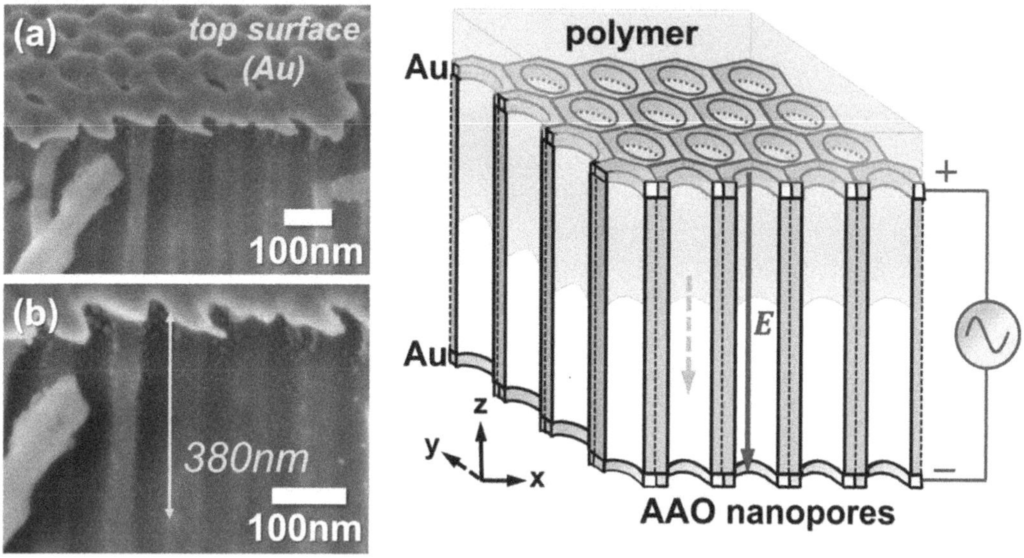

Figure 2. (Left top and bottom) SEM figures showing the top surface and cross section of AAO pores with a dimeter of 65nm after Au sputtering. The left bottom figure is a zoom-in picture. Au particles are distributed at the top of the nanopores within a distance of 380 nm. (Right) Schematic of the experimental setup of in-situ nanodielectric spectroscopy. The blue arrow indicates the direction of the applied electric field, E (along the nanopores, z-axis). The cyan dashed arrow indicates the flow direction of the polymer meniscus within the AAO nanopores. The xyz coordinate represent the laboratory axis.

There are also certain limitations in following the imbibition kinetics by Nanodielectric Spectroscopy. It requires a judicious choice of a polymer with a suitable glass temperature, T_g, and with a bulk viscosity that is neither too low nor too high at the imbibition temperature, so that, the process falls within the experimental frequency window. The frequency window may also be limited by fast kinetics. These are parameters that need to be considered in designing any experiment.

Simulation of the Imbibition Process

Despite the simplicity of the method several factors need to be considered, including the porosity of the templates, the presence of closed ends and the possibility of entrapped air, the background signal from the AAO, the polarity of the polymer and the interactions between the polymer and the AAO walls. The Debye model function is used to simulate the frequency dependence of complex dielectric permittivity ($\varepsilon^* = \varepsilon' - i\varepsilon''$) of a polar and a nonpolar polymer:

$$\varepsilon^*{}_{polymer} = \Delta\varepsilon_{polymer} \frac{1}{1 + i\omega\tau_D} + \varepsilon_{polymer,\infty} \tag{3}$$

with real (ε') and imaginary (ε'') parts

$$\varepsilon'{}_{polymer} = \frac{\Delta\varepsilon_{polymer}}{1 + (\omega\tau_D)^2} + \varepsilon_{polymer,\infty} \tag{4}$$

$$\varepsilon''{}_{polymer} = \Delta\varepsilon_{polymer} \frac{\omega\tau_D}{1 + (\omega\tau_D)^2} \tag{5}$$

Here, $\Delta\varepsilon_{polymer}$ is the dielectric strength of the polymer, $\varepsilon_{polymer,\infty}$ is the dielectric permittivity at very high frequencies, and τ_D is a characteristic relaxation time (here set to 1s). For the polar (nonpolar) polymer cases, the following exemplary values were employed: $\Delta\varepsilon_{polymer}$=1 (0.1), and $\varepsilon^*_{polymer,\infty}$=2.25 for both cases. To provide a comprehensive analysis the equivalent capacitance representation is necessary. Here, two models are considered with respect to the interaction of the polymer/AAO (e.g. by "interaction" we mean the area of contact and the strength of polymer/AAO interaction at the interface: (i) a parallel model with weak interactions in the horizontal direction (e.g. the direction perpendicular to the nanotubes), and (ii) a series model with strong interactions in the horizontal direction. In the parallel model, the capacitance of the polymer (C_1) and air (C_2) are connected in series, and the combinational capacitance of polymer/air (C_{12}) connects in parallel with AAO (C_3) (Figure 3a). In the series model, the combinatorial capacitance of the polymer (C_1) and AAO (C_3), C_{13} connect in series with the combined air/AAO capacitance (C_{23}) (Figure 3b).

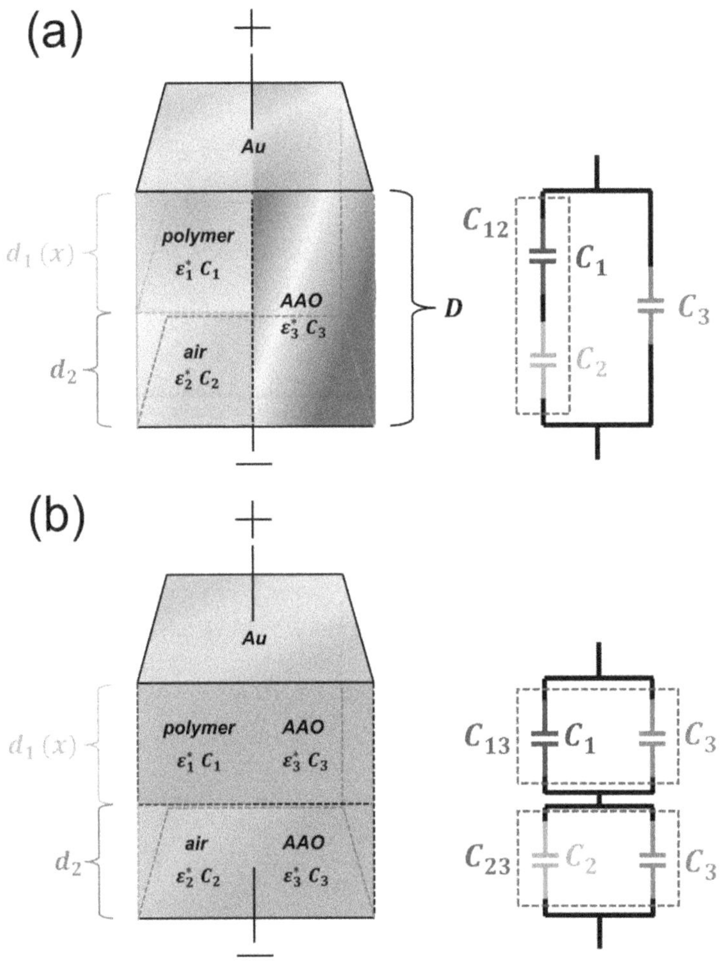

Figure 3. Schematics illustrating the equivalent capacitance of the (a) parallel and (b) series models. Parameters are defined as follows: ε_1^*, ε_2^*, ε_3^*, ε_{12}^*, ε_{13}^*, and ε_{23} are, respectively, the permittivity of the polymer layer, air, AAO, polymer/air composite, the polymer/AAO composite, and air/AAO composite; C_1, C_2, C_3, C_{12}, C_{13}, and C_{23} are the respective capacitances; d_1, and d_2 are the lengths of the pores comprising the polymer and air. D is the full length of nanopores.

Parallel Model: Weak Interaction between Polymer and AAO in the Horizontal Direction

In terms of the equivalent circuit, we have the following equations,

$$\frac{1}{C_{12}} = \frac{1}{C_1} + \frac{1}{C_2} \tag{6}$$

$$C_{123} = C_{12} + C_3 \tag{7}$$

From the definition of capacitance (i.e., $C = \varepsilon A/d$) it follows that,

$$\varepsilon_{12}^* = \frac{D\varepsilon_1^*}{d_1 + d_2\varepsilon_1^*} \tag{8}$$

$$\varepsilon_{123}^* = \varepsilon_{12}^*\varphi_{12} + \varepsilon_3^*\varphi_3 \tag{9}$$

φ_{12} represents the porosity of AAO templates, $\varphi_3 (= 1 - \varphi_{12})$ is the remaining area of the template, and D is the full length of AAO nanopores. After some arrangements, we can obtain the following equation for the complex dielectric permittivity:

$$\varepsilon_{123}^* = \left[\frac{D\varepsilon_1^*}{d_1 + d_2\varepsilon_1^*}\right]\varphi_{12} + \varepsilon_3^*\varphi_3 \tag{10}$$

Therefore, the imaginary part is

$$\varepsilon_{123}'' = \frac{d_1\varepsilon_1''\varphi_{12}}{(d_1 + d_2\varepsilon_1')^2 + (\varepsilon_1''d_2)^2} \tag{11}$$

Series Model: Strong Interaction between Polymer and AAO in the Horizontal Direction

In terms of the equivalent circuit,

$$\frac{1}{C_{123}} = \frac{1}{C_{13}} + \frac{1}{C_{23}} \tag{12}$$

$$C_{13} = C_1 + C_3 \tag{13}$$

$$C_{23} = C_2 + C_3 \tag{14}$$

From the definition of capacitance (i.e., $C = \varepsilon A/d$),

$$\frac{D}{\varepsilon_{123}^*} = \frac{d_1}{\varepsilon_{13}^*} + \frac{d_2}{\varepsilon_{23}^*} \tag{15}$$

$$\varepsilon_{13}^* = \varepsilon_1^*\varphi + \varepsilon_3^*(1 - \varphi) \tag{16}$$

$$\varepsilon_{23}^* = \varepsilon_2^*\varphi + \varepsilon_3^*(1 - \varphi) \tag{17}$$

After some arrangements, we can obtain the following equation for the complex dielectric permittivity:

$$\varepsilon_{123}^{*} = \frac{[\varepsilon_1^{*}\varphi + \varepsilon_3^{*}(1-\varphi)][\varepsilon_2^{*}\varphi + \varepsilon_3^{*}(1-\varphi)]}{[\varepsilon_1^{*}\varphi + \varepsilon_3^{*}(1-\varphi)] + x\varphi(\varepsilon_2^{*} - \varepsilon_1^{*})} \tag{18}$$

Here, x=d_1 (imbibition length of polymer). The imaginary part is

$$\varepsilon_{123}'' = \frac{BC - AG}{C^2 + G^2} \tag{19}$$

$$B \to \varepsilon_1''\varphi^2 + a\varphi\varepsilon_1'' \tag{20}$$

$$C \to \varepsilon_1'\varphi(1-x) + a + x\varphi \tag{21}$$

$$A \to \varepsilon_1'\varphi^2 + a\varphi(\varepsilon_1' + 1) + a^2 \tag{22}$$

$$G \to \varepsilon_1''\varphi(1-x) \tag{23}$$

$$a \to \varepsilon_3^{*}(1-\varphi) \tag{24}$$

For the real and imaginary parts of dielectric constant of AAO we have employed typical values ε_3'=ε_{AAO}'=2.6 and ε_3''=ε_{AAO}''=0 (*14*). The latter involves the subtraction of the empty cell (e.g. background) signal.

Three issues are addressed for two polymers (polar and apolar): (i) the effect of porosity (φ) on the evolution of ε'', (ii) the variation of the dielectric strength ($\Delta\varepsilon$) with the imbibition length (d_1 or x), and (iii) the variation of the relaxation time during imbibition. The results of the parallel and the series model for the nonpolar and polar polymers (Figures 4-7) are discussed with respect to the growing dielectric loss versus the imbibition length. Independent of polarity, at low porosities (i.e., φ=0.1), $\Delta\varepsilon$ shows a linear increase with d_1 in the series model (Figure 4b and 5b). In contrast, within the parallel model a linearity applies only at the initial stages of the imbibition process turning rapidly nonlinear at late stages. At intermediate (i.e., φ=0.5), and high (i.e., φ=0.9) porosities even the series model (Figures 4, 5) deviates from linearity.

On the other hand, polarity affects mainly the relaxation times *during* imbibition. For the nonpolar polymer (e.g. *cis*-1,4-polyisoprene), the position of the relaxation peak stays nearly constant during imbibition and this is independent from the model; by contrast, for the polar polymer [e.g. poly(*n*-butyl methacrylate)] a faster relaxation process at shorter imbibition lengths is followed by a gradual slowing-down at longer imbibition lengths (Figure 6). Notice that the molecular dynamics obtained from the series model are always slower than the parallel model and closer to bulk behavior. This phenomenon reflects on the extensive interactions between polymer and AAO in the series model.

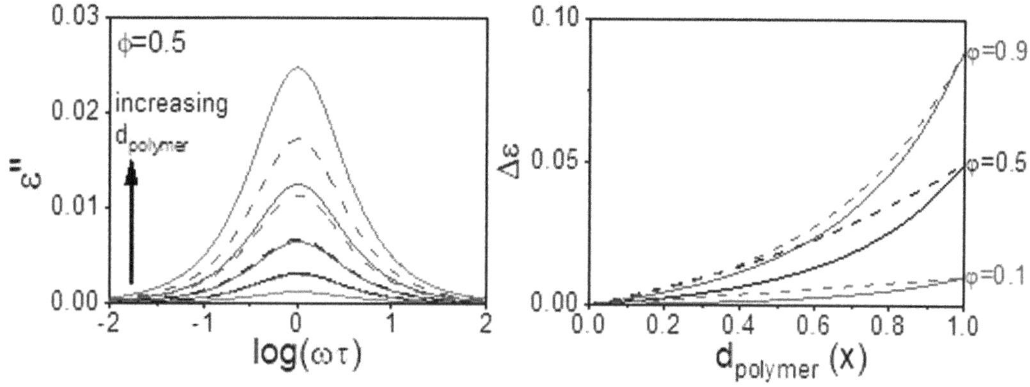

Figure 4. (Left) Comparison in simulated growing ε" of a nonpolar polymer (Δε=0.1) as a function of imbibition length (x) via two different models (solid lines: parallel model; dashed lines: series model) for porosity (φ) equal to 0.5. Color codes are as following: (red) x=0.2, (blue) x=0.4, (magenta) x=0.6, (orange) x=0.8, and (olive) x=1. (Right) Comparison in extracted dielectric intensity (Δε) as a function of x for different porosity: (red) φ=0.1, (black) φ=0.5, and (blue) φ=0.9. Solid (dashed) lines represent the parallel (series) model.

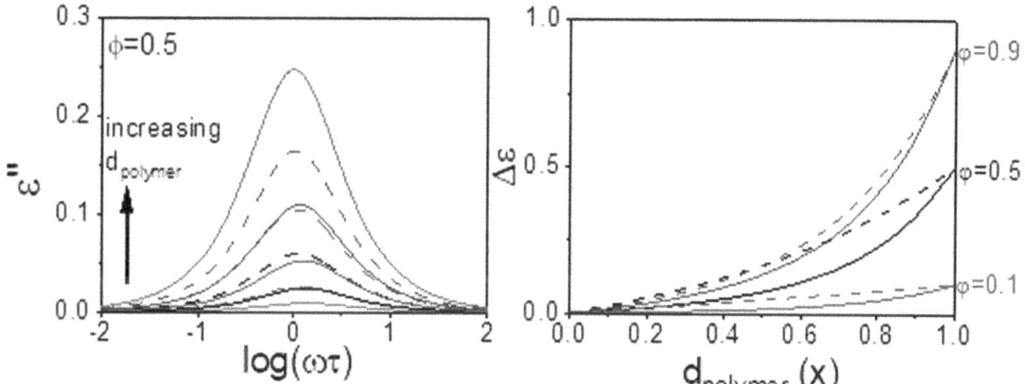

Figure 5. (Left) Simulated ε" of a polar polymer (Δε=1) as a function of imbibition length (x) via two different models (solid lines: parallel model; dashed lines: series model) for porosity (φ) equal to 0.5. Color codes are as following: (red) x=0.2, (blue) x=0.4, (magenta) x=0.6, (orange) x=0.8, and (olive) x=1. (Right) Comparison in extracted dielectric intensity (Δε) as a function of x for different porosity: (red) φ=0.1, (black) φ=0.5, and (blue) φ=0.9. Solid (dashed) lines represent the parallel (serial) model.

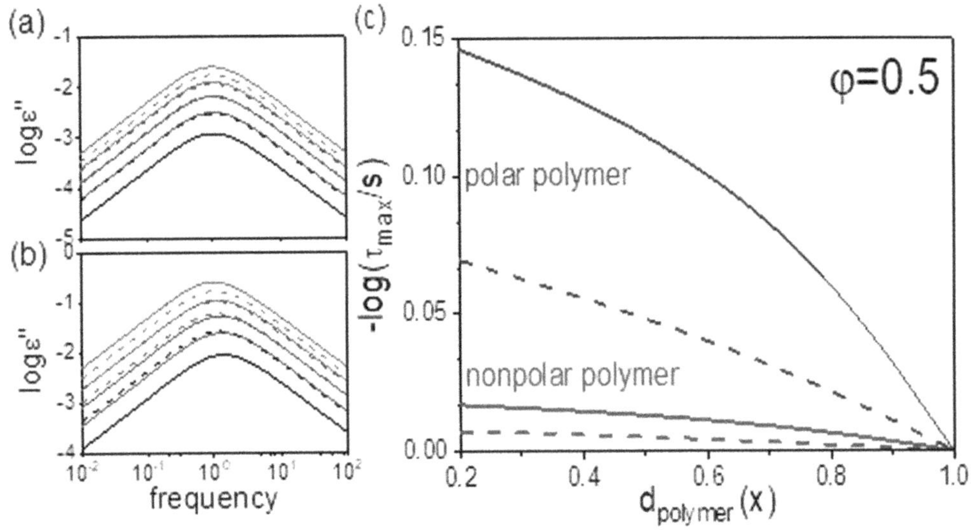

Figure 6. Representative simulated DS loss curves during imbibition for (a) nonpolar and (b) polar polymers. (c) Comparison in extracted relaxation times as a function of imbibition length (x) is plotted. Solid (dashed) lines represent the parallel (series) model. Color codes in (a) and (b): (black) x=0.2, (red) x=0.4, (blue) x=0.6, (magenta) x=0.8, and (olive) x=1.0. Color codes in (c): (blue) polar polymer and (red) nonpolar polymer. The selected porosity (φ) for all cases here is 0.5.

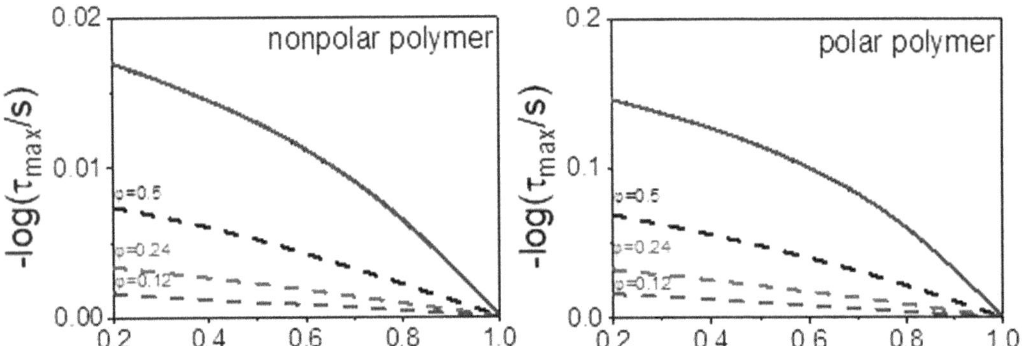

Figure 7. Comparison of the relaxation times for a (left) nonpolar and (right) a polar polymer during imbibition. Porosity: (blue) 0.12, (red) 0.24, and (black) 0.5. Simulation models: (dashed lines) series model and (solid lines) parallel model.

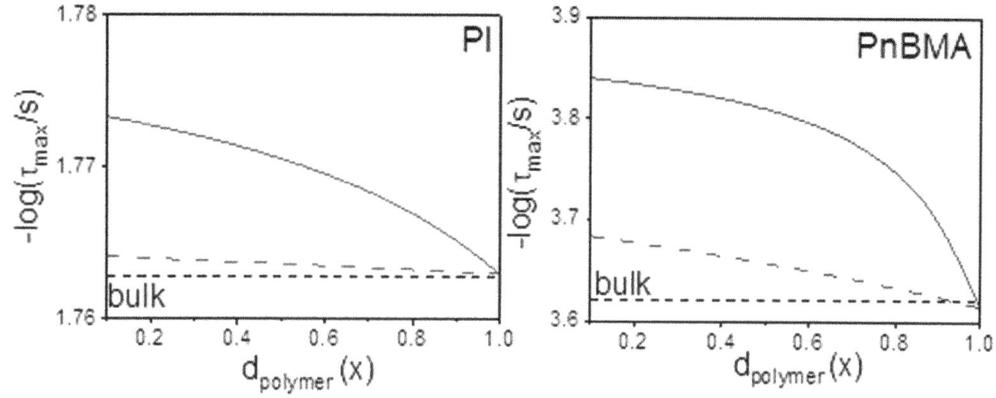

Figure 8. Comparison of the relaxation times of (left) polyisoprene (PI) and (right) poly(n-butyl methacrylate) (PnBMA) during imbibition at temperatures of 303 K and 333 K, respectively. Simulation models: (dashed lines) serial model and (solid lines) parallel model. Porosity: 0.12 (Left, AAO with pore dimeter of 40 nm) and 0.175 (Right, AAO with pore diameter of 65 nm). Black dashed line in the left (right) plot indicates the longest normal mode relaxation (α-relaxation) time in bulk polymers.

The effects of porosity on the relaxation times during imbibition are shown in Figure 7 for the polar and nonpolar polymer. Within the series model increasing porosity results in faster times whereas porosity plays no role within the parallel model. As expected, these changes are more prominent in the more polar polymer and rather insignificant for the nonpolar polymer. More realistic applications of the models can be made for the two particular polymers to be examined here: PI and PnBMA. The results of the two models for the actual experimental porosities are provided in Figure 8. The figure depicts a minor effect on the PI relaxation times and some effect (faster dynamics at the initial stages of imbibition) for the more polar PnBMA. Additional examples are discussed in the appendix (Figures A1-A4).

Overall simulation results show stronger effects on the evolution of the characteristic times and the dielectric strength for the polar polymers and nearly no effect for the nonpolar polymer. Hence, such effects should be considered at the initial stages of imbibition for polar polymers.

In Situ DS Measurements of a Polar (PnBMA) and a Non-polar (PI) Polymer

Herein, in an effort to explore the different possibilities and limitations of the nanofluidic method we compare two polymers with different dielectric properties (i.e., polarity and dielectric relaxation processes)–PnBMA and PI. PnBMA, as a member of alkyl methacrylate polymers has a high polarity due to strong electric dipole located in the side group. For this reason, different processes can be detected dielectrically: From the faster β-process in the glassy state, to the segmental α-process at and above T_g to the merged $\alpha\beta$-processes at higher temperatures (*14, 16, 17*). Because of the high viscosity near T_g, the segmental process (practically the merged $\alpha\beta$-process) can only be followed at elevated temperatures (a temperature $T=T_g+58$ K is shown in Figure 9).

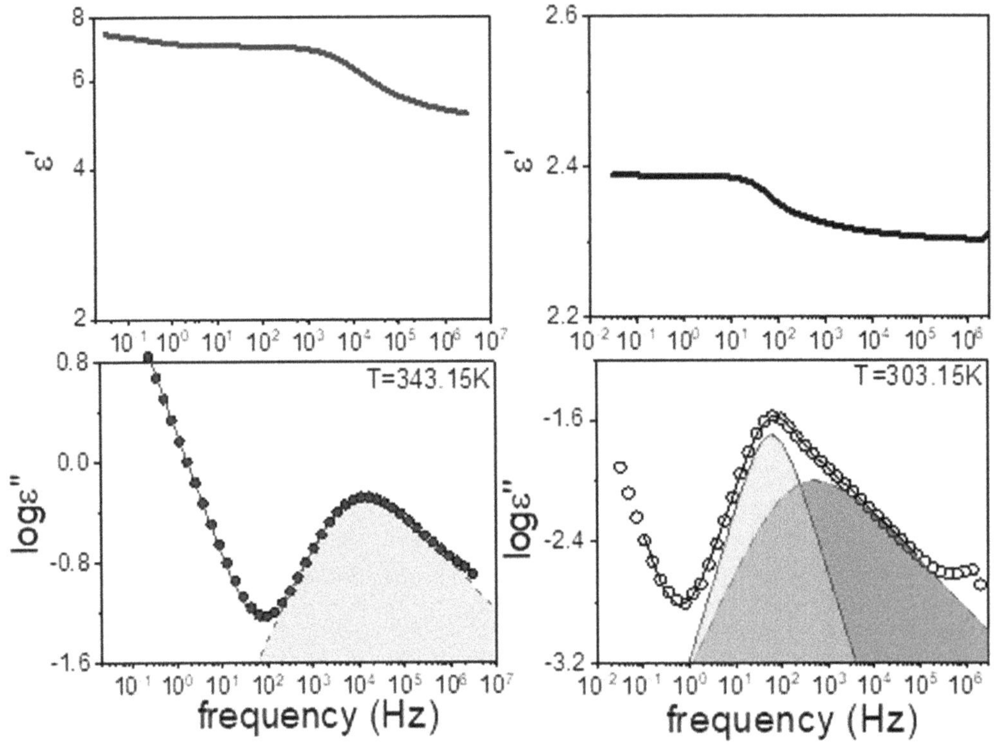

Figure 9. Real (ε') and imaginary (ε") part of complex dielectric function of (Left column) PnBMA 5.5k and (Right column) PI 42k at selected temperatures. The shaded area in left plot represents the segmental relaxation. The cyan (red) shaded area in right plot represents the longest normal mode relaxation (higher normal modes).

PI is a type-A polymer [according to the classification by Stockmayer (*18*)], and as such has components of the dipole moment both perpendicular and parallel to the contour of the polymer chain. The former contributes to the local segmental relaxation, and the latter provides the dynamics of the end-to-end vector as well as of the internal modes (normal mode relaxation). Because of experimental difficulties in observing the fast kinetics of imbibition–during the cooling process–we have followed the imbibition kinetics at a temperature of 303 K (i.e., at T_g+96 K) and at the lengthscale of the normal modes. Figure 9 depicts typical dielectric loss curves for the two polymers together with the respective fits to a single (PnBMA) or dual (PI) Havriliak-Negami processes. With respect to the latter, it is known that a single HN cannot capture the "spectrum" of normal modes in type-A polymers and that a summation of two HN functions is necessary.

Based on the dynamics shown in Figure 9, we can probe the way that PnBMA penetrates AAO nanopores by following the segmental process whereas in the case of PI we can follow the imbibition at the length scale of the chain (normal) modes. The results of the *in situ* imbibition are discussed with respect to Figures 10-12.

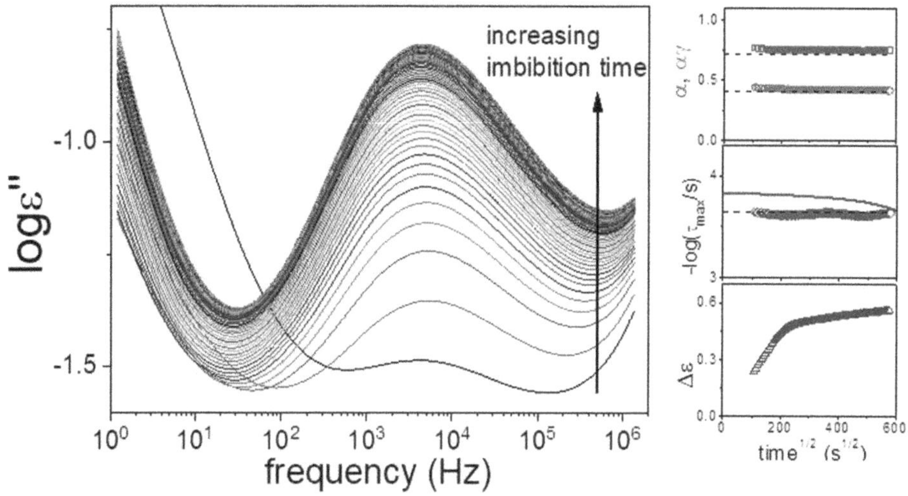

Figure 10. (Left) Evolution of the dielectric loss curves for PnBMA 5.5k entering AAO nanopores with pore diameter of 400 nm at 333 K. The vertical arrow indicates increasing time of imbibition. (Right) Evolution of the HN shape parameters (top), of the relaxation time at maximum loss (middle) and of the dielectric strength (bottom) during the experiment plotted as a function of the square root of imbibition time. The solid line for the characteristic times gives the parallel model predictions. Dashed lines give the parameters corresponding to the bulk polymer.

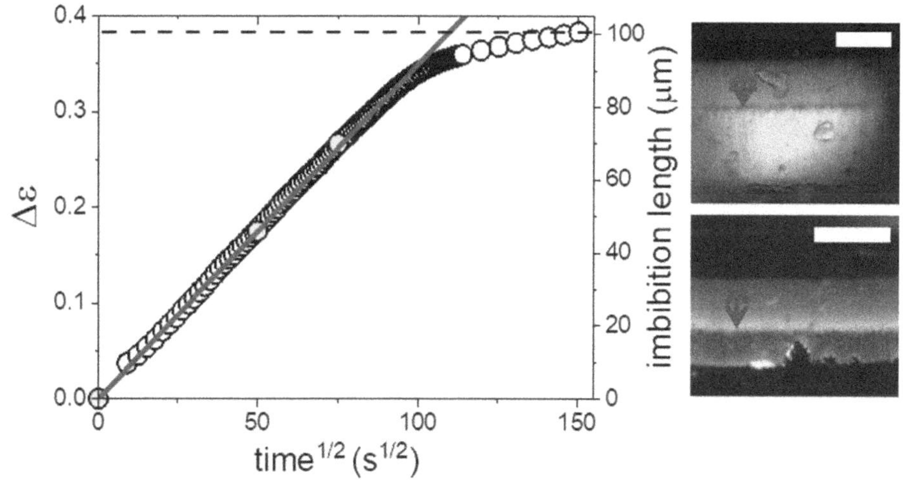

Figure 11. Imbibition length d_1 and dielectric intensity versus imbibition time $t^{1/2}$ for the PnBMA (5.5k) within AAO pores with a diameter of 400 nm at 343 K extracted from the parallel model. Black dashed line indicates the thickness of AAO template. Red line represents the theoretical prediction based on Lucas-Washburn equation. The yellow points represent the calibration points via reflection optical microscopy (ROM). The insets are ROM pictures for selected imbibition periods. The white scale bar indicates 100 μm (top) and 50 μm (bottom).

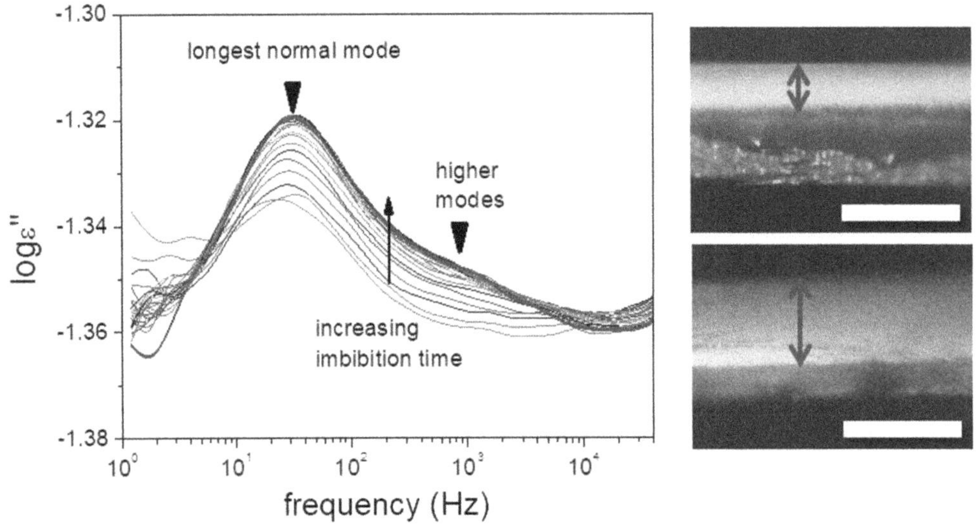

Figure 12. (Left) Evolution of DS curves of PI 42k within AAO 40 nm at temperature of 303 K. The dielectric loss curve with the thick blue color corresponds to imbibition time of 200s$^{0.5}$. (Right) Representative ROM images for imbibition times of 150s$^{0.5}$ (top) and 199s$^{0.5}$ (bottom) respectively. The areas specified by red arrows indicate the filled regions. The white scale bar indicates 100 μm.

Figure 10 depicts the growing dielectric loss curves as PnBMA 5.5k penetrates the pores. It provides details with respect to the dielectric strength, the distribution of relaxation times and the relaxation times at maximum loss. The low- and high- frequency slopes of the segmental process (i.e., α and $\alpha\gamma$) and relaxation times during imbibition are found to be nearly identical to the bulk value (except for the somewhat faster relaxation at the early times, discussed in the simulations part). This observation may sound contradictory to literature data revealing faster dynamics (compared with bulk) under confinement (*19*). However, there is *no* contradiction if we consider the fact that the selected imbibition temperature is much above T_g where these effects are diminished ($T=T_g+96$ K).

Interestingly, the dielectric intensity shows a two stage growth. In the initial part the dielectric strength grows as $t^{1/2}$, whereas at longer times it bends over. The extremely slow imbibition at late stage results from the trapped-air at the nanopore bottom (AAO templates with one close end were employed in the PnBMA case). This is a general feature of all templates with one closed end. To further examine the imbibition kinetics, the information on the effective viscosity that the polymer experiences during flow needs to be extracted. This can be made by *ex situ* measurements of the actual imbibition length by reflection optical microscopy. As an example, the imbibition kinetics of PnBMA within the larger pores (i.e., 400 nm) is well captured by Lucas-Washburn equation (Figure 11). However, the LWE fails to predict the effective viscosity in the smaller nanopores as documented by measurements reported in (*14*). In addition, the imbibition length can be obtained from the dielectric strength and the use of the parallel model discussed earlier (Figure 11). The PnBMA results clearly show the potential of the method to extract precise information of the polymer dynamics at the segmental length scale as well as the potential to extract the effective viscosity during flow.

The case of PI where we probe the imbibition at the longer lengthscales (i.e., the chain dynamics) is substantially different than in the segmental case of PnBMA (Figure 12). Herein, the initial continuous growth of normal modes is followed by a longer time regime where the dielectric loss curves remain unchanged or even reduced in dielectric strength. In parallel, the exact imbibition lengths were measured ex situ by optical reflection microscopy. A direct comparison of the two experiments provides the following scenario: while at the beginning of the process PI penetrates the pores resulting in the increase of the dielectric strength for the chain modes, at longer times the polymer penetrates the pores but at constant or even decreasing dielectric strength for the chain modes. Note that none of proposed models (i.e., parallel and series model) can explain this behavior. The results suggest increasing adsorption of chain segments at the pore walls. Such adsorbed segments give rise to loops that are de-activated dielectrically as their chain-ends are fixed at the pore walls and cannot fluctuate in space and time. Increasing adsorption sites result to the decrease of the intensity of normal modes as observed experimentally (*15*).

Conclusions

Emerged Physical Picture of Polymer Imbibition in Nanopores

In-situ nanodielectric spectroscopy has the capacity to explore not only confinement effects but also polymer/substrate interactions during polymer flow. Following some theoretical considerations, we provided here two examples: one for a polar polymer where the imbibition was followed at the segmental process and for a nonpolar polymer where imbibition was followed at the lengthscale of normal modes. Results demonstrate the possibility to extract important information on the relaxation times and distribution of relaxation times during flow for both length scales. At the same time, and within the context of the LWE, the effective viscosity during polymer flow can be obtained and compared to the bulk. We are not aware of another technique that can access these dynamic properties *during* polymer flow. In addition, imbibition of type A polymers is expected to provide quantitative information on the role of the adsorption both during and at the end of the filling process (including the kinetics). These results can be compared with the adsorption in polymer films where confinement is 1D (*20*).

Several additional polymer physics problems can be addressed with the method. While in-situ monitoring the imbibition process of PI, we observed that the (rather low) ionic conductivity was a decreasing function of imbibition time. It implies that certain ions are entering preferentially at the early stage of imbibition process followed by the polymer segments. This point requires further studies that include systems of higher ion content (ionic liquids, polymer electrolytes) and to further explore the role of the charged AAO surface (*21*). Experiments in this direction are currently in progress in our laboratory. On another issue, we can employ the same method to study the imbibition process of a biomacromolecule possessing an intrinsic secondary structure [e.g. poly(γ-benzyl-L-glutamate), PBLG (*22*)]. With PBLG several issues can be addressed: (i) Can we infiltrate an ordered α-helical secondary structure directly into the nanopores? (ii) Can we follow in-situ the segmental and slower dynamics of polypeptides and proteins during imbibition? It is noted that the slower process in bulk PBLG resembles the normal modes of polyisoprene owing to its dipoles arranged along the α-helical structure (PBLG is also a type - A polymer) (*22*). If successful, these studies will have implications to the problem of protein translocation, e.g. to the motion of a protein through narrow pores.

Appendix

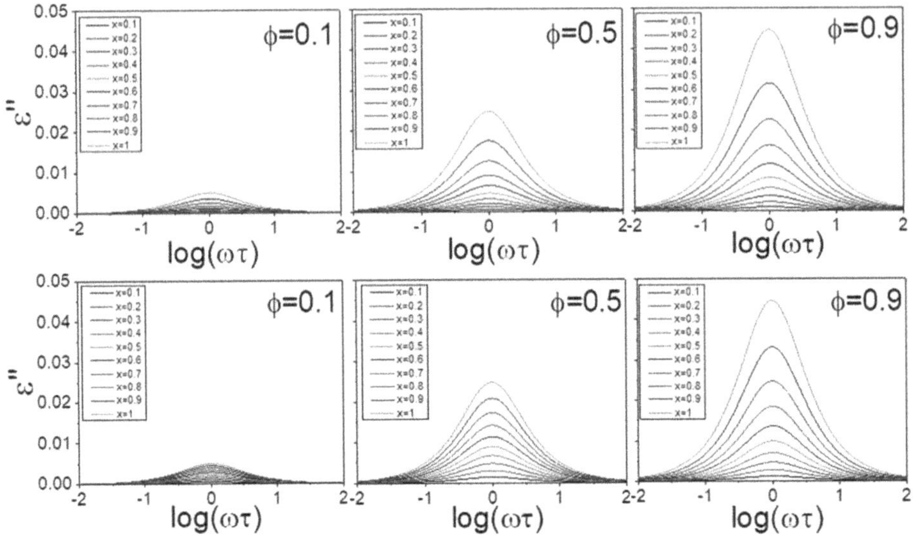

Figure A1. Simulated growing imaginary part of dielectric permittivity of a nonpolar polymer ($\Delta\varepsilon=0.1$) as a function of imbibition length (x) based on the parallel model (top row) and on the series model (bottom row) with different porosities: (left) 0.1, (middle) 0.5, and (right) 0.9.

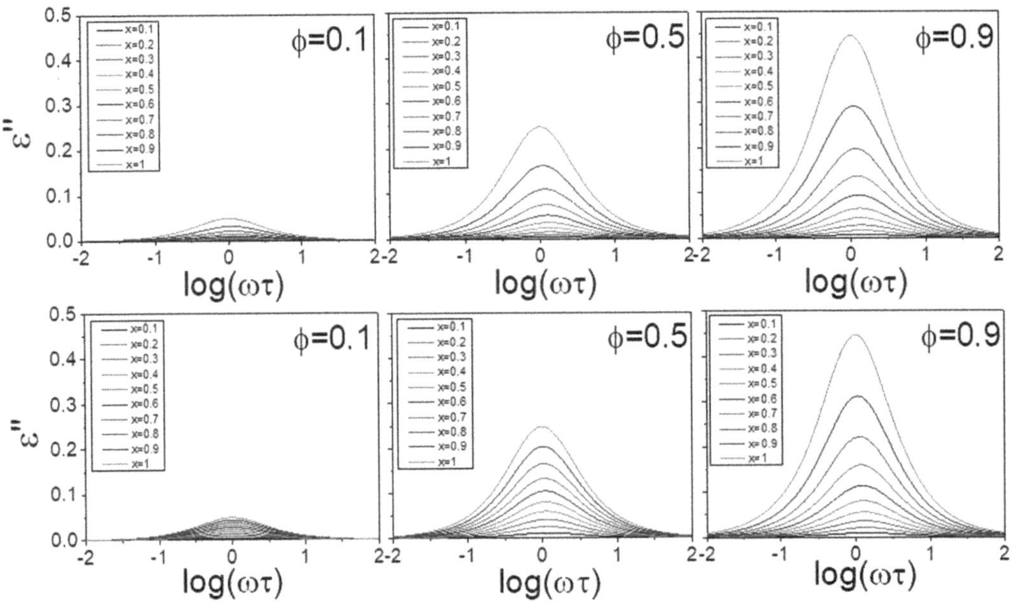

Figure A2. Simulated growing imaginary part of dielectric permittivity of a polar polymer ($\Delta\varepsilon=1$) as a function of imbibition length (x) based on the parallel model (top row) and the series model (bottom row) with different porosities: (left) 0.1, (middle) 0.5, and (right) 0.9.

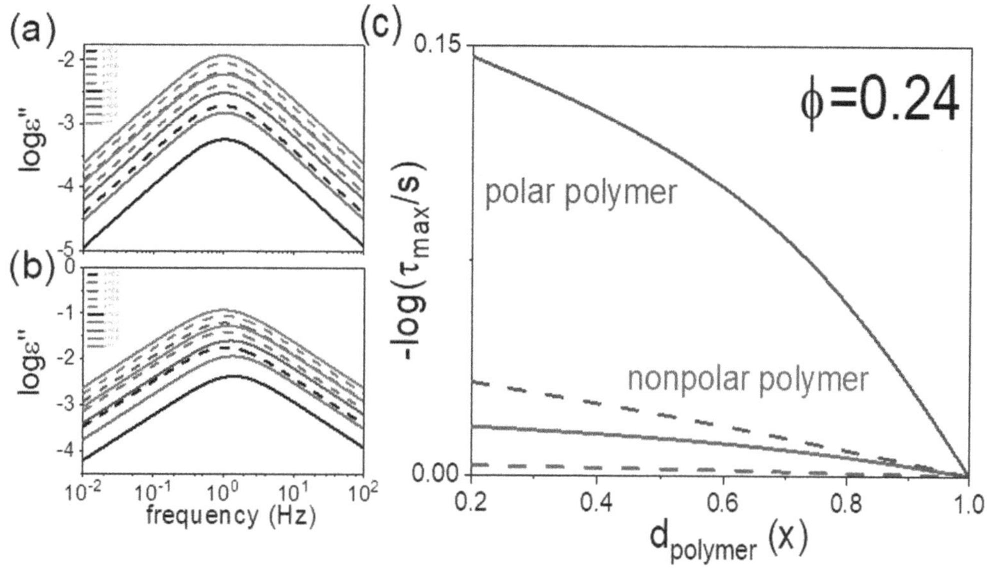

Figure A3. Representative simulated DS curves during imbibition for (a) nonpolar and (b) polar polymers. (c) Comparison in extracted relaxation times as a function of imbibition length (x) is plotted. Real (dashed) lines represent the parallel (serial) model. Color codes in (a) and (b): (black) x=0.2, (red) x=0.4, (blue) x=0.6, (magenta) x=0.8, and (olive) x=1.0. Color codes in (c): (blue) polar polymer and (red) nonpolar polymer. The selected porosity (φ) for all case here is 0.24 (same as InRedox AAO template with pore size of 100 nm).

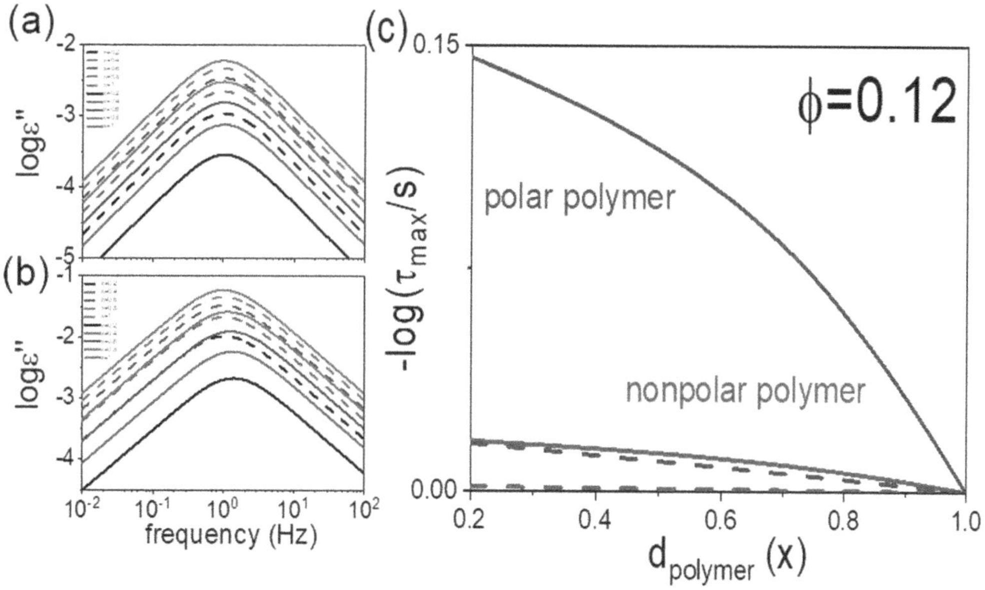

Figure A4. Representative simulated DS curves during imbibition for (a) nonpolar and (b) polar polymers. (c) Comparison in extracted relaxation times as a function of imbibition length (x) is plotted. Real (dashed) lines represent the parallel (serial) model. Color codes in (a) and (b): (black) x=0.2, (red) x=0.4, (blue) x=0.6, (magenta) x=0.8, and (olive) x=1.0. Color codes in (c): (blue) polar polymer and (red) nonpolar polymer. The selected porosity (φ) for all case here is 0.12 (same as InRedox AAO template with pore size of 40 nm).

References

1. Kremer, F., Ed. *Dynamics in Geometrical Confinement*; Springer: Berlin, 2014.
2. Masuda, H.; Hasegawa, F.; Ono, S. Self-ordering of cell arrangement of anodic porous alumina formed in sulfuric acid solution. *J. Electrochem. Soc.* **1997**, *144*, L127–L130.
3. Steinhart, M. Supramolecular organization of polymeric materials in nanoporous hard templates. *Adv. Polym. Sci.* **2008**, *220*, 123–187.
4. Lucas, R. Ueber das zeitgesetz des kapillaren aufstiegs von flüssigkeiten. *Colloid Polym. Sci.* **1918**, *23*, 15–22.
5. Washburn, E. W. The dynamics of capillary flow. *Phys. Rev.* **1921**, *17* (3), 273–283.
6. Yao, Y.; Butt, H. J.; Floudas, G.; Zhou, J.; Doi, M. Theory on capillary filling of polymer melts in nanopores. *Macromol. Rapid Commun.* **2018**, *39* (14), 1800087.
7. Yao, Y.; Alexandris, S.; Henrich, F.; Auernhammer, G.; Steinhart, M.; Butt, H.-J.; Floudas, G. Complex dynamics of capillary imbibition of poly(ethylene oxide) melts in nanoporous alumina. *J. Chem. Phys.* **2017**, *146* (20), 203320.
8. Cencha, L. G.; Urteaga, R.; Berli, C. L. *Macromolecules* **2018**, *51* (21), 8721–8728.
9. Serghei, A.; Chen, D.; Lee, D. H.; Russell, T. P. Segmental dynamics of polymers during capillary flow into nanopores. *Soft Matter* **2010**, *6*, 1111–1113.
10. Serghei, A.; Lutkenhaus, J. L.; Miranda, D. F.; McEnnis, K.; Kremer, F.; Russell, T. P. Density fluctuations and phase transitions of ferroelectric polymer nanowires. *Small* **2010**, *6* (16), 1822–1826.
11. Serghei, A.; Zhao, W.; Wei, X.; Chen, D.; Russell, T. P. Nanofluidics with phase separated block-copolymers: Glassy dynamics during capillary flow. *Eur. Phys. J.: Spec. Top.* **2010**, *189*, 95–101.
12. Serghei, A.; Zhao, W.; Miranda, D.; Russell, T. P. Curie transitions for attograms of ferroelectric polymers. *Nano Lett.* **2013**, *13*, 577–580.
13. Houachtia, A.; Alcouffe, P.; Boiteux, G.; Seytre, G.; Gérard, J.-F.; Serghei, A. Nanofluidics approach to separate between static and kinetic nanoconfinement effects on the crystallization of polymers. *Nano Lett.* **2015**, *15*, 4311–4316.
14. Tu, C.-H.; Steinhart, M.; Butt, H.-J.; Floudas, G. In situ monitoring of the imbibition of poly (n-butyl methacrylates) in nanoporous alumina by dielectric spectroscopy. *Macromolecules* **2019**, *52* (21), 8167–8176.
15. Tu, C.-H.; Zhou, J.; Doi, M.; Butt, H.-J.; Floudas, G. Interfacial interactions during in situ polymer imbibition in nanopores. *Phys. Rev. Lett.* **2020**, *125*, 127802.
16. Mpoukouvalas, K.; Floudas, G.; Williams, G. Origin of the α, β, (βα), and "slow" dielectric processes in poly(ethyl methacrylate). *Macromolecules* **2009**, *42*, 4690–4700.
17. Deres, A.; Floudas, G. A.; Müllen, K.; Van der Auweraer, M.; De Schryver, F.; Enderlein, J.; Uji-i, H.; Hofkens, J. The origin of heterogeneity of polymer dynamics near the glass temperature as probed by defocused imaging. *Macromolecules* **2011**, *44*, 9703–9709.
18. Stockmayer, W. H. Dielectric dispersion in solutions of flexible polymers. *Pure Appl. Chem.* **1967**, *15* (3–4), 539–554.

19. Alexandris, S.; Papadopoulos, P.; Sakellariou, G.; Steinhart, M.; Butt, H.-J.; Floudas, G. Interfacial energy and glass temperature of polymers confined to nanoporous alumina. *Macromolecules* **2016**, *49*, 7400.
20. Napolitano, S.; Wübbenhorst, M. The lifetime of the deviations from bulk behaviour in polymers confined at the nanoscale. *Nat. Commun.* **2011**, *2*, 260.
21. Zardalidis, G.; Ioannou, E.; Pispas, S.; Floudas, G. Relating Structure, Viscoelasticity and local Mobility to Conductivity in PEO/LiTf Electrolytes. *Macromolecules* **2013**, *46*, 2705–2714.
22. Floudas, G.; Spiess, H. W. Self-assembly and dynamics of polypeptides. *Macromol. Rapid Commun.* **2009**, *30*, 278.

Chapter 10

Polymer Dynamics in Nanostructured Environments: Structure-Property Relations Unraveled by Dielectric Spectroscopy

Martin Tress,[1,*] Maximillian Vielhauer,[2] Pierre Lutz,[3] Rolf Mülhaupt,[4] Friedrich Kremer,[1] Kunyue Xing,[5] Sirui Ge,[6] Pengfei Cao,[7] Tomonori Saito,[7] and Alexei Sokolov[5,7]

[1]Peter-Debye-Institute, University Leipzig, Linnestr. 5, D-04103 Leipzig, Germany

[2]Freiburg Materials Research Center FMF, University of Freiburg, Stefan-Meier-Str. 21, D-79104 Freiburg, Germany

[3]Institute Charles Sadron, CNRS UPR 22, University of Strasbourg, 23 rue du Loess, F-67034 Strasbourg, France

[4]Institute for Macromolecular Chemistry, University of Freiburg, Stefan-Meier-Str. 31, D-79104 Freiburg, Germany

[5]Department of Chemistry, University of Tennessee, Knoxville, 1420 Circle Dr., Knoxville, Tennessee 37996, United States

[6]Department of Materials Science, University of Tennessee, Knoxville, 1420 Circle Dr., Knoxville, Tennessee 37996, United States

[7]Chemical Sciences Division, Oak Ridge National Laboratory, Oak Ridge, Tennessee 37831, United States

[*]Email: martin.tress@uni-leipzig.de

Many polymers exhibit complex levels of molecular and supramolecular organization, particularly if their chemical structure shows an increased complexity itself. This internal nanoscale-structuring is often accompanied by altered dynamics which in turn affects the macroscopic performance. Using this to design highly desired properties, e.g., mechanical stability, transport functions or self-healing, requires a detailed understanding of the interrelation between chemical structure and the corresponding change in dynamics. A versatile tool to explore polymer dynamics is dielectric spectroscopy, as demonstrated in two examples. The first case is a semi-crystalline star-polymer with a polyhedral oligomeric silesquioxane core; due to the unusual architecture of this molecule, the partial crystallization constrains the chain conformations in the amorphous regions and results in an accelerated segmental relaxation–in stark contrast to the commonly observed slower dynamics in semi-crystalline polymers. A detailed analysis of the extracted relaxation time distribution (RTD) reveals three separate amorphous regions with different dynamics. In a second example, the impact of meso-phase separation in associating telechelic polymers is studied. The hydrogen-bonding end groups in these polydimethylsiloxane-based materials phase separate, which is accompanied by a tremendously broadened RTD indicating severe constraints on the segmental dynamics. Again, altered chain conformations are proposed which offer a qualitative explanation for the tremendously enhanced rubbery plateau found in complimentary shear modulus measurements. In both examples, beyond the severe impact of nanostructured environments on the dynamics, dielectric spectroscopy also reveals subtle structural details leading to an improved macroscopic mechanical performance.

Introduction: Nanostructured Environments and Polymers

Miniaturization is an ongoing trend in both fundamental research as well as applied technology. The interest has grown particularly with the ability to reproducibly fabricate nanostructures with unprecedented physical effects. In many cases such effects offer great potential for novel applications, and hence also polymer science often focuses on systems where nanostructures of similar dimensions as the polymer chain are present. Traditional systems are block copolymers (*1*, *2*) and nanocomposites (*3–8*); in the former, the chemical incompatibility drives the formation of segregated domains which often result in a highly regular morphology on nanometric length scales while in the latter the presence of a nanometer-sized particle represents a structure which itself affects the surrounding chains. In both cases, e.g., the mechanical properties are considerably altered (*1*, *6*) which has been used to fabricate materials of strongly enhanced durability (*1*). Clearly, the details of the improved performance depend on several parameters of the structures like their amount and size, the chemical dissimilarity of the components, etc. In other words, depending on the characteristics of this environment, the embedded polymer may exhibit altered properties.

One characteristic of polymers is their large molecule size which causes a hierarchy of structures and dynamics on different length and time scales (*9*). Many technologically relevant properties, e.g., glass transition temperature (T_g), mechanical strength, conductivity, processability, etc. depend on the dynamics of the polymer. Consequently, it is of paramount interest how the dynamics can be manipulated by the characteristics of the nanostructured environment. Dielectric spectroscopy is a well-established method to study polymer dynamics in the melt (*9*) and is therefore also suitable to examine alterations due to nanostructured environments (*10*). In the following this is demonstrated in two examples where polymer dynamics have been altered in an unexpected or excessive way by internal structures of nanometric size. The first example is a semi-crystalline polymer where, due to the star-shape of the molecules, severe constraints are exerted on the segments unexpectedly resulting in faster dynamics. In the second example, the impact of phase separation in end-functionalized telechelic polymers is shown to slow down the dynamics which strongly impacts the mechanical properties. It is demonstrated that in both cases the common focus on mean fluctuation rates is insufficient to understand the details of the dynamical changes caused by the nanoscale structure formation. Only an extensive analysis of their distributions reveals the complex alterations in these nanostructured environments which is essential to gain insight into the structure-property relations.

Enhanced Dynamics in a Semi-crystalline Star Polymer

Most polymers have the tendency to partially crystallize; the chains arrange in lamellae forming a highly ordered polymer crystal (*11*). A complete crystallization is usually not achieved due to conformational structures like loops or entanglements which cannot be incorporated into the crystal. Hence, in between the crystallites a certain fraction of the polymer remains in its amorphous phase (Figure 1a). However, even though these amorphous domains do not undergo a phase transition, the adjacent crystallites have a strong impact on and alter the properties of these amorphous regions. It has been shown for several materials that around the crystallites the segments form a so-called rigid amorphous fraction (RAF) which is characterized by slower dynamics than in the purely amorphous phase (*11, 12*). The reason for this is that the chains protruding from a crystallite are limited in their mobility due to the constraints in the highly ordered phase. As is typical for interfacial effects in polymers, the range of these alterations is on nanometer to sub-nanometer scale (*10, 11*); at greater distance from the crystallites the dynamics are unchanged in the so-called mobile amorphous fraction (MAF) (*12*).

Dielectric spectroscopy is a method which is suitable to study the details of dynamics of polymers on the molecular scale; for a detailed description of the fundamentals the reader is directed to basic text books (*9, 10*). In short, it relies on the molecular dipole moments present in most materials due to non-uniform electron density. The fluctuations of these dipole moments directly reflect the motion of the molecules (or parts thereof). Applying an external oscillating electrical field, e.g., in a parallel-plate capacitor filled with the sample material, leads to a coupling of the molecular dipoles to the external field and dissipates energy from the latter. This resembles an energy loss in the capacitor which is frequency dependent and reaches a maximum when the oscillation frequency of the external field matches the fluctuation rate of the molecular dipoles. The result is an absorption curve in the dielectric loss, i.e., the imaginary part of the complex permittivity, which is called a relaxation process. Its intensity is proportional to the effective dipole moment and the concentration of the participating species of molecular structures. Note that only fluctuating dipoles contribute to the signal which means that immobile dipoles or motions without change in the net dipole moment are invisible to this technique. The high spectral resolution and wide spectral range achieved with this technique also enables the detailed examination of the spectral shape of the relaxation peak. This provides access to the distribution of relaxation times which is typically polydisperse reflecting heterogeneous dynamics. Consequently, in polymers, be it in the bulk, in confinement or under the constraints of crystal formation, the dynamics in general as well as its heterogeneity are often investigated using dielectric spectroscopy (*9, 10, 12, 13*).

A counterintuitive and surprising change in dynamics has recently been found in a star-shaped semi-crystalline polymer based on a polyhedral oligomeric silesquioxane (POSS) molecule as a star center connected to 6-8 isotactic polystyrene (iPS) arms (Figure 1b) (*14*). Each iPS arm has a molecular weight (MW) of $M_{n,a} = 5500$ g/mol; details of the synthesis can be found elsewhere (*14*). A thin film of the material was spun-cast from a toluene solution and immediately covered with a silicon counter electrode (a specialized sample arrangement was employed which relies on insulating silica nanostructures on the electrode surface as spacers to control the capacitor thickness; details on this preparation method can be found elsewhere) (*13, 15*). Subsequently, the sample was annealed at 425 K (ca. $T_g + 75$ K) for 12 h in high vacuum (10^{-6} mbar) in order to remove residual solvent. The dielectric measurements were conducted using a Novocontrol Alpha-A Analyzer connected to a Quatro Cryosystem in dry nitrogen atmosphere in the temperature range from 350 to 480 K at frequencies between 0.1 Hz and 1 MHz. Prior to the measurement, the sample was annealed in the cryostat of the spectrometer at 484 K (the melting point of the POSS-iPS has been determined as T_m = 479 K) for 1 h in order to melt the sample and erase the thermal history. To study the sample in its amorphous state it was cooled to 350 K at a rate of 20 K/min which was sufficiently fast considering the quite slow crystallization of iPS (*12*, *13*). To obtain the semi-crystalline state, the sample was annealed at 460 K for 20 h resulting in a crystallinity of 10 %. In a subsequent experiment after another complete melting, measurements were also carried out applying shorter crystallization times (obtained crystallinities were 4.6 % and 0.6 %). These crystallinities were deduced from DSC. Despite these relatively low degrees of crystallinity, the fraction of immobilized segments is considerably bigger as determined by the reduction in the dielectric relaxation strength (see below) to 34 %, 31 %, and 20 %, respectively.

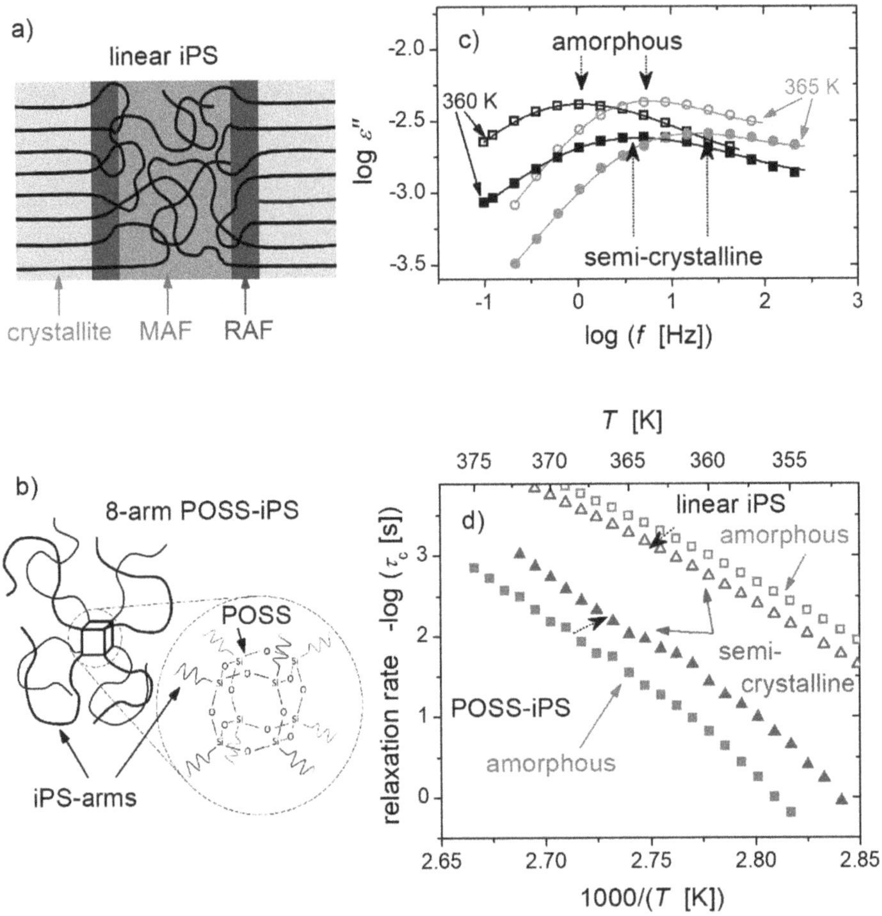

Figure 1. (a) Schematic chain conformations in a semi-crystalline polymer like isotactic polystyrene (iPS) indicating the lamellar organized crystallites, the mobile amorphous fraction (MAF), and the intermediate rigid amorphous fraction (RAF). (b) Molecular structure of the star-shaped POSS-iPS. (c) Dielectric loss spectra of POSS-iPS in the purely amorphous and in the semi-crystalline state at two different temperatures as indicated; the solid lines are fits to the Havriliak Negami function and the mean relaxation rates indicated by the peak positions are indicated by dotted arrows. (d) Activation plot (logarithmic mean relaxation rate vs inverse temperature) of POSS-iPS (solid symbols) in the purely amorphous and in the semi-crystalline state as indicated; as reference, also the mean relaxation rates of linear iPS with a molecular weight of 10 kg/mol is depicted (open symbols). Data in part from reference (13).

The α-relaxation peak in the dielectric loss spectra which reflects the segmental motion of the iPS chains is clearly visible and appears at significantly higher frequencies when the material is semi-crystalline compared to its purely amorphous state (Figure 1c). This indicates higher mobility of the segments and opposes the commonly observed slower segmental dynamics in semi-crystalline polymers, e.g., as found in linear iPS (Figure 1d) (*11, 12*). In order to analyze the complex permittivity $\varepsilon^* = \varepsilon' - i\varepsilon''$ in detail, the loss spectra (i.e., the imaginary part ε'') have been fitted to the Havriliak Negami (HN) function (*16*) (first term) combined with a conductivity contribution (*9*) (second term)

$$\varepsilon''(\omega) = -\mathrm{Im}\left[\frac{\Delta\varepsilon}{(1+(i\omega\tau_{HN})^{\beta})^{\gamma}}\right] + \frac{\sigma_{DC}}{\varepsilon_0\omega} \quad (1)$$

where $\Delta\varepsilon$ is the relaxation strength, τ_{HN} the characteristic relaxation time, and β and γ are the symmetric and asymmetric shape parameters, respectively. The relaxation strength corresponds to the area under the relaxation peak and it is proportional to the number of fluctuating dipoles, i.e., the mobile segments. Thus, its reduction upon crystallization reflects the immobilization of segments in and around the growing crystallites. It is important to note that also a considerable fraction of the RAF does not contribute to the segmental relaxation (*14*). Consequently, the crystallinity cannot be calculated from the ratio of the relaxation strength after crystallization and its value in the purely amorphous sate; this ratio rather reflects the number of immobilized segments. Instead, crystallinity must be determined from complementary measurements, e.g., by calorimetry. The mean relaxation time τ_c corresponding to the maximum position of the peak is given by (*9*)

$$\tau_c = \tau_{HN}\left[\frac{\sin\left(\frac{\pi\beta\gamma}{2+2\gamma}\right)}{\sin\left(\frac{\pi\beta}{2+2\gamma}\right)}\right]^{1/\beta} \quad (2)$$

From the shape parameters, the relaxation time distribution can be calculated (Figure 2) according to (*9, 17*)

$$L(\tau) = \frac{\sin\left(\gamma\left[\frac{\pi}{2} - \arctan\left(\frac{\left(\frac{\tau}{\tau_{HN}}\right)^{\beta} + \cos(\pi\beta)}{\sin(\pi\beta)}\right)\right]\right)}{\pi\left(1 + 2\cos(\pi\beta)\left(\frac{\tau_{HN}}{\tau}\right)^{\beta} + \left(\frac{\tau_{HN}}{\tau}\right)^{2\beta}\right)^{\gamma/2}} \quad (3)$$

This quantity allows one to subtract the contribution of segments which exhibit the same dynamics as in the purely amorphous state. For that, the distribution obtained from the purely amorphous sample is downscaled to just fit in the distribution deduced in the semi-crystalline state (Figure 2b). This corresponds to an estimate of the upper limit of the MAF. The result are two remaining fractions, a relatively small one at longer relaxation times, assigned to the RAF (It should be noted that the RAF does not only comprise of this fraction of slower mobility but also contains segments which are completely immobilized while still being amorphous (*14*). Occasionally. it is even defined as just the amorphous segments which do not contribute to relaxation; however, the slower but still mobile fraction we find here is clearly caused by the constraints from the crystallites. Hence, it seems appropriate to merge it with the RAF.), and a larger one at short times. The latter is unexpected in this context and resembles the altered dynamics in confinement due to packing frustration (*10, 18*); hence, we refer to these segments as the confined amorphous fraction (CAF).

The assignment of the different dynamic fractions becomes more evident from their evolution during crystallization. The area of the slow fraction assigned to the RAF increases with progressing crystal growth as one would expect (Figure 2d). Additionally, its mean relaxation time is stable throughout this process indicating that the dynamic constraints remain the same (since these depend only on the local environment but not on the size of the crystallite). In contrast, the area of the fast fraction reduces with increasing crystallinity (Figure 2c) which can be clearly referred to the process of mobile segments getting immobilized in the growing crystallites. At the same time, the mean relaxation time reduces in the course of crystallization; a sign of an increasing degree of packing frustration.

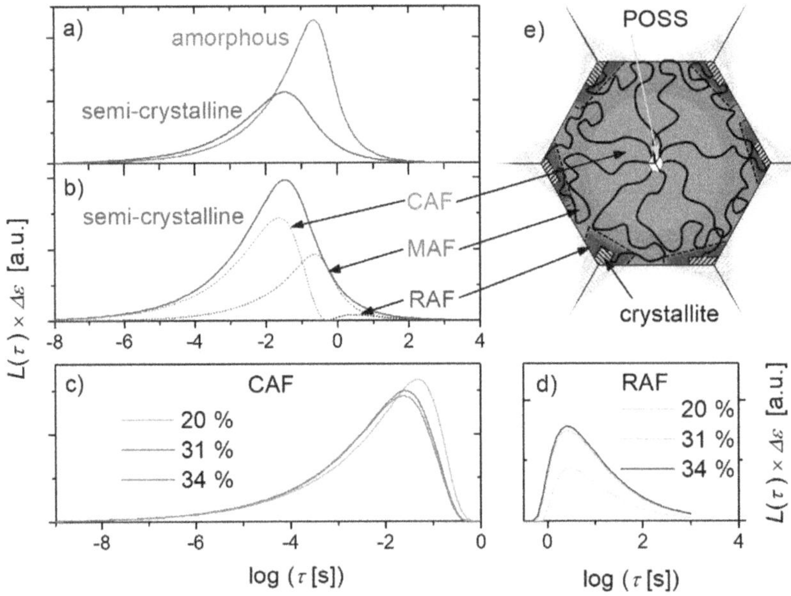

Figure 2. (a) Relaxation time distribution (RTD) $L(\tau)$ scaled with the dielectric relaxation strength $\Delta\varepsilon$ of POSS-iPS in the amorphous and semi-crystalline state. (b) RTD of the semi-crystalline state of POSS-iPS and its dissection into different dynamical fractions (dotted lines) using the downscaled amorphous RTD to approximate the mobile amorphous fraction (MAF). (c) and (d) Evolution of the extracted RTD portion of the CAF and the RAF, respectively at different extents of crystallinity as indicated. (e) Scheme of a single POSS-iPS molecule and its proposed dynamical domains as indicated; the white cube symbolizes the central POSS and the solid black lines sketch the conformation of the iPS-arms. Data from reference (13).

This complex dynamic signature suggests quite heterogeneous organization of the material. The basis of this appears to be the star-shape of the molecule which, due to the rather high number of arms (6-8) and their short length, behave like soft spheres that slightly interpenetrate their neighbors. Hence, the arrangement of the molecules may be approximated by a close packing of spheres with the POSS in the centers of the spheres. Although the random close packing of spheres may capture the situation more accurately, the cubic close packing offers the simpler theoretical treatment. The crystal formation of the iPS arms most likely takes place distant from the star centers consisting of the POSS molecules since POSS is known to suppress crystallization in other systems (*19–22*). Consequently, the regular distribution of the POSS cores will force a scaffold-like pattern of crystallites at the intersection planes of adjacent molecules (Figure 2e). Apparently, this nanoscale structure prevents a large-scale contraction of the material during the (partial) densification upon crystallization. Hence, to compensate for the increased density in the crystallites, domains of reduced density are formed. For reasons of continuity, such less dense domains are expected to be around the POSS cores with their faster dynamics (CAF). These dynamics gradually shift to normal amorphous dynamics (MAF) as radial distance from the POSS increases until close to the crystallites (in greatest distance from the POSS) the slower dynamics (RAF) develops.

The area of the different proportions of the RTD curve can be utilized as measure for the number of mobile segments assigned to each dynamic fraction (assuming comparable relaxation strengths). Combined with the inferred spatial arrangement of these fractions (Figure 2e) and the volume of a single POSS-iPS molecule, an estimate for their size can be given. Based on the bulk density of amorphous iPS, $\rho_{\mathrm{am}} = 1.06\ \mathrm{g/cm^3}$, (the volume of the POSS core (*23*) is negligible), a single molecule volume $V_{\mathrm{m}} = 60\ (\pm 9)\ \mathrm{nm^3}$ is obtained from

$$V_m = n_a \frac{M_{n,a}}{\rho_{am} N_A} \quad (4)$$

Here N_{A} is Avogadro's number; the uncertainty of the value given for V_{m} reflects the variation of the number of iPS arms n_{a} between 6 and 8. Assuming that the POSS cores are arranged in a face-centered cubic lattice (corresponding to a cubic close packing of spheres), the respective Voronoi cell corresponds to V_{m}. This yields a distance of the POSS cores of adjacent molecules of 4.4 (±0.2) nm which also approximates the extension of a single molecule.

In order to estimate the size of the different dynamic regions, first, the reduction of the relaxation strength upon crystallization must be considered, i.e., the reduction of the area under the overall scaled RTD curve $L(\tau)\times\Delta\varepsilon$ (Figure 2a). Its drop by 34 % means that in the semi-crystalline state about 34 % of the segments are immobilized in the crystallites (or as part of a completely immobilized sub-fraction of the RAF) (*13*). This comprises about 18 (±3) $\mathrm{nm^3}$ of the volume of a single POSS-iPS molecule corresponding to about 95 to 127 iPS repeat units (i.e., 12 to 16 repeat units per arm on average). Then, the area fractions of the dissected RTD can be utilized to divide the remaining volume, into the RAF (3 % of the mobile portion), the MAF (34 %), and the CAF (63 %) which yields 1.2 (±0.2) $\mathrm{nm^3}$, 14.5 (±2) $\mathrm{nm^3}$, 26.5 (±4) $\mathrm{nm^3}$, respectively. Following the proposed spatial arrangement of the different dynamic fractions, i.e., the CAF occupies the central region of each POSS-iPS molecule surrounded by a shell formed by the MAF which contains the crystallites wrapped in the RAF (Figure 2e), the obtained volumes suggest a spherical CAF with a radius of 1.85 (±0.1) nm surrounded by a MAF layer of about 0.29 (±0.02) nm thickness.

This demonstrates that a considerable proportion of the polymer is affected by the structures in the semi-crystalline state. While about one third of all segments are immobilized due to crystal formation (either in the crystallites directly or as part of the RAF), almost another half of them exhibits faster segmental dynamics (compared to the purely amorphous state) most likely as a result of altered chain conformations. This quite unusual observation is assigned to the interplay of the complex chain architecture and the crystallization at the chain ends which creates a structured environment on the nanoscale to the iPS chains. We want to emphasize that our analysis is based on the assumption that the individual contribution of each mobile segment to the dielectric relaxation strength of a relaxation process is identical; only if this condition is fulfilled, the dielectric response reflects the number of fluctuating dipoles. However, recently it was proposed that in polymer nanocomposites and thin polymer films, a certain fraction of segments (those firmly attached to a solid interface) might have strongly reduced fluctuation amplitudes which leads to a decrease of their dielectric response, although these segments are not completely immobilized (and in some cases might even fluctuate faster) (*24*). Such effects were not considered here.

Slower Dynamics in Phase Separating Associating Polymers

Supramolecular associations are mediated by attractively interacting atom groups which form dynamic links between the molecules (*25, 26*). This often generates additional levels of organization on the length scale of a single or a few molecules, i.e., at nanometric size (*27*). Materials which contain such associations have been the subject of scientific research for decades (*25, 28–30*) yet their scientific attraction is still increasing because they are promising candidates to realize properties such as self-healing (*31–36*), recyclability (*37–39*), and extraordinary mechanical performances (*36–40*). To some extent they can even mimic the extraordinary functions of biological materials (*36*). Thus, it is highly desirable to understand the details of the association mechanisms and how its mobility reduction impacts the surrounding matter on the molecular scale. A crucial factor thereby is the distribution of the dynamically associating groups on nanometric length scales which creates a nanostructured environment for the matrix material.

Among the existing types of dynamic bonds such as ionic interactions (*41, 42*), π-π-stacking (*43*), and others (*44*), hydrogen (H) bonds (*25–32, 35, 37, 38, 40, 45–48*) may be the most studied and also most abundant in biological matter (*49*). Recent investigations on telechelic polymers terminated with H-bonding groups, which form dimeric transient bonds, have successfully employed the bond lifetime renormalization model (*50*) to quantitatively link the observed lifetimes of the transient bonds and the mechanical stress relaxation as probed by dielectric spectroscopy and oscillatory shear rheology, respectively (*45–47*). It is expected that slower dynamics in such dimeric aggregates also reduces the mobility of adjacent segments of the matrix polymer. Even more constraining, however, should be the case where the associating groups form large aggregates because the dynamics of an additional segregated phase of H-bonding motifs will be even slower (*48*).

To investigate this, a series of telechelic polydimethylsiloxane (PDMS) of different degrees of polymerization (*DP*) terminated with 4-(propylamino)-4-oxobutanoic acid (POA) were synthesized (Figure 3a) and have been studied by dielectric spectroscopy, oscillatory shear rheology, and differential scanning calorimetry (*27, 48*).

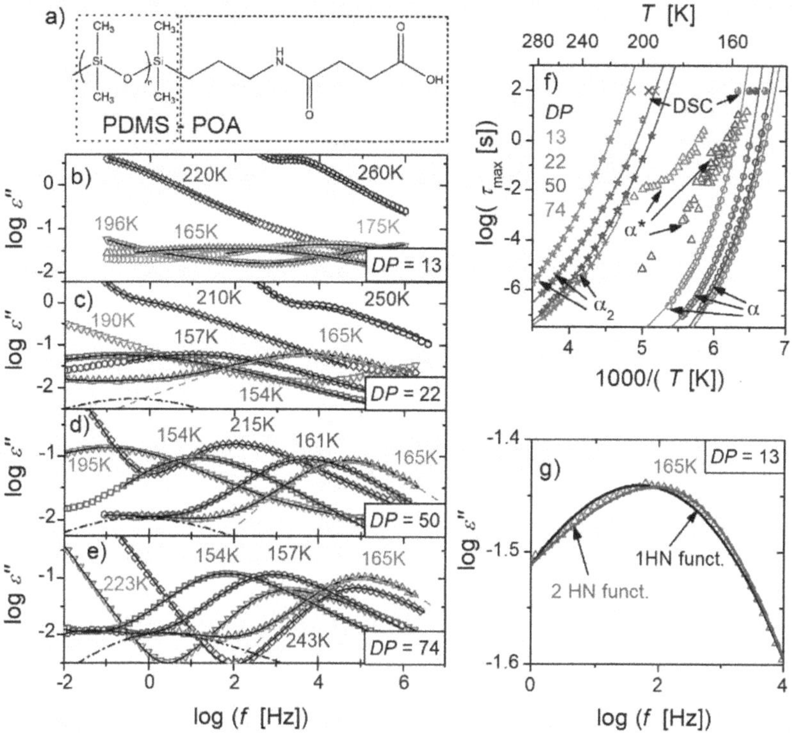

Figure 3. (a) Chemical structure of PDMS-POA; for clarity only one of the two POA end groups is depicted. (b-e) Dielectric loss spectra recorded at different temperatures as indicated for PDMS-POA with degree of polymerization (DP) of 13, 22, 50, and 74, respectively. The solid lines are fits to eq. 1; the dashed and dash-dotted lines show the individual contributions of the HN functions describing the α- and α-relaxation, respectively. (f) Activation plot showing the mean relaxation times of the α-, α*- and α_2-relaxations of PDMS-POA for different DP as indicated. Additionally, the glass transition temperatures of the PDMS domains (circles) and of the POA end group aggregates (crosses) as determined by DSC are depicted at the conventional relaxation time of 100 s. (g) Dielectric loss spectrum of PDMS-POA-13 at a temperature of 165 K and the corresponding fits consisting of one (black line) and two HN functions (red line), respectively. Data from reference (48).*

For the dielectric measurements, the samples were filled in a measurement cell made of stainless steel with a sapphire disc separating the electrodes; its active sample area has a diameter of 10 mm and an electrode separation of 50 μm resulting in a capacity of ~20 pF of the empty cell (more details on this type of cell can be found elsewhere) (*51*). Prior to the dielectric measurement, the samples were kept in a vacuum oven (~1 mbar) for 5 days at 313 K to remove residual solvents. The dielectric loss spectra of all these materials exhibit a clear α-relaxation peak assigned to the segmental motion of the PDMS main chain (Figure 3b-e). Additionally, at about 2-3 decades lower in frequency, another process identified as the α*-relaxation is visible which indicates the H-bond dissociation (*27, 45–48*). A third relaxation process is found at considerably higher temperatures which stems from the phase separated POA groups, hence it is referred to as α_2-relaxation (*27, 48*). This assignment is corroborated by the DSC trace which reveals two distinct glass transition steps (*27*), each of which coincides with the extrapolated mean relaxation times of the α-relaxation and the α_2-relaxation, respectively, in the activation plot (Figure 3f). The mean relaxation times were deduced from HN fits to the data. Since most spectra show contributions of several relaxation processes, a sum of up to three HN functions (one for each of the above-mentioned relaxation processes) has been used (*9, 16*)

$$\varepsilon''(\omega) = \sum_n -\mathrm{Im}\left[\frac{\Delta\varepsilon_n}{\left(1 + \left(i\omega\tau_{HN,n}\right)^{\beta_n}\right)^{\gamma_n}}\right] + \frac{\sigma_{DC}}{\varepsilon_0\omega} \quad (5)$$

Here, the same parameter assignment as in eq. 1 is used with n as index for the nth process. The extrapolation of temperature dependence of the thus obtained mean relaxation times are based on the Vogel-Fulcher-Tammann equation (*52–54*).

$$\tau_c(T) = \tau_0 \exp\left(\frac{DT_0}{T - T_0}\right) \quad (6)$$

With the constant D, the so-called Vogel temperature T_0, and the limiting relaxation time τ_0. Conventionally it is proposed that at T_g the mean relaxation time of the α-relaxation corresponds to 100 s, which enables cross-checks with DSC experiments.

A closer inspection of the α-relaxation spectra indicates an increased broadening in the shorter PDMS chains. This suggests already that the segmental dynamics of the PDMS may be split into different fractions. The respective RTD of the α-relaxation is calculated from the fit parameters according to eq. 3. Since in the sample with lowest *DP* (13) the shape of the broad process is not properly captured by a single HN function, two of these were combined to describe the α-relaxation (Figure 3g). Despite employing two processes, a separate depiction seems unjustified. Consequently, the two RTD obtained from two such processes are added to reproduce the RTD of the whole broadened relaxation.

Comparing the RTD of PDMS-POA of different *DP* against one another reveals not only a considerable shift of the maximum position towards longer relaxation times but also a tremendous increase in broadening of the distribution (Figure 4a). Further, the relatively narrow RTD of PDMS-POA-74 is very similar in shape to that of PDMS without aggregating end groups; however, their maxima are still about one decade in relaxation time apart. This indicates that the POA aggregates impact the segmental mobility of the PDMS chains already significantly at rather small concentrations. In contrast, the severe broadening found at lower *DP* may suggest the emergence of a different effect as the POA fraction increases.

A comparison of the RTD scaled with respect to the position and level of their maxima reveals that the wing towards short times begins to broaden not until the *DP* is well below 50 while on the other wing towards long times the RTD deviates from the non-aggregating reference PDMS-H (the shape of the reference applies to all examined *DP* since it is stable in a wide MW range) (*55*) for even the largest $DP = 74$ under study (Figure 4b). Although a dissection of these RTD utilizing the non-functionalized reference to approximate the fraction of segments exhibiting unperturbed dynamics (similar to the analysis of POSS-iPS shown above) may be desirable, it is not unequivocally executable in this case. The reason for that is the yet unresolved origin of the two alterations, i.e., the shift which is present already at high *DP*, and the broadening setting in at much lower *DP*; it is not clear whether the subtraction of the reference shifted to match the maximum positions yields a meaningful dissection, or whether the unscaled curves should be analyzed, or whether a smaller shift to compensate the slowing observed for the higher *DP* is the correct choice. Nonetheless, even without means to decide for a proper continuation of this analysis, it is evident that particularly at longer times additional relaxation modes arise.

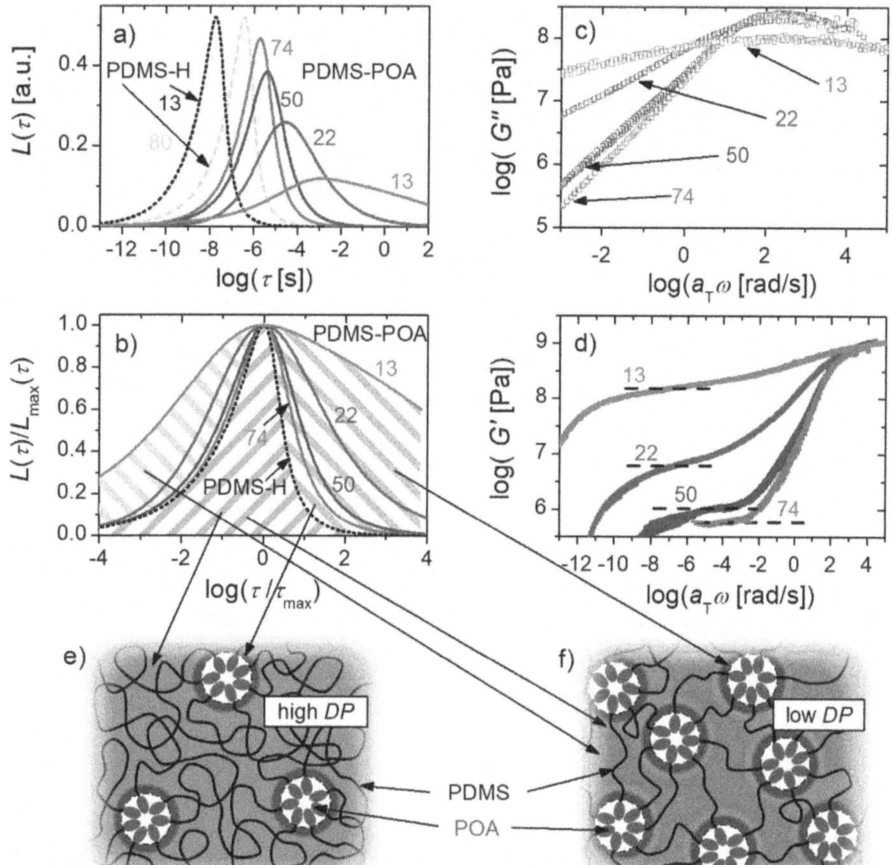

Figure 4. (a) Relaxation time distributions (RTD) of PDMS-POA and of non-aggregating PDMS-H of different degree of polymerization (DP) as indicated at a temperature of 165 K. (b) The same RTD as shown in (a) normalized with respect their maximum (for PDMS-H only one curve is shown since the shape is stable in a wide MW range); the striped areas mark different dynamic fractions with respect to the non-aggregating reference. (c) and (d) Shear modulus loss part G" and storage part G', respectively, of different DP as indicated; the dashed vertical lines in (d) mark the rubbery modulus plateau. (e) and (f) Scheme of the PDMS-POA materials segregating into POA end group aggregates embedded in and connected by the PDMS matrix for high and low DP, respectively; emphasized is the suspected change in chain conformation from rather unperturbed to relatively stretched, as well as regions of respective dynamics colored corresponding to the striped areas in (b).

Such a modification of the relaxation time spectrum is expected to alter also the mechanical performance of the system as was examined for the presented samples in shear modulus measurements (details of linear oscillatory shear rheology can be found elsewhere) (*27, 48*). Indeed, the shear modulus loss part G'' also exhibits a severe broadening of the relaxation peak assigned to the segmental dynamics (Figure 4c). However, this affects not only the segmental dynamics; particularly at low frequencies the G'' curves exhibit a severe change of slope. Since the slope of this part of the curve reflects the Rouse dynamics of the chains, severe alterations of larger-scale dynamics of the chains and hence their conformations can be suspected. In fact, the storage part G' of the shear modulus is dominated by a tremendous increase of the rubbery plateau level by more than 2 decades (Figure 4d) which is another indication for altered chain conformations counteracting shear deformation.

Since the rubbery plateau reflects dynamics well above T_g, it is not directly governed by segmental relaxation but by processes on larger length and longer time scales, typically the motion of whole chains and their entanglements. In the investigated PDMS-POA, however, the chains are too short to form a plateau through entanglement. However, the POA aggregates crosslink the PDMS chains which forms the plateau. In that case, the plateau level should be inversely proportional to the chain length, i.e., $\sim 1/DP$. The experimental results, in contrast, show that the plateau value increases by a factor of ~100 while the *DP* is reduced only by a factor of ~5 times. This suggests that the separation between POA aggregates does not decrease accordingly as *DP* is reduced and the following picture is proposed: While for the higher *DP* samples the chain conformation seems largely unperturbed (as suggested by the narrow RTD) meaning the linking PDMS chains between the aggregates allow for considerable stretching (Figure 4e), in the lower *DP* samples the connecting PDMS chains are already considerably stretched even without external mechanical load (reflected in the broad RTD) leaving little extensibility and generating much higher forces (Figure 4f). It is worthwhile to note that the effect of mechanical enhancement is not primarily based on the contribution of a hard filler (the POA aggregates with their increased T_g will certainly act like those); instead, the mechanical response of the matrix polymer (possibly located in an interfacial layer around the POA aggregates) is altered on the molecular scale.

Conclusion

For the cases of semi-crystalline star-shaped iPS and a series of telechelic PDMS terminated with H-bonding POA groups it has been shown that complex structural morphologies of the nanometric length scale strongly alter the structural relaxation in the polymer component. Thereby, two different effects were observed. In the former, the interplay between crystallizing iPS chains and crystallization-suppressing star-centers creates regions of increased mobility, most likely due to a reduced density. In contrast, in the latter case the H-bonding POA groups phase separate from the PDMS main chains and form clusters of nanometric size and separation which stretches the interconnecting chains. As a result, the segments surrounding the aggregates exhibit a far slower dynamics with an extremely broadened relaxation time distribution; this suggests drastically altered chain conformations which strongly enhances the rubbery shear modulus. Employing dielectric spectroscopy in these studies has revealed detailed characteristics of the dynamical changes; particularly extracting the relaxation time distribution has proven useful.

These examples demonstrate the versatility of dielectric spectroscopy in the characterization of complex and functional materials in order to unravel molecular details. Particularly the gained insight in the impact of the confinement of nanostructured environments on the dynamics of the polymer matrix is extremely valuable to understand the underlying mechanisms.

Acknowledgments

The authors thank the German Science Foundation, DFG, (IRTG Soft Matter Science-Freiburg-Strasbourg-Basel-Mulhouse) for financial support for the POSS-iPS study. The work on H-bonding telechelic PDMS was funded by the NSF Polymer program under grant DMR-1904657 with partial financial support for polymer synthesis by the U.S. Department of Energy, Office of Science, Basic Energy Sciences, Materials Science & Engineering Division. M.T. and F.K. are grateful for the support of the (DFG) within the SFB-TRR 102.

References

1. Hadjichristidis, N.; Pispas, S.; Floudas, G. *Block Copolymers: Synthetic Strategies, Physical Properties, and Applications*; John Wiley & Sons: New York, 2002.
2. Segalman, R. A. Patterning with block copolymer thin films. *Mater. Sci. Eng., R* **2005**, *48*, 191–226.
3. Holt, A. P.; Sangoro, J. R.; Wang, Y.; Agapov, A. L.; Sokolov, A. P. Chain and Segmental Dynamics of Poly(2-vinylpyridine) Nanocomposites. *Macromolecules* **2013**, *46*, 4168–4173.
4. Holt, A. P.; Griffin, P. J.; Bocharova, V.; Agapov, A. L.; Imel, A. E.; Dadmun, M. D.; Sangoro, J. R.; Sokolov, A. P. Dynamics at the Polymer/Nanoparticle Interface in Poly(2-vinylpyridine)/Silica Nanocomposites. *Macromolecules* **2014**, *47*, 1837–1843.
5. Cheng, S.; Mirigian, S.; Carrillo, J.-M. Y.; Bocharova, V.; Sumpter, B. G.; Schweizer, K. S.; Sokolov, A. P. Revealing spatially heterogeneous relaxation in a model nanocomposite. *J. Chem. Phys.* **2015**, *143*, 194704.
6. Cheng, S.; Bocharova, V.; Belianinov, A.; Xiong, S.; Kisliuk, A.; Somnath, S.; Holt, A. P.; Ovchinnikova, O. S.; Jesse, S.; Martin, H.; Etampawala, T.; Dadmun, M.; Sokolov, A. P. Unraveling the Mechanism of Nanoscale Mechanical Reinforcement in Glassy Polymer Nanocomposites. *Nano Lett.* **2016**, *16*, 3630–3637.
7. Cheng, S.; Holt, A. P.; Wang, H.; Fan, F.; Bocharova, V.; Martin, H.; Etampawala, T.; White, B. T.; Saito, T.; Kang, N.-G.; Dadmun, M. D.; Mays, J. W.; Sokolov, A. P. Unexpected Molecular Weight Effect in Polymer Nanocomposites. *Phys. Rev. Lett.* **2016**, *116*, 038302.
8. Carroll, B.; Cheng, S.; Sokolov, A. P. Analyzing the Interfacial Layer Properties in Polymer Nanocomposites by Broadband Dielectric Spectroscopy. *Macromolecules* **2017**, *50*, 6149–6163.
9. Kremer, F.; Schönhals, A. *Broadband Dielectric Spectroscopy*; Springer Science & Business Media: 2002.
10. Kremer, F., Ed. *Dynamics in Geometrical Confinement*; Springer International Publishing: 2014.
11. Strobl, G. *The Physics of Polymers*; Springer: Berlin, Heidelberg, 2007.
12. Natesan, B.; Xu, H.; Seyhan Ince, B.; Cebe, P. Molecular relaxation of isotactic polystyrene: Real-time dielectric spectroscopy and X-ray scattering studies. *J. Polym. Sci., Part B: Polym. Phys.* **2004**, *42*, 777–789.
13. Tress, M.; Vielhauer, M.; Lutz, P. J.; Mülhaupt, R.; Kremer, F. Crystallization-Induced Confinement Enhances Glassy Dynamics in Star-Shaped Polyhedral Oligomeric

Polysilesquioxane-Isotactic Polystyrene (POSS-iPS) Hybrid Material. *Macromolecules* **2018**, *51*, 504–511.

14. Vielhauer, M.; Lutz, P. J.; Reiter, G.; Mülhaupt, R. Linear and star-shaped POSS hybrid materials containing crystalline isotactic polystyrene chains. *J Polym. Sci., Part A: Polym. Chem.* **2013**, *51*, 947–953.
15. Tress, M..; Mapesa, E. U.; Kossack, W.; Kipnusu, W. K.; Reiche, M.; Kremer, F. Glassy Dynamics in Condensed Isolated Polymer Chains. *Science* **2013**, *341*, 1371–1374.
16. Havriliak, S.; Negami, S. A complex plane representation of dielectric and mechanical relaxation processes in some polymers. *Polymer* **1967**, *8*, 161–210.
17. Zorn, R. Applicability of distribution functions for the Havriliak-Negami spectral function. *J. Polym. Sci., Part B: Polym. Phys.* **1999**, *37*, 1043–1044.
18. Kipnusu, W. K.; Elsayed, M.; Kossack, W.; Pawlus, S.; Adrjanowicz, K.; Tress, M.; Mapesa, E. U.; Krause-Rehberg, R.; Kaminski, K.; Kremer, F. Confinement for More Space: A Larger Free Volume and Enhanced Glassy Dynamics of 2-Ethyl-1-hexanol in Nanopores. *J. Phys. Chem. Lett.* **2015**, *6*, 3708–3712.
19. Zheng, Z.; Kuang, F.; Zhao, J. Direct observation of rotational motion of fluorophores chemically attached to polystyrene in its thin films. *Macromolecules* **2010**, *43*, 3165–3168.
20. Maitra, P.; Wunder, S. L. Oligomeric Poly(ethylene oxide)-Functionalized Silsesquioxanes: Interfacial Effects on T_g, T_m, and ΔH_m. *Chem. Mater.* **2002**, *14*, 4494–4497.
21. Núñez, E.; Ferrando, C.; Malmström, E.; Claesson, H.; Gedde, U. W. Crystallization Behavior and Morphology of Star Polyesters with Poly(ε-Caprolactone) Arms. *J. Macromol. Sci., Part B: Phys.* **2004**, *43*, 1143–1160.
22. Kuo, S.-W.; Chang, F.-C. POSS related polymer nanocomposites. *Prog. Polym. Sci.* **2011**, *36*, 1649–1696.
23. Yani, Y.; Lamm, M. H. Molecular dynamics simulation of mixed matrix nanocomposites containing polyimide and polyhedral oligomeric silsesquioxane (POSS). *Polymer* **2009**, *50*, 1324–1332.
24. Popov, I.; Carroll, B.; Bocharova, V.; Genix, A.-C.; Cheng, S.; Khamzin, A.; Kilsiuk, A.; Sokolov, A. P. Strong Reduction in Amplitude of the Interfacial Segmental Dynamics in Polymer Nanocomposites. *Macromolecules* **2020**, *53*, 4126–4135.
25. Brunsveld, L.; Folmer, B.; Meijer, E. W.; Sijbesma, R. Supramolecular polymers. *Chem. Rev.* **2001**, *101*, 4071–4098.
26. Monkenbusch, M.; Krutyeva, M.; Pyckhout-Hintzen, W.; Antonius, W.; Hovelmann, C. H.; Allgaier, J.; Bras, A.; Farago, B.; Wischnewski, A.; Richter, D. Molecular View on Supramolecular Chain and Association Dynamics. *Phys. Rev. Lett.* **2016**, *117*, 147802.
27. Xing, K.; Tress, M.; Cao, P.; Cheng, S.; Saito, T.; Novikov, V. N.; Sokolov, A. P. Hydrogen-bond strength changes network dynamics in associating telechelic PDMS. *Soft Matter* **2018**, *14*, 1235–1246.
28. Hackelbusch, S.; Rossow, T.; van Assenbergh, P.; Seiffert, S. Chain dynamics in supramolecular polymer networks. *Macromolecules* **2013**, *46*, 6273–6286.
29. Binder W. H.; Zirbs R. *Hydrogen bonded polymers*; Springer: 2006; pp 1–78.
30. Lewis, C. L.; Stewart, K.; Anthamatten, M. The influence of hydrogen bonding side-groups on viscoelastic behavior of linear and network polymers. *Macromolecules* **2014**, *47*, 729–740.

31. Sijbesma, R. P.; Beijer, F. H.; Brunsveld, L.; Folmer, B. J. B.; Hirschberg, J. H. K. K.; Lange, R. F. M.; Lowe, J. K. L.; Meijer, E. W. Reversible polymers formed from self-complementary monomers using quadruple hydrogen bonding. *Science* **1997**, *278*, 1601–1604.
32. Cordier, P.; Tournilhac, F.; Soulie-Ziakovic, C.; Leibler, L. Self-healing and thermoreversible rubber from supramolecular assembly. *Nature* **2008**, *451*, 977–980.
33. Burnworth, M.; Tang, L.; Kumpfer, J. R.; Duncan, A. J.; Beyer, F. L.; Fiore, G. L.; Rowan, S. J.; Weder, C. Optically healable supramolecular polymers. *Nature* **2011**, *472*, 334–337.
34. Kang, J.; Miyajima, D.; Mori, T.; Inoue, Y.; Itoh, Y.; Aida, T. A rational strategy for the realization of chain-growth supramolecular polymerization. *Science* **2015**, *347*, 646–651.
35. Yanagisawa, Y.; Nan, Y.; Okuro, K.; Aida, T. Mechanically robust, readily repairable polymers via tailored noncovalent cross-linking. *Science* **2018**, *359*, 72–76.
36. Filippidi, E.; Cristiani, T. R.; Eisenbach, C. D.; Waite, J. H.; Israelachvili, J. N.; Ahn, B. K.; Valentine, M. T. Toughening elastomers using mussel-inspired iron-catechol complexes. *Science* **2017**, *358*, 502–505.
37. Folmer, B. J. B.; Sijbesma, R. P.; Versteegen, R. M.; van der Rijt, J. A. J.; Meijer, E. W. Supramolecular polymer materials: chain extension of telechelic polymers using a reactive hydrogen-bonding synthon. *Adv. Mater.* **2000**, *12*, 874–878.
38. Feldman, K. E.; Kade, M. J.; Meijer, E. W.; Hawker, C. J.; Kramer, E. J. Model transient networks from strongly hydrogen-bonded polymers. *Macromolecules* **2009**, *42*, 9072.
39. Zhang, Z. P.; Rong, M. Z.; Zhang, M. Q. Polymer engineering based on reversible covalent chemistry: A promising innovative pathway towards new materials and new functionalities. *Prog. Polym. Sci.* **2018**, *80*, 39–93.
40. Cao, P.-F.; Li, B.; Hong, T.; Townsend, J.; Qiang, Z.; Xing, K.; Vogiatzis, K. D.; Wang, Y.; Mays, J. W.; Sokolov, A. P.; Saito, T. Superstretchable, self-healing polymeric elastomers with tunable properties. *Adv. Funct. Mater.* **2018**, *28*, 1800741.
41. Kulkarni, A.; Lele, A.; Sivaram, S.; Rajamohanan, P. R.; Velankar, S.; Chatterji, A. Star Telechelic Poly(l-lactide) Ionomers. *Macromolecules* **2015**, *48*, 6580–6588.
42. Lundberg, R. D.; Makowski, H. S.; Westerman, L. *Ions in Polymers*; American Chemical Society: Washington, DC, 1980; pp 67−76.
43. Li, S.-L.; Xiao, T.; Lin, C.; Wang, L. Advanced supramolecular polymers constructed by orthogonal self-assembly. *Chem. Soc. Rev.* **2012**, *41*, 5950–5968.
44. Tang, S.; Olsen, B. D. Relaxation Processes in Supramolecular Metallogels Based on Histidine-Nickel Coordination Bonds. *Macromolecules* **2016**, *49*, 9163–9175.
45. Gold, B. J.; Hövelmann, C. H.; Lühmann, N.; Székely, N. K.; Pyckhout-Hintzen, W.; Wischnewski, A.; Richter, D. Importance of Compact Random Walks for the Rheology of Transient Networks. *ACS Macro Lett.* **2017**, *6*, 73–77.
46. Gold, B. J.; Hövelmann, C. H.; Lühmann, N.; Pyckhout-Hintzen, W.; Wischnewski, A.; Richter, D. The microscopic origin of the rheology in supramolecular entangled polymer networks. *J. Rheol.* **2017**, *61*, 1211–1226.
47. Ge, S.; Tress, M.; Xing, K.; Cao, P.; Saito, T.; Sokolov, A. P. Viscoelasticity in associating oligomers and polymers: experimental test of the bond lifetime renormalization model. *Soft Matter* **2020**, *16*, 390–401.

48. Tress, M.; Xing, K.; Ge, S.; Cao, P.; Saito, T.; Sokolov, A. P. What dielectric spectroscopy can tell us about supramolecular networks. *Eur. Phys. J. E* **2019**, *42*, 133.
49. Desiraju, G. R.; Steiner, T. *The weak hydrogen bond*; Oxford University Press: 1999.
50. Stukalin, E. B.; Cai, L. H.; Kumar, N. A.; Leibler, L.; Rubinstein, M. Self-Healing of Unentangled Polymer Networks with Reversible Bonds. *Macromolecules* **2013**, *46*, 7525–7541.
51. Wagner, H.; Richert, R. Equilibrium and non-equilibrium type β-relaxations: d-sorbitol versus o-terphenyl. *J. Phys. Chem. B* **1999**, *103*, 4071–4077.
52. Vogel, H. Das Temperaturabhängigkeitsgesetz der Viskosität von Flüssigkeiten. *Phys. Z.* **1921**, *22*, 645–646.
53. Fulcher, G. S. Analysis of recent measurements of the viscosity of glasses. *J. Am. Ceram. Soc.* **1925**, *8*, 339–355.
54. Tammann, G.; Hesse, W. Die Abhängigkeit der Viscosität von der Temperatur bei unterkühlten Flüssigkeiten. *Z. Anorg. Allg. Chem.* **1926**, *156*, 245–257.
55. Hintermeyer, J.; Herrmann, A.; Kahlau, R.; Goiceanu, C.; Rössler, E. A. Molecular Weight Dependence of Glassy Dynamics in Linear Polymers Revisited. *Macromolecules* **2008**, *41*, 9335–9344.

Chapter 11

Impacts of Crosslinking and Degassing on the Conductivity, Dielectric Loss, and Morphology of Low-Density Polyethylene and Crosslinked Polyethylene

Roger Craig Walker II,[1,*] Hossein Hamedi,[2] William H. Hunter Woodward,[3] Ramakrishnan Rajagopalan,[4] and Michael Lanagan[1,4]

[1]Department of Materials Science and Engineering, Penn State, University Park, Pennsylvania 16802, United States

[2]Department of Chemical Engineering, Penn State, University Park, Pennsylvania 16802, United States

[3]Dow, Analytical Sciences, Core R&D, Midland, Michigan 48667, United States

[4]Department of Engineering Science and Mechanics, Penn State, University Park, Pennsylvania 16802, United States

[*]Email: rcw202@psu.edu

Crosslinked polyethylene (XLPE) is a key material used for power cable insulation due to its electrical properties and structure. Here, it was shown that the technique of broadband dielectric spectroscopy can be used to analyze both by extracting the temperature coefficient of capacitance (TCC), dielectric loss, and AC conductivity. TCC for low-density polyethylene (LDPE) can be directly compared to its linear thermal expansion coefficient and used as a tool to compare different types of polyethylene. Degassing, a thermal treatment to remove volatile species and increase crystallinity, can be used to control TCC magnitude. In general, changes in the TCC magnitude are reflective of changes in the crystallinity and byproduct concentrations. Degassing also reduced the dielectric loss for both due to morphological changes and byproduct removal. However, it only reduces conductivity at elevated temperatures such as 90 °C for XLPE. Other processing parameters, as well as excessive byproduct concentrations, were also seen to influence TCC, dielectric loss, and conductivity.

Introduction

Crosslinked polyethylene (XLPE) is the main material used for high voltage alternating current (AC) power cable insulation. It is typically made by using dicumyl peroxide (DCP) to generate a three-dimensional network from the branched polymer low-density polyethylene (LDPE). In this process, LDPE branches and chains are connected by crosslinks that add strength and stability. There is evidence suggesting that crosslinks also enhance insulating properties (*1–4*). DCP leaves behind byproducts such as acetophenone (ACP), cumyl alcohol, and methylstyrene. They introduce excess conductivity and shallow traps, leading to space charge buildup and early failure (*5, 6*). Methane and water are also produced. Methane is a notable safety hazard due to its flammability, while water has synergistic effects with other byproducts. Notably, it reacts with ACP to produce protons and space charge (*7–9*). While there has been work regarding power cable insulation materials that do not have to deal with these byproducts (*10, 11*), XLPE is currently still in wide use and so the impacts of the DCP byproducts remain important. As such, their detrimental effects must be mitigated, and the current method of choice is to remove them is degassing (*12, 13*). XLPE becomes more insulating and its resistivity surpasses LDPE when the byproducts are removed (*3, 4*), making it the preferred choice for power cable insulation.

Polymeric insulators are characterized for use in this application by their dielectric loss and AC conductivity. Dielectric loss is defined as the energy lost during AC operation relative to the total energy stored by a dielectric during the polarization cycle. It is also referred to as the loss tangent, dissipation factor, tan δ, and so on. Loss and the AC conductivity are related by the following equation:

$$\sigma = \varepsilon' * \tan\delta * 2\pi f \quad (1)$$

where σ is the AC conductivity, ε′ is the real component of the complex permittivity, tan δ is the dielectric loss, and f is the frequency. This value is often measured at power frequency (50 or 60 Hz), and a table of literature values for the dielectric loss in LDPE is provided in Table A1. Low frequency dielectric loss is proportional to the leakage current (*14, 15*), and large currents are detrimental to the XLPE insulation. Low loss thus leads to negligible conductivity in AC conditions, and also in the case of direct current (DC) as the frequency approaches zero. Such a state can be achieved by removing sources of trapped charge such as DCP byproducts (*13*).

Degassing to remove byproducts is typically done in the temperature range from 50 °C to 80 °C due to constraints from cable manufacturing (*13*). Recrystallization is expected upon cooling from temperatures within the melting range of polyethylene (50°C to 110°C) (*16*). This process alters the morphology (*17*) and potentially impacts the dielectric loss. The changes in structure that occur when transforming LDPE into XLPE, which consist of crosslinking and annealing, should also alter the loss and conductivity. There have been several notable studies relating structural changes to modification of the electrical properties (*1, 3, 18*). It was shown by Sato and Yashiro that the secondary structure (crosslinks, branches, etc.) can be as important in determining the dielectric loss as polar impurities, though this effect tends to be seen only at very high frequencies such as MHz. Preventing oxidation and reducing branching concentration was recommended for reducing the loss (*1*). This was also reflected in current density measurements (*3*). One study demonstrated that annealing LDPE at temperatures of 45°C or greater under applied electric field led to reduced activation energy (*18*). This was correlated with differential scanning calorimetry and discharge current measurements to show that a structural transformation occurs that reduces space charge susceptibility (*18*). In that work, follow-up studies were recommended for determining the impact of crosslinks and antioxidants, as well as separating the effects of field and temperature. It is clear both that significant progress has been made and that more work is needed to fully understand how the structure and electrical properties of polyethylene relate to each other.

As such, those findings motivated this study on the influence of thermal treatments on XLPE electrical properties under AC conditions down to low frequencies and a correlation to changes in structure. In particular, broadband dielectric spectroscopy (BDS) was used as the primary tool of this study. It allows for direct measurement of changes in dielectric properties and indirect measurement of changes in structure by obtaining the temperature coefficient of capacitance (TCC). TCC values for non-polar polymers are typically small and negative due to their structure and chemistry, and can be used as approximations of the linear thermal expansion coefficient (LTEC) due to this (*19*). Significant changes from this can be attributed to either extrinsic factors such as impurities or intrinsic factors such as changes in the polymer structure. Thus, variations in the TCC can be used to understand structural and chemical differences between samples.

Experimental

LDPE pellets were provided by Dow and DCP was obtained from Sigma-Aldrich. DCP was added to LDPE pellets by soaking at 70 °C. LDPE and XLPE thin films were generated via melt pressing of polymer beads between two Teflon sheets. The addition of DCP allowed for crosslinking. The pressure was fixed at 17.24 MPa (2500 psi). Standard conditions for LDPE and XLPE processing were 180 °C for the crosslinking temperature (T_c) and 12 minutes for the crosslinking time (t_c). Temperature and time for crosslinking were varied for XLPE. Samples were cooled in air while under pressure. Some LDPE and XLPE films were degassed in an Across VO Series 300°C Vacuum Drying Oven at 65 °C for 72 hours. In this oven, a vacuum of 500 to 600 mmHg was applied, reducing the atmosphere to the range of 34.7 kPa to 21.3 kPa. Film thicknesses were in the range of 100-300 μm. A table listing the samples analyzed is displayed in Table 1.

Aluminum electrodes were applied using electron beam evaporation via a Kurt J. Lesker Lab-18 unit. Inside the evaporation chamber, aluminum metal is exposed to an electron beam, and the ablated material travels 11 inches across the chamber and coats the samples. The samples are attached to a six inch metal plate during this process to keep them in place. Deposition was done in high vacuum (i.e. ~10^{-4} Pa) at a rate of 2.5 Å s^{-1} to a thickness of 50 nm. To ensure that all electrodes had the same area, a mask made of 0.005" thick blue polyester plastic was used with holes cut in it having diameters of 3 cm. To ensure a uniform coating, the plate holding the samples was rotated at 20 rpm. Under these conditions, electrodes were able to be applied to a wide variety of samples.

LDPE and XLPE samples were electrically characterized using broadband dielectric spectroscopy (BDS) via a Solartron analyzer, specifically a Modulab unit, allowing for the extraction of loss, conductivity, and capacitance values over a frequency range from 0.1 to 1000 Hz. A 1 V rms AC signal was applied during measurement. The impedance was measured and converted to complex capacitance and dielectric loss values, allowing for extraction of real capacitance and conductivity. Loss and conductivity values from room temperature and 90 °C were examined. The average low frequency values for these two parameters were extracted by measuring from 0.1 to 10 Hz and then averaging the obtained values. Analysis of the capacitance in 10 °C steps allowed for estimation of the temperature coefficient of capacitance for LDPE and XLPE, and the 90 °C data were obtained during this sweep. A Delta Design 9023 Oven was used for temperature control. A custom-built door fixture was used to allow for measurement of the dielectric properties inside the oven. The samples are held in place inside the oven using spring contact probes from Smiths Interconnect, which are themselves held in place by the custom-built anodized aluminum fixture. Only real capacitance data taken at 1 kHz was used for the TCC data displayed here in order to avoid space charge or conduction contributions, which become prominent at lower frequencies.

Table 1. List of the types of polyethylene samples examined.

XLPE or LDPE	*%DCP*	*XL Temperature (°C)*	*Degas?*	*Soak w/ ACP?*
LDPE	0	180	No	No
LDPE	0	180	Yes	No
XLPE	1.78	180	No	No
XLPE	1.78	200	No	No
XLPE	2.06	120	No	No
XLPE	2.06	140	No	No
XLPE	2.06	160	No	No
XLPE	2.06	180	No	No
XLPE	2.10	180	No	No
XLPE	2.12	120	No	No
XLPE	2.12	120	Yes	No
XLPE	2.12	140	No	No
XLPE	2.12	140	Yes	No
XLPE	2.12	160	No	No
XLPE	2.12	160	Yes	No
XLPE	2.12	180	No	No
XLPE	2.12	180	Yes	No
XLPE	2.14	180	Yes	No
XLPE	2.2	180	Yes	Yes
XLPE	2.4	180	Yes	No
XLPE	3.2	180	Yes	No
XLPE	4.9	180	Yes	No
XLPE	6.7	180	Yes	No
XLPE	8.2	180	Yes	No
XLPE	10	180	Yes	No

Further characterization was done using thermogravimetric analysis (TGA), wide-angle X-ray scattering (WAXS), and Fourier transform infrared spectroscopy (FTIR). TGA was used to analyze potential changes in sample mass with heating. A TA Instruments Discovery Series TGA Q5500 analyzed samples with an initial mass of around 2 mg. Samples were heated in an alumina pan at a rate of 10 °C/min. from 25 °C to 300 °C in ambient air with a nitrogen purge flow of 100 mL/min. WAXS was used to calculate the crystalline fraction via a Xenocs Xeuss 2.0 beamline in transmission mode. Samples with 100 μm thickness were used for measurement, and areas of 1 cm^2 were mapped with a resolution of 1 mm^2. FTIR was used to obtain the concentration of byproducts in the samples. It was performed via the Bruker HYPERION 3000 FTIR microscope in transmission mode on samples with approximately 100 μm thickness. To detect ACP in the XLPE, the C=O stretching peak at 1695 cm^{-1} was measured in mapping areas of around 1 cm^2 with a resolution of 100 μm^2. To compare samples, the measured peak was normalized with the corresponding polyethylene backbone peak.

Temperature Coefficient of Capacitance

TCC is the fractional change in the capacitance (C; C′) of a dielectric during a unit change of temperature (T) at constant pressure. In practice, a low TCC value is desirable for insulating materials in order to have stable capacitance values over a wide temperature range. Excessive variation in the capacitance as a function of temperature leads to degradation of the insulating material, much the same as would occur for excessive variation in conductivity. Polymers such as polyethylene fulfill this requirement due to the factors influencing their TCC value, which will be elaborated on in this section. The TCC is defined as:

$$TCC = \frac{1}{C'}\left(\frac{\partial C'}{\partial T}\right)_P = \frac{1}{\varepsilon'}\left(\frac{\partial \varepsilon'}{\partial T}\right)_P + \alpha_T \quad (2)$$

It can be seen in Equation 2 that TCC is related to the LTEC (α_T), which in a semicrystalline material will be influenced by the crystallinity as both the amorphous and crystalline phases contribute to the total TCC. By using the Clausius-Mossotti relation, the term involving the permittivity can be expanded further (*19*). This is shown in Equations 3 and 4:

$$\frac{\varepsilon-1}{\varepsilon+2} = \frac{\alpha_m}{3\bar{V}\varepsilon_0} \quad (3)$$

$$\frac{\partial\varepsilon}{\partial T} = (\varepsilon+2)(\varepsilon-1)\left(\frac{1}{3\alpha_m}\frac{\partial\alpha_m}{\partial T} - \alpha_T\right) \quad (4)$$

$$TCC = \frac{(\varepsilon+2)(\varepsilon-1)}{\varepsilon}\left(\frac{1}{3\alpha_m}\frac{\partial\alpha_m}{\partial T} - \alpha_T\right) + \alpha_T \quad (5)$$

Equation 4 was obtained by differentiating Equation 3 with respect to temperature. In these equations, α_m is the molecular polarizability, V is the molar volume, and ε_0 is the vacuum permittivity. The thermal expansion coefficient that appears in Equation 4 comes from changes in the molar volume of the sample. Substituting Equation 4 into Equation 2 yields the full TCC equation, which is shown in Equation 5. TCC values for a given material will therefore be determined by thermal expansion and changes in polarizability. The expression above assumed that the two were independent; if not, then another term must be added (*19*).

By making certain assumptions, the TCC equation can be simplified (*19*). The relative permittivity of the non-magnetic material polyethylene is governed by its refractive index. The dielectric response in this material is solely from polarized electrons, as it lacks any dipolar or ionic species. Space charge polarization is only relevant in conditions with high electrical field and low frequency and will not be considered here. As such, only the polarizability of the electrons in polyethylene will determine the polarizability of the whole molecule. The factors determining this number include the electron density on the atom and the average distance of the electron cloud from the nucleus. These factors are not expected to depend much on the temperature or the molar volume of the material. As such, terms involving changes in polarizability can be set to zero. Following this, the relative permittivity could then be considered constant, having a value around 2 based on the refractive index of polyethylene and directly measured as 2.2 via BDS. This allows the TCC to be simplified as follows, using the value extracted from BDS:

$$TCC = \frac{(2.2+2.2)(2.2-1)}{2.2}(-\alpha_T) + \alpha_T \quad (6)$$

$$TCC = -1.4\alpha_T \quad (7)$$

$$TCC \approx -\alpha_T \quad (8)$$

These assumptions that led to Equation 8 can then be affirmed via measurement of TCC using BDS and comparison to literature values of the LTEC for polyethylene. The results from such a measurement and comparison are shown in Figure 1a. The experimental data cited from literature examined the impact of the crystalline fraction on thermal expansion of LDPE (*20*). Two samples of low-density polyethylene were prepared with crystallinity around 42% which matched that of one of the literature samples. The values of negative TCC extracted from the LDPE samples were good matches for LTEC data from LDPE with similar crystallinity, with the exception of one sample having anomalous behavior around 0°C (Figure 1a). The cause of said anomalous behavior could not be determined at this time. In general, there was good confidence not only that TCC data can be used to approximate the LTEC but also that comparisons between samples may reveal information regarding the structure of polyethylene samples. This is due to the relation between increased crystallinity and reduced linear thermal expansion coefficient for PE (*20*, *21*), which can then be related to differences in structure.

Degassing Effect on TCC, Loss, and Conductivity

Following this, a comparison of TCC values was made between different PE samples: as-received XLPE, as-received LDPE, degassed XLPE, and degassed LDPE. Due to the necessity of keeping the magnitude of the TCC low, it is useful to be able to show how the crosslinking process – which introduces both crosslinks and byproducts – can influence its value. The same is true for the degassing process. The results are shown in Figure 1b.

Around room temperature and below, the as-received XLPE and LDPE are similar. The magnitude of the TCC increases significantly for XLPE at elevated temperatures when compared to LDPE starting at 40°C. By analyzing these samples via TGA as shown in Figure A1, it became clear that this difference was due to the DCP byproducts left behind in the XLPE. The as-received XLPE lost mass upon heating at a rate that accelerated near 40 °C. That process ceased near 90 °C with a total mass loss of approximately 1.5%, a value congruent with the loss of residual DCP byproducts (*13*). The LDPE, meanwhile, had no significant change in mass below 90 °C. We therefore concluded that the increase in the magnitude of the TCC above 40 °C was caused by an increase in thermal expansion arising from trace concentrations of the DCP byproducts. Degassing generally reduced the magnitude of the TCC for both LDPE and XLPE above 0 °C, with the differences becoming more apparent starting at 20 °C. The crystallinity of as-received and degassed XLPE were also measured and are shown in Figure A2. In XLPE, it can be seen that degassing leads to increased crystallinity and loss of byproducts, which led to the altered TCC behavior. The same appeared to be case for LDPE, although the change was not as large due to a lack of DCP byproducts, meaning that the change came primarily from the increased crystallinity. It can be seen thus that both impurities and inherent structure impact the TCC of this polymer.

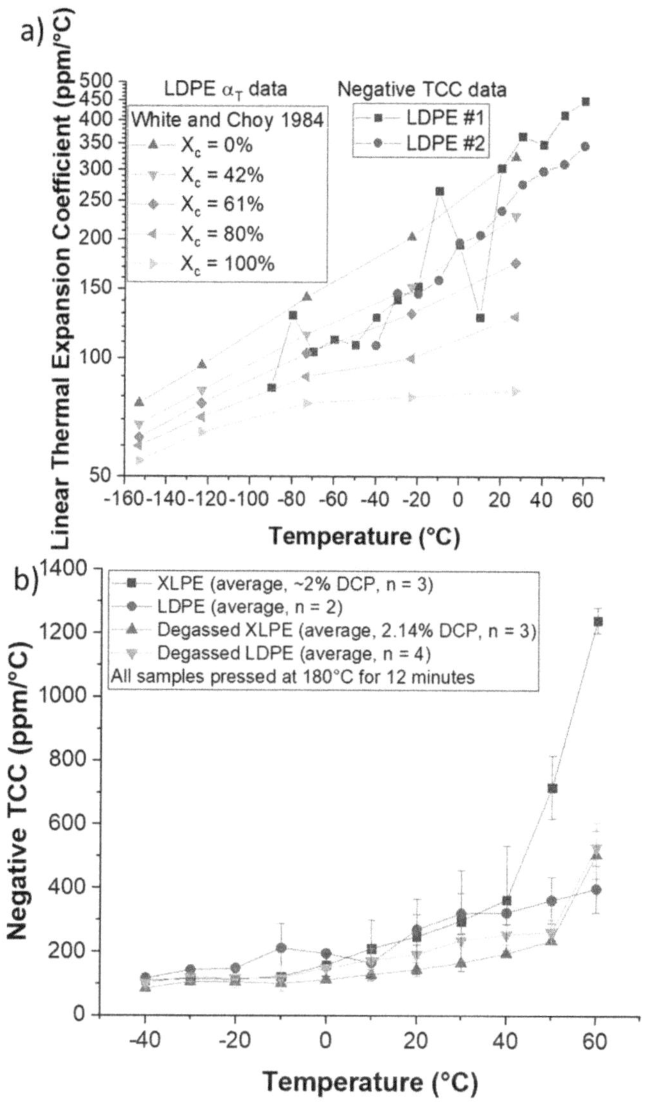

Figure 1. a) TCC measurement of two samples of low-density polyethylene (LDPE) and comparison to literature data for the LTEC of the same polymer. X_c represents the crystallinity of polyethylene. LTEC data from Reference (20). b) Average TCC comparison between as-received XLPE, as-received LDPE, degassed XLPE, and degassed LDPE samples. N is the number of samples analyzed.

To determine if changes in morphology due to degassing impact the dielectric loss, a comparison was made between XLPE (which contains DCP byproducts in the as-received state) and LDPE (which would not). The result of a comparison made at room-temperature is presented in Figure 2, with error bars showing the standard error of the calculated average. The raw broadband dielectric spectra for these samples are shown in Figure A3. The loss values were similar between the as-received samples of these two materials, which agrees with prior investigations (22). Degassing reduced the dielectric loss for both XLPE and LDPE. The choice of crosslinking temperature also did not lead to systematic variations in the dielectric loss for samples in either condition. Such differences could not be explained by variations in the content of DCP byproducts, as the LDPE films lacked them and their concentration should have systematically varied with the choice of crosslinking temperature following their consumption to create crosslinks (23). As such, this difference was attributed to morphological changes induced by degassing, in particular the changes to the crystallinity.

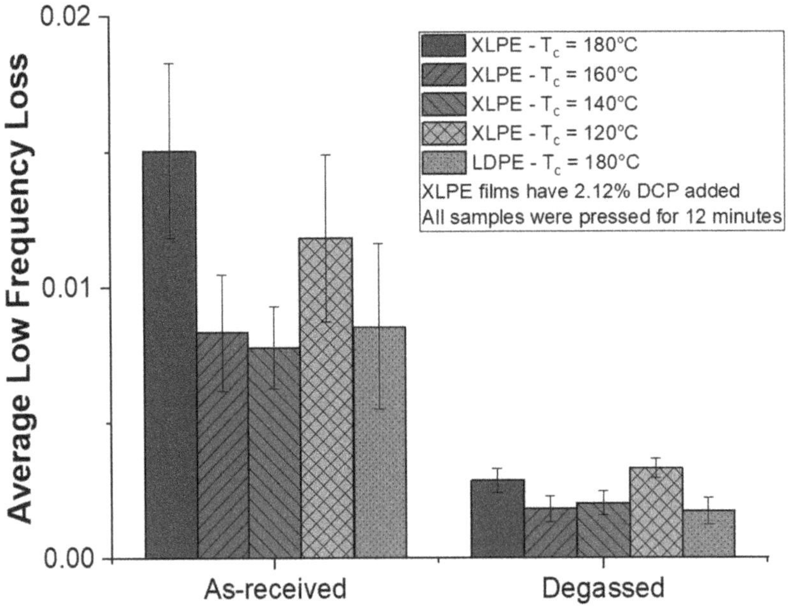

Figure 2. Comparison of the room temperature dielectric loss values measured at 10 Hz and below for XLPE and LDPE samples crosslinked at various temperatures.

As polyethylene exhibits thermally activated conductivity, it was useful to determine if the reduction in dielectric loss due to degassing would be maintained at elevated temperatures. That experiment began with measurement of the dielectric loss at elevated temperatures for as-received samples, which is shown in Figure 3a for a temperature of 90 °C. At high frequencies such as 1 kHz, the loss was roughly constant. At low frequencies such as 10 Hz and below, the loss and frequency were inversely proportional due to contributions from conductive losses. This meant that a constant value for the AC conductivity could be extracted, based on Equation 1. When this is true, a material may be considered to be undergoing DC conduction (24, 25), and the results of applying Equation 1 are shown in Figures 3b and 3c. The as-received XLPE samples were less conductive than the as-received LDPE samples on average.

Similarly, by comparing degassed and as-received samples in Figure 4, it became possible to see distinct differences between XLPE and LDPE samples. In Figure 4a, it can be seen that the degassed XLPE samples were less conductive than their as-received counterparts by roughly an order of magnitude. On the other hand, the degassed and as-received LDPE samples were comparable, as shown in Figure 4b. The differences in electrical behavior between these samples can be attributed to the differences in structure between these samples and how that structure responds to elevated temperatures. In the as-received condition, XLPE was less conductive than LDPE due to its crosslinks. There was a slightly greater reduction of conductivity than loss upon degassing, which may be a sign that further changes occurred during measurement at elevated temperatures. The removal of crosslinking byproducts, which contribute significantly to the conductivity (26), was likely responsible for this observation. Meanwhile, the LDPE conductivity was roughly unchanged.

As the measurement temperature of 90 °C is a significant fraction of the nominal melting temperature of LDPE (~105 °C), the benefits of degassing appear to have been lost.

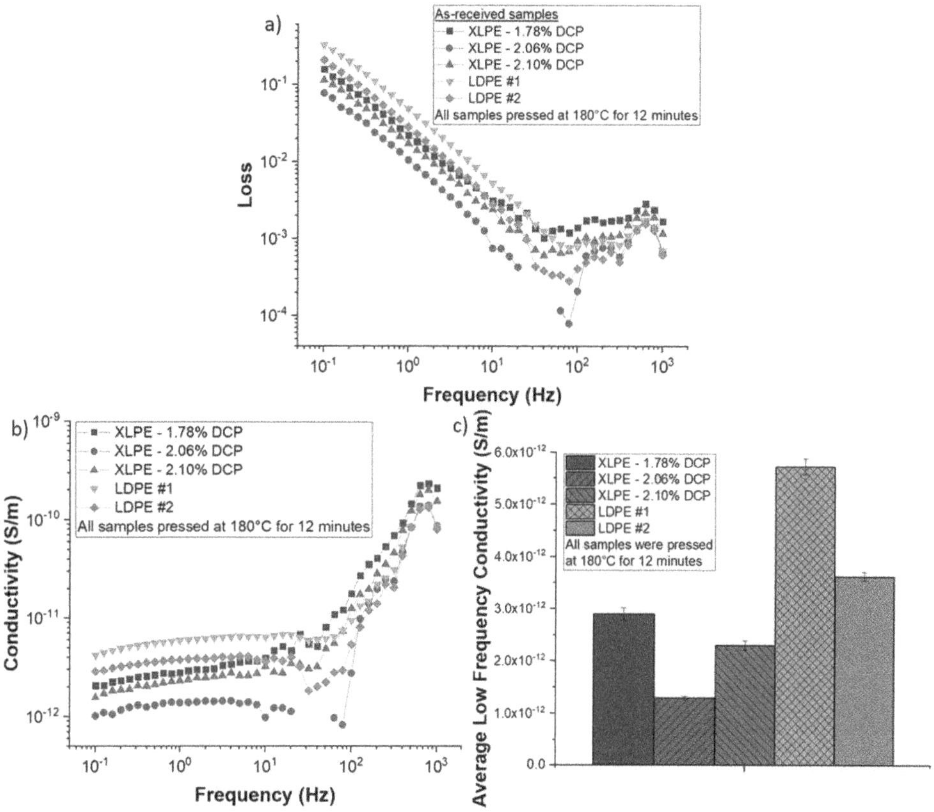

Figure 3. a) Dielectric loss of as-received XLPE and LDPE samples measured at 90°C; b) Conductivity of as-received XLPE and LDPE samples measured at 90°C; c) Average low frequency conductivity of the same samples from frequencies of 10 Hz and below.

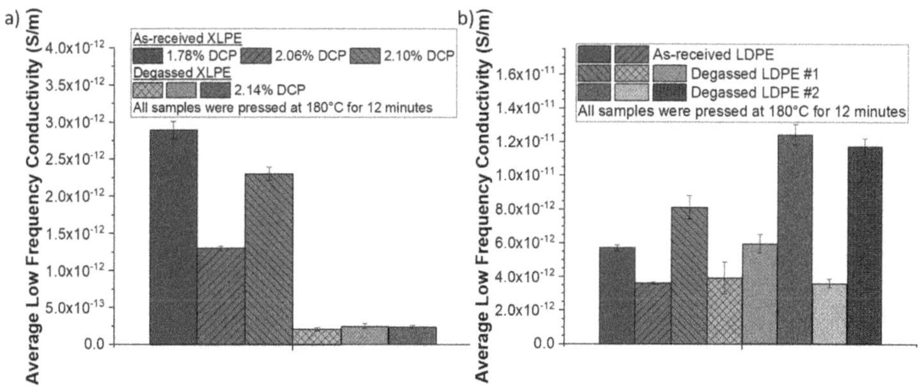

Figure 4. a) Average low frequency 90°C conductivity comparison between as-received and degassed XLPE samples; b) Average low frequency 90°C conductivity comparison between as-received and degassed LDPE samples.

Crosslinking Temperature Effect on TCC, Loss, and Conductivity

As far as the impact of crosslinking temperature goes, a 90 °C conductivity analysis shown in Figure 5 revealed no clear dependence between the two with one major exception. The sample with the lowest crosslinking temperature used in this study had a significantly higher conductivity when compared to the other samples. An analysis of the TCC and crystallinity of these samples helps to show why this occurred. The variation in TCC was slight, as shown in Figure 6, with an increased crosslinking temperature on average leading to a reduced TCC magnitude. This was due to a reduction in the concentration of residual byproducts with higher crosslinking temperature. Meanwhile, the sample crosslinked at 120 °C had the highest crystallinity, with decreased extent of crystallinity corresponding to increased temperature for 140 °C and 160 °C, and above this point the crystallinity appeared to be invariant with respect to temperature (27). This corresponded with literature data showing that the crosslinking reaction is highly temperature dependent (23). As such, the XLPE sample "crosslinked" at 120°C would have had the smallest extent of crosslinking reaction occurring and thus had both crystallinity most similar to LDPE, and conductivity much higher than other DCP samples due to the presence of unreacted DCP and other impurities.

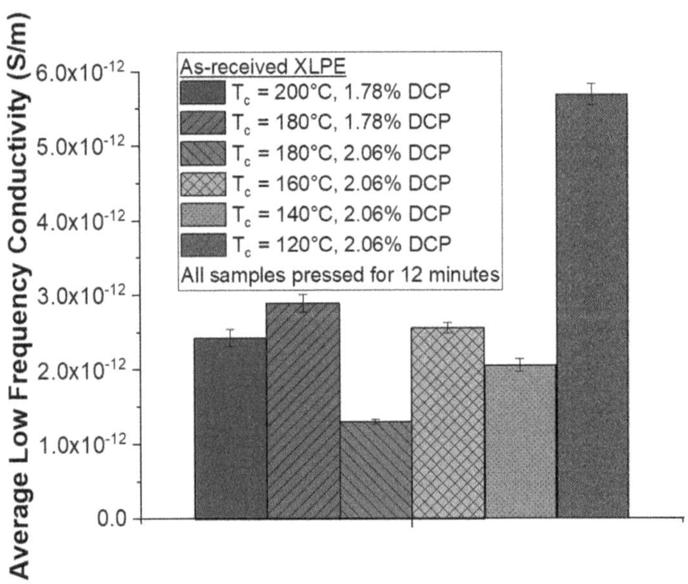

Figure 5. Average low frequency 90°C conductivity comparison between as-received XLPE samples crosslinked at different temperatures.

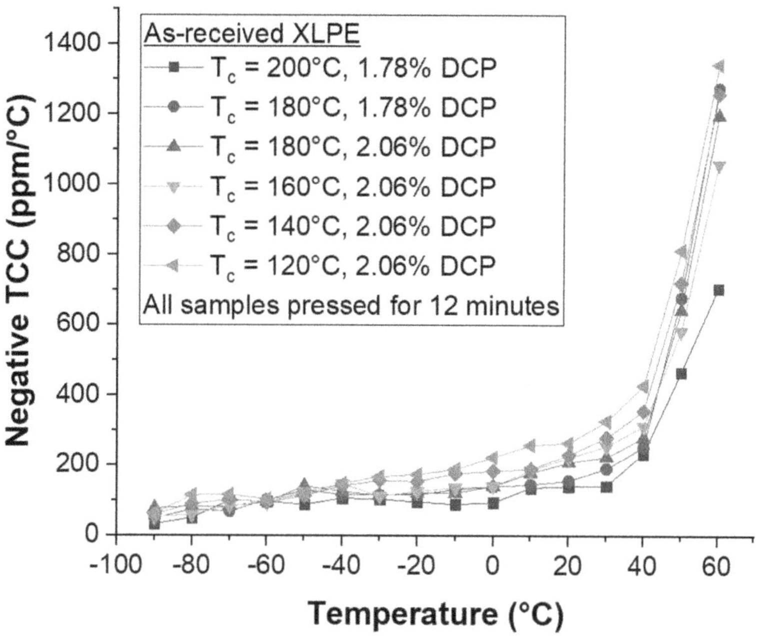

Figure 6. TCC comparison between as-received XLPE samples crosslinked at different temperatures.

Byproduct Effect on TCC, Loss, and Conductivity

Besides the crosslinking temperature, the amount of DCP added prior to crosslinking has a significant impact on the polyethylene structure and residual byproduct concentrations (*23*, *28*). As such, the impact of the amount of DCP on the resulting TCC of XLPE was explored, and the results from degassed samples are shown in Figure 7. The data from the as-received samples in Figure 1a is also plotted here. It can be seen that increasing the amount of DCP added led to an increase in the magnitude of the TCC that met and then exceeded what was observed in the as-received samples. A reduction in the crystalline fraction was also observed (*27*). It is known that increased DCP concentration leads to increased extent of crosslinking and increased concentration of residual DCP byproducts. The excess crosslinks led to further disruption of the crystal structure, particularly at 7% DCP added and above where the spherulitic structure is destroyed and random assortments of lamellae are seen instead (*28*). As such, the increase in TCC magnitude and reduction of crystallinity arose from a combination of two factors. One was the increased extent of crosslinks which heavily limited the formation and growth of crystals, and the other was the increased concentration of residual DCP and its byproducts that promoted the formation of voids and amorphous regions in order to accommodate them.

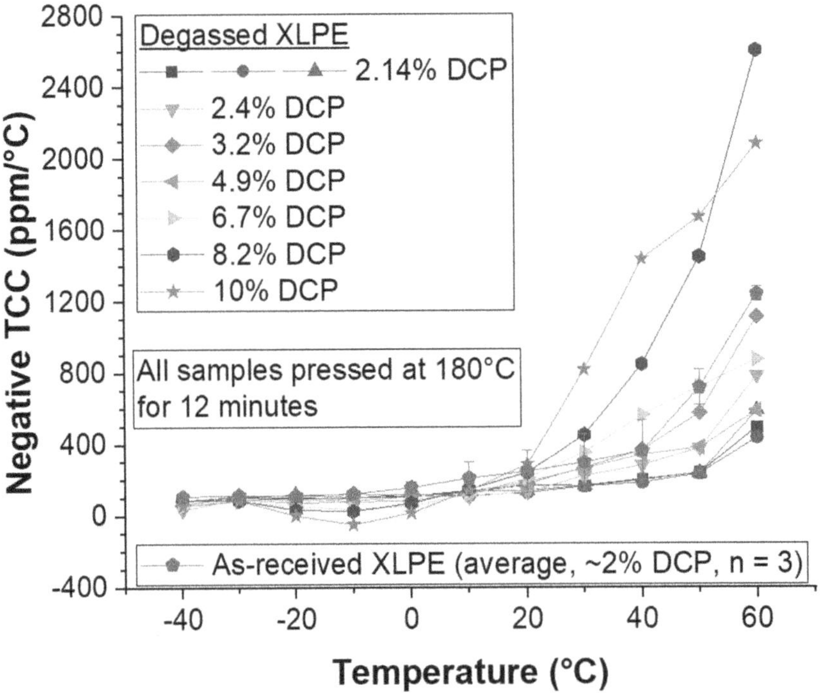

Figure 7. TCC comparison between degassed XLPE samples crosslinked with different concentrations of DCP added. A comparison to as-received samples has been included.

To confirm that these changes could be correlated to chemical residues, samples crosslinked with various amounts of DCP were also analyzed using FTIR, with the main results presented in other work (*27*). Being one of the most common byproducts, the concentration of ACP was used as a proxy for them all (*12*). Increasing the initial amount of DCP used led to gradual increases in the amount of residual byproducts, until a DCP concentration of around 8%. Beyond that point, the concentration of ACP was roughly constant. This matches up with the threshold for increased dielectric loss at room temperature (see Figure 8). As such, we determined that surpassing a certain threshold in DCP concentration is necessary for increases in dielectric loss to be from remnant byproducts in degassed XLPE.

Further analysis using TGA and FTIR was done on XLPE samples crosslinked with 13% DCP, far in excess of what would be used commercially, where one set of samples was degassed and the other was not. The results are shown in Figure A4. The TGA results showed that significant concentrations of byproducts could be removed via degassing, yet the FTIR results showed the retention of ACP. This was interpreted as there being a population of permanently trapped byproducts, a result that has been observed by other researchers (*29*). Based on the results in Figure 7, the population scales with the initial DCP concentration and so is best minimized by limiting the amount of DCP introduced in the first place.

This increase in the byproduct concentration would also be expected to increase the low frequency dielectric loss, and the data shown in Figure 8 clearly indicates that this occurred for DCP concentrations above 7-8%. Values for these and all samples discussed in the text are presented in Table A2. This also explains why XLPE made with low amounts of DCP appeared to be indistinguishable from LDPE, as the loss was governed primarily by factors intrinsic to the polymer and not by the byproduct concentrations.

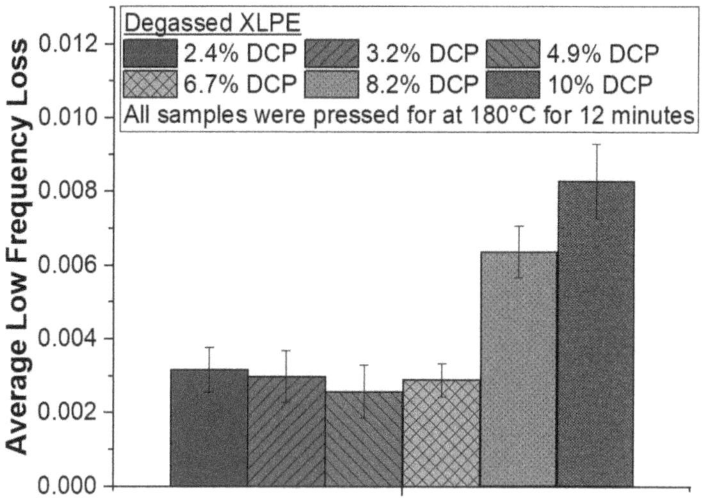

Figure 8. Average low frequency room temperature dielectric loss in degassed XLPE samples crosslinked with different DCP concentrations.

An analysis of these samples at higher temperatures was then done to extract the conductivity. The result is shown in Figure 9. Low frequency conductivity values could not be extracted from the sample crosslinked with 8.2% DCP. Here, the addition of more DCP, which was confirmed to lead to a greater concentration of byproducts, did not lead to enhanced conductivity. It may be tentatively surmised that the slightly reduced conductivity with greater initial DCP concentration was due to the extra crosslinks. Excess concentrations of DCP byproducts may not contribute much to quasi-DC conduction when compared to the changes in the AC dielectric loss discussed previously.

A comparison was also made between the as-received and degassed XLPE samples discussed previously and degassed XLPE exposed to excess concentrations of ACP. The excess ACP was added by soaking films in a bath of liquid ACP at a temperature of 50°C for two hours prior to adding electrodes. The capacitance and dielectric loss of these samples was measured around room temperature (nominally 20°C) and is displayed in Figure 10. The capacitance of the samples relative to the value obtained at 1 kHz is presented in order to allow a comparison between samples.

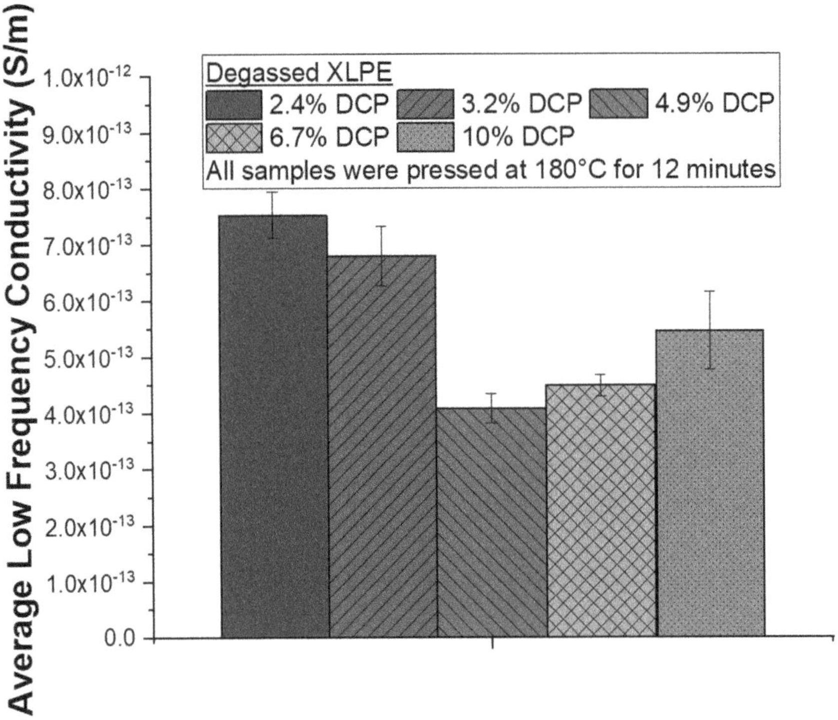

Figure 9. a) Average low frequency 90°C conductivity comparison between degassed XLPE samples crosslinked with different concentrations of DCP added.

It is well known that ACP can engage in a solvation reaction with water or antioxidants with oxygen or sulfur groups, producing ions and space charge in DC conditions (*7–9*). It was thus expected that intentionally retaining these ions and molecules should lead to significant changes not only in the dielectric loss and conductivity under AC conditions, but also in the capacitance. This is exactly what was observed. There were significant increases in both capacitance and loss around room temperature that can be correlated to the addition of ACP, a small polar molecule with a melting point of 20°C and a dielectric constant of 17-18, much larger than the 2.2 of polyethylene. The higher permittivity comes from its ketone group, which gives it the ability to have hydrogen bonds and electron pair interactions with solvents like water. Water has trace concentrations in polyethylene samples (*30*). As such, these results can be explained by the solvation reaction occurring within the XLPE samples. ACP will form hydrogen bonds with water (*31*), and the water molecules will congregate into a shell surrounding the ACP molecule. The formation of complexes of ACP and water led to a significant increase in sample polarizability, and thus the capacitance. The dielectric loss was also enhanced by these defect complexes and their ionic contributions to the AC conductivity. As for the TCC, the large values of the dielectric loss in those samples would be expected to provide a positive extrinsic contribution (*19*). Significant increases in the TCC magnitude would also be expected based solely on the addition of these small molecules based on results such as Figure 8, but that contribution would be negative rather than positive. It would thus be interesting future work to determine the temperature-dependence of these results for the ACP-soaked samples and the role that annealing procedures and temperature swings play in mitigating or enhancing conductivity losses caused by the presence of ACP and water.

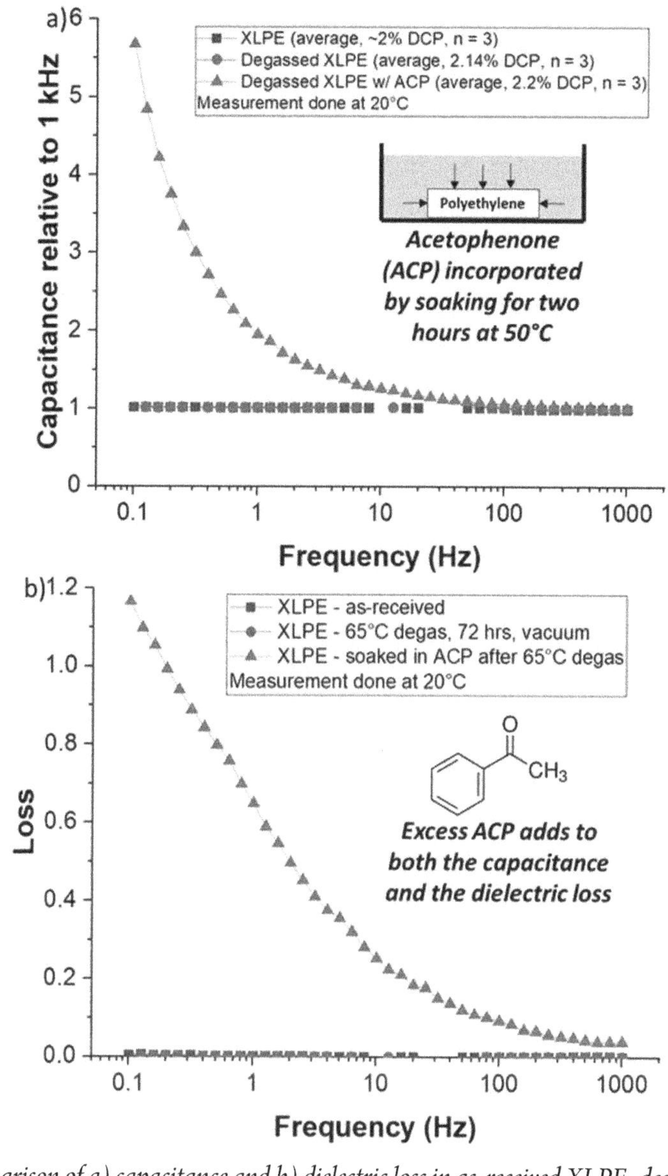

Figure 10. Comparison of a) capacitance and b) dielectric loss in as-received XLPE, degassed XLPE, and degassed XLPE after soaking ACP for two hours at 50 °C. The capacitance is normalized to the value obtained at an applied frequency of 1 kHz.

Conclusions

In summary, it is shown here that BDS of polyethylene can be used to examine changes in both the structure and the electrical properties. This is through the use of the TCC, the dielectric loss, and the AC conductivity. TCC for LDPE matches the linear thermal expansion coefficient of LDPE, as expected. XLPE sees an increase in the TCC magnitude, while degassing reduces it for both. This matches well with the relative changes in crystallinity and impurity concentrations for these samples. Thus, minimizing the TCC of XLPE and thus improving its temperature stability requires degassing. XLPE and LDPE have similar dielectric loss values at room temperature, which are reduced by similar extents with degassing. Broadly speaking, degassing continued to show itself as an effective tool in controlling the electrical properties. Meanwhile, XLPE is less conductive than

LDPE at elevated temperatures and is improved by degassing while LDPE sees no benefit. The choices of crosslinking temperature and DCP added prior to processing were seen to modify the structure as well as the dielectric and electrical properties of XLPE, and so should be chosen with care in practical work. Crosslinking below 140°C and DCP concentrations in excess of 8% lead to increased loss and morphological changes that were deemed not desirable for power cable operation. This is despite the reduced conductivity at elevated temperatures for high DCP concentrations. For this crosslinking agent, it will be important to find the optimum concentration – enough so that the reaction completes, but not so much that excess impurities result. For crosslinking temperature, it will be important to balance having a sufficient temperature with the energy costs associated elevated temperatures. There is likely also an optimum temperature, which could be determined through careful experiment. Additionally, an examination of capacitance and loss parameters in samples soaked with ACP showed a negative impact at room temperature, due to the buildup of mobile polar groups from a solvation reaction between it and water. Because of their generation of excess conductive loss, mobile polar groups – whether from impurities or from the material itself – must be removed from polymers meant for insulating applications. Future work was then suggested to determine the temperature-dependence of the conductive losses added by the solvation reaction, which is important to know for power cable operation. More generally, it will also be important to analyze the temperature coefficient of conductivity and the field dependence of conductivity, which also determine the electric fields within the insulator and shape its long-term stability.

Appendix

Table A1. Reference dielectric loss values for LDPE samples measured at 50 Hz.

Sample	*Dielectric Loss*	*Electric Field (kV/mm)*	*Temp.*	*Ref.*
LDPE	≈0.01	1	<50°C	(32)
LDPE	≈0.01	0.5	n/a	(33)
LDPE coated with ACP	≈0.3	0.5	n/a	(33)
LDPE	≈0.01	5	25°C	(26)
LDPE	≈0.01	40	25°C	(26)
LDPE with ACP (0.3 wt%)	≈0.01	5	25°C	(26)
LDPE with ACP (0.3 wt%)	≈0.2	40	25°C	(26)

Table A2. Room temperature dielectric loss for samples discussed in the main text.

Sample	*Dielectric Loss*
As-received LDPE and XLPE (Figure 2)	0.0103 ± 0.0028
Degassed LDPE and XLPE w/ ~2% DCP (Figure 2)	0.0023 ± 0.0006
Degassed XLPE w/ up to 6.7% DCP (Figure 8)	0.0029 ± 0.0002
Degassed XLPE w/ 8.2% DCP (Figure 8)	0.0064
Degassed XLPE w/ 10% DCP (Figure 8)	0.0083
Degassed XLPE soaked in acetophenone	0.4236 ± 0.0957

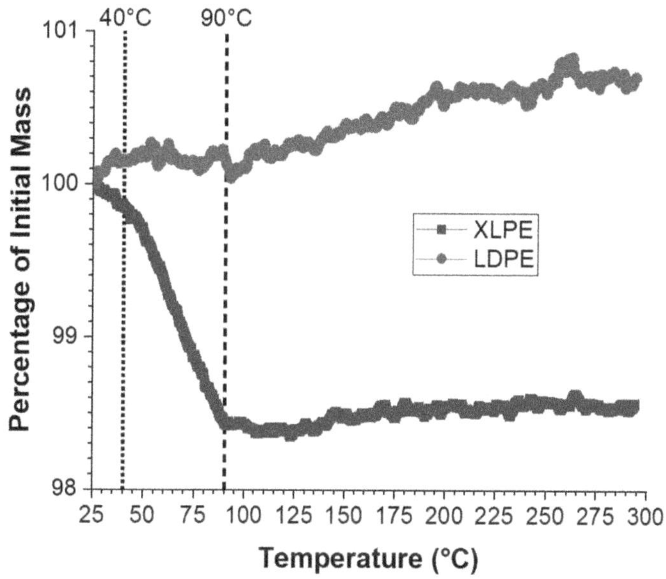

Figure A1. Mass change in as-received XLPE and LDPE as measured via TGA. DCP concentration used to generate the XLPE sample was approximately 2%.

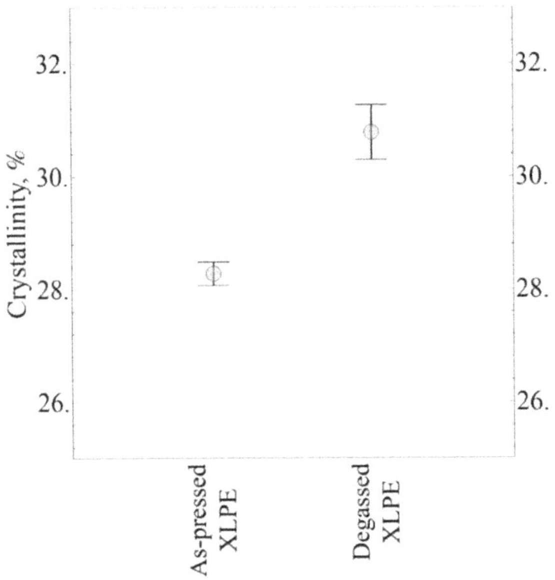

Figure A2. XLPE crystallinity comparison between as-pressed and degassed samples when both were pressed for 12 minutes at 180°C.

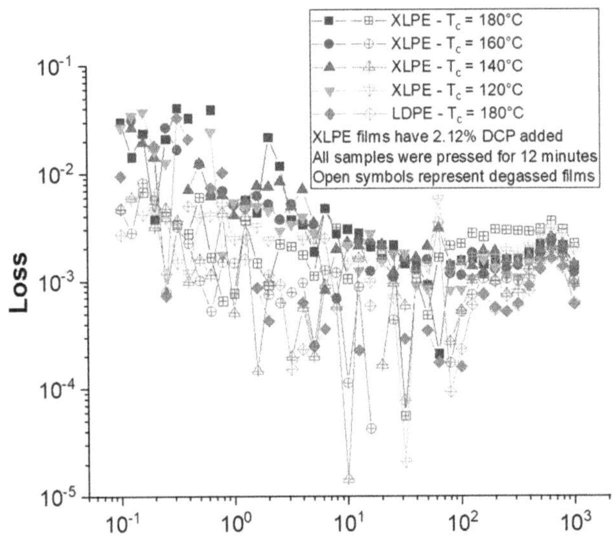

Figure A3. Comparison of raw broadband dielectric spectra for XLPE and LDPE samples crosslinked at various temperatures.

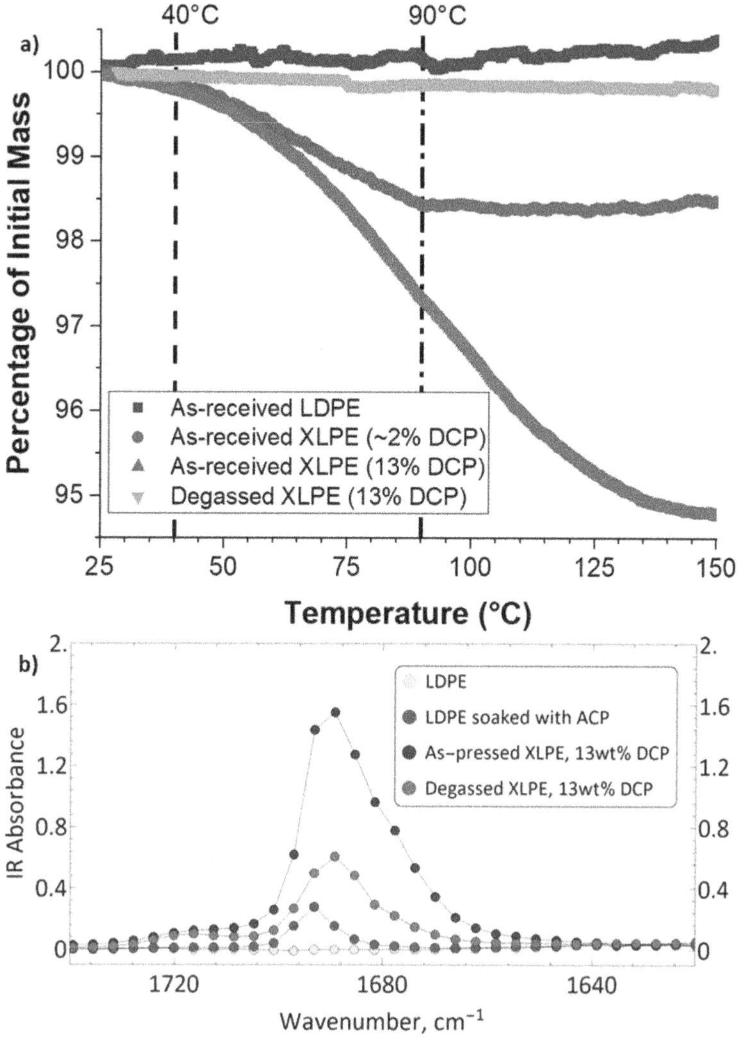

Figure A4. a) Mass change in as-received XLPE with different amounts of DCP as measured via TGA. An LDPE and a degassed XLPE sample were also prepared for comparative purposes; b) IR spectra comparison of concentration of ACP in XLPE (13% DCP) samples with and without degassing. Acetophenone IR peak at 1690 cm^{-1} (carbonyl bond stretching mode). A peak at 1720 cm^{-1} is related to the XLPE in-chain carbonyl bond (oxidation).

References

1. Sato, Y.; Yashiro, T. The Effect of Polar Groups on the Dielectric Loss of Polyethylene. *J. Appl. Polym. Sci.* **1978**, *22* (8), 2141–2153. https://doi.org/10.1002/app.1978.070220807.
2. Vahedy, V. Polymer Insulated High Voltage Cables. *IEEE Electr. Insul. Mag.* **2006**, *22* (3), 13–18. https://doi.org/10.1109/MEI.2006.1639025.
3. Boudou, L.; Guastavino, J. Influence of Temperature on Low-Density Polyethylene Films through Conduction Measurement. *J. Phys. D: Appl. Phys.* **2002**, 35 (13), 317. https://doi.org/10.1088/0022-3727/35/13/317.

4. Montanari, G. C.; Mazzanti, G.; Palmieri, F.; Motori, A.; Perego, G.; Serra, S. Space-Charge Trapping and Conduction in LDPE, HDPE and XLPE. *J. Phys. D: Appl. Phys.* **2001**, *34* (18), 2902–2911. https://doi.org/10.1088/0022-3727/34/18/325.
5. Hirai, N.; Minarni, R.; Tanaka, T.; Ohki, Y.; Okashita, M.; Maeno, T. Chemical Group in Crosslinking Byproducts Responsible for Charge Trapping in Polyethylene. *IEEE Trans. Dielectr. Electr. Insul.* **2003**, *10* (2), 320–330. https://doi.org/10.1109/TDEI.2003.1194118.
6. Wang, X.; Yoshimura, N.; Murata, K.; Tanaka, Y.; Takada, T. Space-Charge Characteristics in Polyethylene. *J. Appl. Phys.* **1998**, *84* (3), 1546–1550. https://doi.org/10.1063/1.368221.
7. Sekii, Y.; Taya, A.; Maeno, T. Effect of Antioxidants on Space Charge Generation in Cross-Linked Polyethylene and EPR. In *Annual Report - Conference on Electrical Insulation and Dielectric Phenomena, CEIDP*; IEEE: 2006. https://doi.org/10.1109/CEIDP.2006.312079.
8. Sekii, Y.; Ohbayashi, T.; Uchimura, T.; Hukuyama, T.; Maeno, T. A Study on the Space Charge Formation in XLPE. In *2001 Annual Report Conference on Electrical Insulation and Dielectric Phenomena (Cat. No.01CH37225)*; IEEE: 2002; pp 469–472. https://doi.org/10.1109/ceidp.2001.963583.
9. Sekii, Y.; Ohbayashi, T.; Uchimura, T.; Mochizuki, U.; Maeno, T. The Effects of Material Properties and Inclusions on the Space Charge Profiles of LDPE and XLPE. In *Annual Report Conference on Electrical Insulation and Dielectric Phenomena*; IEEE: 2002; pp 635–639. https://doi.org/10.1109/CEIDP.2002.1048876.
10. Tefferi, M.; Li, Z.; Cao, Y.; Uehara, H.; Chen, Q. Novel EPR-Insulated DC Cables for Future Multi-Terminal MVDC Integration. *IEEE Electr. Insul. Mag.* **2019**, *35*, 20–27. https://doi.org/10.1109/MEI.2019.8804331.
11. Mauri, M.; Peterson, A.; Senol, A.; Elamin, K.; Gitsas, A.; Hjertberg, T.; Matic, A.; Gkourmpis, T.; Prieto, O.; Müller, C. Byproduct-Free Curing of a Highly Insulating Polyethylene Copolymer Blend: An Alternative to Peroxide Crosslinking. *J. Mater. Chem. C* **2018**, *6*, 11292–11302. https://doi.org/10.1039/c8tc04494e.
12. Sahyoun, J.; Crepet, A.; Gouanve, F.; Keromnes, L.; Espuche, E. Diffusion Mechanism of Byproducts Resulting from the Peroxide Crosslinking of Polyethylene. *J. Appl. Polym. Sci.* **2017**, *134* (9), 44525. https://doi.org/10.1002/app.44525.
13. Andrews, T.; Hampton, R. N.; Smedberg, A.; Wald, D.; Waschk, V.; Weissenberg, W. The Role of Degassing in XLPE Power Cable Manufacture. *IEEE Electr. Insul. Mag.* **2006**, *22* (6), 5–16. https://doi.org/10.1109/MEI.2006.253416.
14. Liu, T.; Fothergill, J.; Dodd, S.; Nilsson, U. Dielectric Spectroscopy Measurements on Very Low Loss Cross-Linked Polyethylene Power Cables. *J. Phys.: Conf. Ser.* **2009**, *183*, 012002. https://doi.org/10.1088/1742-6596/183/1/012002.
15. Fothergill, J. C.; Dodd, S. J.; Dissado, L. A.; Liu, T.; Nilsson, U. H. The Measurement of Very Low Conductivity and Dielectric Loss in XLPE Cables: A Possible Method to Detect Degradation Due to Thermal Aging. *IEEE Trans. Dielectr. Electr. Insul.* **2011**, *18* (5), 1544–1553. https://doi.org/10.1109/TDEI.2011.6032823.
16. Tamayo, I.; Belana, J.; Mudarra, M.; Diego, J. A.; Sellarés, J. Thermally Stimulated Depolarization Currents of Crosslinked Polyethylene Relaxations in the Fusion Range of Temperatures. *J. Polym. Sci., Part B: Polym. Phys.* **2003**, *41* (12), 1412–1421. https://doi.org/10.1002/polb.10489.

17. Diego, J. A.; Belana, J.; Òrrit, J.; Sellarès, J.; Mudarra, M.; Cañadas, J. C. TSDC Study of XLPE Recrystallization Effects in the Melting Range of Temperatures. *J. Phys. D: Appl. Phys.* **2006**, *39* (9), 1932–1938. https://doi.org/10.1088/0022-3727/39/9/031.
18. Boudou, L.; Guastavino, J. Influence of Temperature Treatment on the Electrical Properties of Low-Density Polyethylene. *J. Phys. D: Appl. Phys.* **2000**, *33* (21), L129–L131. https://doi.org/10.1088/0022-3727/33/21/102.
19. Cockbain, A. G.; Harrop, P. J. The Temperature Coefficient of Capacitance. *J. Appl. Phys.* **1968**, *1* (2), 1109–1115. https://doi.org/10.1085/jgp.1.3.295.
20. White, G. K.; Choy, C. L. Thermal Expansion and Grüneisen Parameters of Isotropic and Oriented Polyethylene. *J. Polym. Sci., Polym. Phys. Ed.* **1984**, *22* (5), 835–846. https://doi.org/10.1002/pol.1984.180220505.
21. Kardos, J. L.; Raisoni, J.; Piccarolo, S.; Halpin, J. C. Prediction and Measurement of the Thermal Expansion Coefficient of Crystalline Polymers. *Polym. Eng. Sci.* **1979**, *19* (14), 1000–1009. https://doi.org/10.1002/pen.760191407.
22. Aida, F.; Wang, S.; Fujita, M.; Tanimoto, G. Space Charge Behavior in Polyethylene: Influence of Moisture and Temperature. *IEEJ Trans. Fundam. Mater.* **1997**, *117* (9), 922–929. https://doi.org/10.1541/ieejfms1990.117.9_922.
23. Gulmine, J. V.; Akcelrud, L. Correlations between the Processing Variables and Morphology of Crosslinked Polyethylene. *J. Appl. Polym. Sci.* **2004**, *94* (1), 222–230. https://doi.org/10.1002/app.20863.
24. Jonscher, A. K. A New Understanding of the Dielectric Relaxation of Solids. *J. Mater. Sci.* **1981**, *16* (8), 2037–2060. https://doi.org/10.1007/BF00542364.
25. Greenhoe, B. M.; Hassan, M. K.; Wiggins, J. S.; Mauritz, K. A. Universal Power Law Behavior of the AC Conductivity versus Frequency of Agglomerate Morphologies in Conductive Carbon Nanotube-Reinforced Epoxy Networks. *J. Polym. Sci., Part B: Polym. Phys.* **2016**, *54* (19), 1918–1923. https://doi.org/10.1002/polb.24121.
26. Nakatsuka, T.; Takahashi, T.; Miyata, H.; Yokoyama, A.; Ishikawa, I.; Niwa, T. Effect on Dielectric Loss of Polyethylene Caused by Acetophenone and Cumylalcohol. In *Proceedings of 1994 IEEE International Symposium on Electrical Insulation*; IEEE: 1994; pp 574–577.
27. Akbarian, D.; Hamedi, H.; Damirchi, B.; Yilmaz, D. E.; Penrod, K.; Woodward, W. H. H.; Moore, J.; Lanagan, M. T.; van Duin, A. C. T.; Damirichi, B.; Yilmaz, D. E.; Penrod, K.; Woodward, H.; Moore, J.; Lanagan, M. T.; van Duin, A. C. T. Atomistic-Scale Insights into the Crosslinking of Polyethylene Induced by Peroxides. *Polymer (Guildf).* **2019**, *183*, 121901. https://doi.org/10.1016/j.polymer.2019.121901.
28. Nilsson, S.; Hjertberg, T.; Smedberg, A. Structural Effects on Thermal Properties and Morphology in XLPE. *Eur. Polym. J.* **2010**, *46* (8), 1759–1769. https://doi.org/10.1016/j.eurpolymj.2010.05.003.
29. Chong, Y. L.; Chen, G.; Hosier, I. L.; Vaughan, A. S.; Ho, Y. F. F. Heat Treatment of Cross-Linked Polyethylene and Its Effect on Morphology and Space Charge Evolution. *IEEE Trans. Dielectr. Electr. Insul.* **2005**, *12*, 1209–1221. https://doi.org/10.1109/TDEI.2005.1561801.
30. McCall, D. W.; Douglass, D. C.; Blyler, L. L.; Johnson, G. E.; Jelinski, L. W.; Bair, H. E. Solubility and Diffusion of Water in Low-Density Polyethylene. *Macromolecules* **1984**, *17* (9), 1644–1649. https://doi.org/10.1021/ma00139a001.

31. Lei, J.; Zhang, J.; Feng, G.; Grabow, J. U.; Gou, Q. Conformational Preference Determined by Inequivalent N-Pairs: Rotational Studies on Acetophenone and Its Monohydrate. *Phys. Chem. Chem. Phys.* **2019**, *21* (41), 22888–22894. https://doi.org/10.1039/c9cp03904j.
32. Tanimoto, G.; Okashita, M.; Aida, F.; Fujiwara, Y. Temperature Dependence of Tan Delta in Polyethylene. In *Proceedings of the 3rd International Conference on Properties and Applications of Dielectric Materials*; IEEE: 2002; pp 1068–1071. https://doi.org/10.1109/icpadm.1991.172259.
33. Aida, F.; Tanimoto, G.; Aihara, M.; Hosokawa, E. Influence of Curing By-Products on Dielectric Loss in XLPE Insulation. In *Annual Conference on Electrical Insulation and Dielectric Phenomena*; IEEE: 1990; pp 465–473. https://doi.org/10.1109/CEIDP.1990.201383.

Applied Dielectric Spectroscopy of Aqueous Systems

Chapter 12

Isotope Effect on the Dynamics of Hydrophilic Solutions at Supercooled Temperatures

Jorge H. Melillo[1] and Silvina Cerveny[1,2,*]

[1]Centro de Física de Materiales (CSIC-UPV/EHU)-Material Physics Centre (MPC), Paseo Manuel de Lardizábal 5, 20018 San Sebastián, Spain

[2]Donostia International Physics Center (DIPC), Paseo Manuel de Lardizabal 4, 20018 San Sebastián, Spain

[*]Email: silvina.cerveny@ehu.es

Broadband dielectric spectroscopy (BDS) is a highly versatile technique to analyze the dynamics of any material with a dipolar moment and therefore is very useful to study water solutions. Numerous studies have been performed in the last few years concerning the dynamics of water in hydrophilic solutions. Here, we examine the isotope effects on the dynamics of water in hydrophilic solutions at supercooled temperatures using BDS. The solutes used were tripropylene glycol (3PG) and pentaethylene glycol (5EG) in solution with three isotopes of water (H_2O, D_2O and $H_2{}^{18}O$). We analyze the effects of the mass and moment of inertia as well as the impact of nuclear quantum effects (NQEs) on the dynamics of water. We show that for low water contents, the classical mass effect dominates the response, whereas for high water contents, NQEs drive the difference between the relaxation times of H_2O and D_2O. In addition, comparisons between solutions made with H_2O and $H_2{}^{18}O$ show that the water molecule translational mobility freezes independently of the water content of the solutions.

Introduction

Water, known as the *universal solvent*, can dissolve a wide range of solutes (ions and polar molecules), making it highly relevant in different aspects of life. On a biological level, it can assist the transport of nutrients and minerals in the human body, and it participates in many metabolic processes. On an industrial level, water is used in the manufacturing of semiconductors, food, and paper, among several other products. On an ecosystem level, water boosts plant growth or supplies a temporary habitat for insects or amphibians. It is commonly believed that the physical properties and chemical composition of water make it an exceptional solvent. In particular, water molecules are polar, which allows the formation of electrostatic interactions (such as hydrogen bonds and hydrophobic interactions) with ions and other polar molecules. The existence of life on Earth depends on the ability of water to dissolve polar molecules. In a solution, several water molecules surround the solute molecules. The interactions between water and solute molecules lead to the formation of a three-dimensional network, which is called the solvation shell or hydration shell. The hydration shell is formed by water layers, and in the case of proteins in the human body, this interfacial region comprises up to 6 layers (*1*). Far away from the hydration shell, it is possible to find water molecules interacting with other water molecules, which leads to bulk-like water (see Figure 1). For these water molecules, the influence of the solute is less important than for water involved in hydration shells (*2*, *3*). The hydration shell and bulk-like water are relevant in proteins and peptides since both their structures and functionalities are substantially influenced by solvent molecules. Therefore, the dynamics of these water molecules that interact with different types of solutes (biological or nonbiological) have been the focus of many research scientists for many years with an emphasis on the supercooled temperature region.

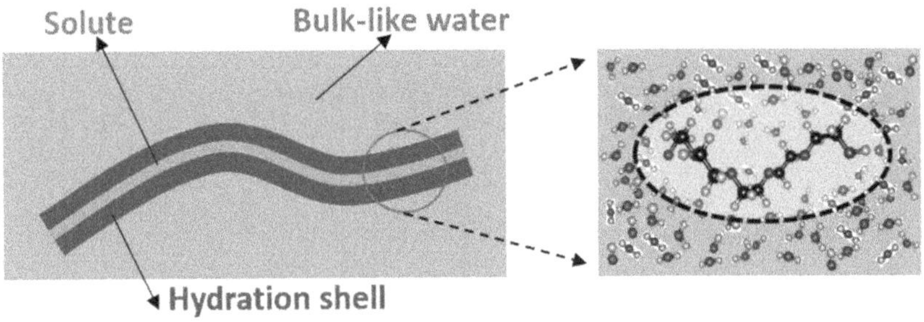

Figure 1. Scheme of a solute molecule (in gray) in solution. The hydration layer formed by some water layers (in red) and the bulk-like water (in blue) can be distinguished.

Supercooling occurs when a liquid does not freeze, although its temperature is below its freezing point. The fact that water can enter a supercooling region (*4*) has important implications for the food industry, for the survival of plants and animals at extremely low temperatures, and for atmospheric phenomena (*5*). In addition, the interest of physicists in the supercooled region is based on the unique physical and chemical properties of water in this temperature range because water exhibits anomalies in dynamical, structural and thermodynamic properties (*6*, *7*). In particular, in the "no man's land" (*8*) (between 150 and 235 K), some of the physical properties (for instance, the value of the glass transition temperature (T_g)) are under debate in the literature (*9*). In this temperature region, it was proposed that bulk water undergoes a fragile-to-strong transition that is observed as a crossover in the temperature dependence of the α-relaxation associated with a liquid-liquid (from high density to low density) phase transition (*10*). However, it is difficult to experimentally access this temperature region because of the homogeneous ice nucleation of water. Therefore, the T_g value of bulk water and the existence of the liquid-liquid transition continue to be debated in the literature (*11–13*).

Water is a small polar molecule consisting of two light hydrogen atoms, each of which is attached to a 16-fold heavier oxygen atom with a high dipole moment (1.85 D in the gas phase and higher in the liquid phase). When water molecules are in the liquid state, the dipoles interact with each other, and therefore, water molecules can form an extended hydrogen-bonded network. This hydrogen-bonded network is assumed to be responsible for the unusual properties of liquid water. In the solid phase (ice Ih), water molecules form a tetrahedral network, whereas in the liquid phase, thermal fluctuations can disturb the hydrogen-bonded network, which fluctuates at picosecond to nanosecond timescales. These fluctuations can be measured by numerous spectroscopic techniques (for instance, broadband dielectric spectroscopy (BDS), nuclear magnetic resonance (NMR) or quasielastic neutron scattering (QENS)). These three mentioned spectroscopic methods can access the molecular mobility of the molecules under study (*11*).

To deeply understand the dynamics of water, isotope substitution can help in determining the origin of the relaxation processes. In fact, isotope effects have been used to obtain evidence on the structure of liquid water (*14, 15*) as well as to understand the nature of its relaxation processes (*16, 17*). Hydrogen (H) and oxygen (O) exist in nature in the form of different isotopes, and the most stable isotopes are ^{16}O, ^{17}O, and ^{18}O and ^{1}H and ^{2}H (D, deuterium). Consequently, water has more than one naturally occurring stable isotope. Light (H_2O) and heavy (D_2O, $H_2{}^{18}O$) water types differ in their physical properties. The extra neutrons in D_2O and $H_2{}^{18}O$ make the density of heavy water approximately eleven percent higher than the density of H_2O (*18*). The melting points are also different: 0 °C for H_2O and $H_2{}^{18}O$ and 3.81 °C for D_2O. However, the viscosities of $H_2{}^{18}O$ and H_2O are very similar to each other, whereas the viscosity of D_2O is 24% higher (*19*). In this regard, Cho et al. (*20*) and Harris and Woolf (*21*) studied the viscosity of D_2O and H_2O as a function of temperature. They proposed that the properties of both isotopes are nearly identical when shifting the temperature by approximately 6.5 degrees. Later, it was observed that the relaxation times of H_2O and D_2O (close to room temperature) in the GHz-THz frequency range (*16*) are identical over a temperature shift of 7.2 K. Moreover, the temperature dependence of the dielectric relaxation times (τ) of low density amorphous ice (LDA) was also studied under isotope substitution (*22*). It was observed that τ strongly increases under the substitution of H $\rightarrow$ D with a shift factor of approximately 10 degrees. In addition to the dynamic properties, the thermodynamic and static properties of water show some differences considering the different isotopes (*23*). Moreover, some controversial questions exist (*24, 25*) related to the mechanism of unfreezing at the glass transition temperature of LDA, and isotopic substitution can help elucidate this topic.

The difference observed in the dynamical properties when considering different isotopes (H_2O, $H_2{}^{18}O$ and D_2O) can be a consequence of the increase in the mass or the moment of inertia of the molecules (*26*). When D_2O is considered, there is a deceleration in the dynamics because of the increase in mass with respect to H_2O. However, all these differences observed in the dynamical properties and in the static and thermodynamical properties under the substitution H$\rightarrow$D can be considered a consequence of many nuclear quantum effects (*23, 27*) (NQEs) produced in water. At low temperatures and due to the small mass of hydrogen atoms, tunneling and proton delocalization (*28, 29*) can be significant, whereas at room temperature, zero-point energy (small H atoms have a greater frequency of vibration and therefore higher zero-point energy) effects can take place. Substantial computational, theoretical and experimental studies have been performed on the NQEs of water (see the reviews (*23, 27, 30*) for a complete list of references), but in comparison, experimental studies of the NQEs on the dynamic properties of water in solutions at supercooled temperatures are limited.

In this chapter, we first review some early works in the field of the dynamics of water in solutions at supercooled temperatures as observed by BDS. Then, we provide new experimental results on the dynamics of two hydrophilic liquids, tripropylene glycol (3PG) and pentaethylene glycol (5EG), mixed with the three isotopologues (H_2O, $H_2{}^{18}O$ and D_2O). Using these three solvents, we can analyze the influence of the mass, the effects due to the change in the moment of inertia and the NQEs on the dynamics of amorphous water in solutions of two hydrophilic solutes.

Experimental Techniques

Samples

Water (^{16}O, product 95284), deuterium oxide (D_2O, product 151882) and $H_2{}^{18}O$ (^{18}O, product 329878) were purchased from Sigma Aldrich and used without any other purification. 3PG (M_w = 192 g/mol) and 5EG (M_w = 238 g/mol) were also purchased from Aldrich Chemical and were pretreated with analytical grade mixed-bed resin (AG 501-X8, Bio-Rad Laboratories) for 15 min and then lyophilized. We transferred the cleaned and lyophilized 3PG and 5EG to a glove box for preparation of the solutions where $H_2{}^{16}O$, $H_2{}^{18}O$ and D_2O were added. The mixtures were sealed for one month to achieve a good water distribution. Both 3PG and 5EG solutions with different isotopes were prepared with different water contents.

Calorimetric Experiments

The differential scanning calorimetric (DSC) measurements described here were performed using a DSC Q2000 from TA Instruments in standard mode. Samples weighing approximately 10 to 15 mg were prepared in hermetic pans. DSC experiments were performed at heating and cooling rates of 5 and 10 K/min, as indicated in each experiment. A helium flow rate of 25 mL/min was used throughout the experiments. In addition, the melting point of the three liquids was measured at 1 K/min.

Dielectric Experiments

BDS is a broadband (10^{-6} to 10^{12} Hz) and powerful technique to evaluate the dielectric properties of aqueous solutions and to assess the molecular dynamics at various time and length scales. BDS is based on the interaction between an external electric field and the electric dipoles of the material under study (*31*). Both dielectric dispersion and absorption phenomena occur due to the dipole relaxation arising from the reorientational motions of molecular dipoles and electrical conduction arising from translational motions of electric charges (ions, electrons).

To measure the complex dielectric permittivity, $\varepsilon^*(\omega) = \varepsilon'(\omega) - i\,\varepsilon''(\omega)$, we used a Novocontrol Alpha-Analyzer (0.01 to 10^6 Hz). The solutions were placed between parallel gold-plated electrodes with a diameter of 30 mm. The electrodes gap was adjusted by fused silica spacer fibers. Therefore, the sample thickness for all measurements was 0.1 mm and the sample diameter was 30 mm. The cryostat was pre-cooled at 140 K and the sample was cooling at a rate of approximately of 10-15 K/min. Isothermal frequency scans recording $\varepsilon^*(\omega)$ were performed every 5° over the temperature range of 150–180 K and every 3° over the temperature range of 183-250 K. The sample temperature was controlled with stability better than ±0.1 K.

To analyze the complex permittivity (ε^*), fitting of the imaginary (ε'') component was performed by using the phenomenological Havriliak-Negami (HN) function (*31*)

$$\varepsilon^*(\omega) = \varepsilon'(\omega) + i\varepsilon''(\omega) = \varepsilon_\infty + \frac{\Delta\varepsilon}{[1 + (i\omega\tau)^\alpha]^\beta} \tag{1}$$

where $\Delta\varepsilon$ is the dielectric strength, ε_∞ is the unrelaxed value of the dielectric constant, τ is the relaxation time and $\omega = 2\pi f$ is the angular frequency. In eq. 1, α and β are shape parameters ($0 < \alpha$, $\alpha\beta < 1$) that describe the symmetric and asymmetric broadening of the equivalent relaxation time distribution function. By setting $\alpha = \beta = 1$, a Debye function is obtained, whereas by setting $\beta = 1$, a symmetrical function is obtained (the Cole-Cole (CC) function), which is widely used to describe secondary relaxations in glassy materials (*31*). At low frequencies, the conductivity effects dominate, and to account for that, a power law term was added to the sum of CC and/or HN functions.

Dynamics of Water Solutions at Supercooled Temperatures

In this section, we briefly review the more important results related to the dynamics of aqueous solutions at supercooled temperatures where the solute is hydrophilic. We will discuss only the case in which water does not show any crystallization upon cooling/heating. This condition is normally fulfilled for solutions with water contents lower than 50 wt% (*32, 33*). The dynamics of water in solutions have been extensively studied by BDS. The solutes analyzed encompass glass-forming materials (for instance, alcohols (*34*)), synthetic polymers (*32–34*), biopolymers (amino acids (*35–37*), peptides (*38*), or proteins (*39, 40*)) and more complex solutes such as blood (*41, 42*), food (*43, 44*), tissues (*45, 46*), and cells (*47, 48*).

When analyzing the dynamics of a solution (two miscible components, water + solute), two scenarios can be distinguished. For some solutes, we can find two significant relaxations whereas, for other types of solutes, three relevant relaxations can be found. First, we focus on solutions where two relaxations are observed (*49–51*). The slower relaxation corresponds to the α-relaxation, which is cooperative, and its shape is asymmetric. It exhibits a super-Arrhenius temperature-dependent relaxation time that extrapolates to the glass transition temperature (T_g, determined by calorimetric experiments) at 100 s. The secondary relaxation or β-relaxation is faster than α-relaxation, and its characteristics are different from those found in α-relaxation (*31*). The relaxation time shows an Arrhenius temperature dependence, and the shape of the relaxation is symmetric. For most of the aqueous solutions investigated, α-relaxation is related to the mobility of the solute molecules affected by the surrounding water molecules. The water molecules, in this case, act as a plasticizer (*52*) and accelerate the solute dynamics. Below the glass transition of the solutions, the mobility of the solute molecules is frozen, but water still has some mobility at low temperatures; therefore, the dynamics of amorphous water can be investigated. Because water is analyzed when the solute molecules are frozen, this type of confinement is usually referred to as *soft confinement* (*49*), in contrast to *hard confinement,* such as found in rocks or metals (*53, 54*).

The temperature dependence of the relaxation times for this type of solution can be schematically seen in Figure 2a, where α-relaxation and the fast solvent relaxation are displayed. The origin of α-relaxation is the glass transition-related structural relaxation of the solution due to its Vogel–Fulcher–Tammann (*55, 56*) (VFT) temperature dependence and the fact that it reaches a time scale of approximately 100 s at the calorimetric T_g. Increasing the water content in the mixtures, the α-relaxation becomes faster because water acts as plasticizer of the solute molecules (see Figure 2b). The fast solvent relaxation (called fast relaxation (*57*), ν-relaxation (*58*) or just water relaxation, depending on the authors) shown in Figure 2c is present below T_g and shows an Arrhenius temperature dependence. The intensity of this relaxation increases with increasing water concentration (*59, 60*), so at least at higher water contents, this relaxation is mainly caused by water (at very low water contents, a large contribution may come from the solute). The fast solvent relaxation speed ups with water content, as shown in Figure 3c. In water-rich mixtures (i.e., water content > 30 wt%), the self-association of water molecules is expected, and in this case, we found very similar activation energies (~ 0.5 eV) (*49*). This activation energy is independent of the solute, and therefore, the fast water process is universal. When increasing the temperature (*59*), there is a deviation in the Arrhenius behavior at a temperature that coincides with T_g, where decoupling of the local water relaxation from the structural α-relaxation occurs (*32, 49*).

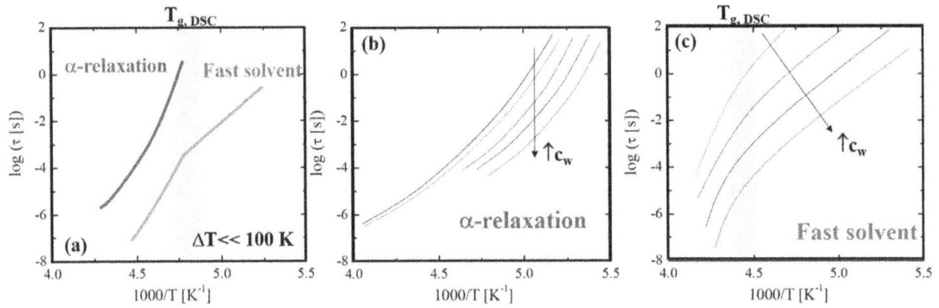

Figure 2. (a) Temperature dependence of the relaxation times for water solutions with ΔT_g << 100 K where a single water relaxation (fast solvent) is observed. (b and c) Concentration dependence of the relaxation times corresponding to α-relaxation and fast solvent relaxation where the same colors correspond to the same water concentration (c_w). Both relaxations accelerate with increasing concentration.

On the other hand, we recently found that certain hydrophilic solutions exhibit two water relaxations (*36–38*) instead of a single relaxation, as explained above. The appearance of this second water relaxation depends on how large the difference is in the value of the glass transition between the dry and wet materials ($\Delta T_g = T_{g,dry} - T_{g,wet}$). If ΔT_g is 100 K or more, two water relaxations can be detected, whereas for ΔT_g << 100 K, a single relaxation is observed (as shown in Figure 2) (*57, 61*). In Figure 3, we show the concentration dependence of T_g, where the two different scenarios for soft materials are displayed. Among the hydrophilic solutes with T_g << 100 K, we can find liquids such as 3PG (*59*) or polymers such as poly (vinyl methyl ether) (*60*) that display a linear dependence of T_g on the water concentration. On the other hand, two water relaxations were previously found in protein solutions (for instance, bovine serum albumin (*58*) or gelatin (*62*)), and it was believed that the occurrence of two water relaxations was exclusively observed in aqueous solutions of biological molecules. However, these two water relaxations were also recently found in lysine peptides (*36, 38*) and synthetic polymers such as poly(vinyl pyrrolidone) (*61*) (PVP) and other hydrophilic solutes under the condition of ΔT_g >> 100 K. It is important to note that the occurrence of two water relaxations was not reported in previous studies on PVP (*63*), in which only a single relaxation was observed; this issue has been revisited in recent papers by us (*57*) and other authors (*64*).

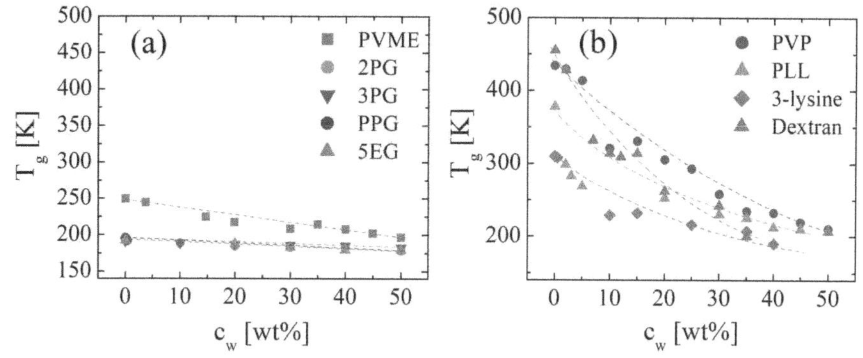

Figure 3. Concentration dependence of the glass transition temperature (T_g) of different solutions. (a) Poly(vinyl methyl ether) (PVME), poly(propylene glycol) (PPG), oligomers of propylene glycol (nPG), and pentaethylene glycol (5EG). (b) Poly(vinyl pyrrolidone), ε-poly(lysine), dextran, and 3-lysine. Adapted with permission from reference (61). Copyright 2019 American Institute of Physics (AIP).

The temperature dependence of the relaxation times for the second type of solutions can be schematically seen in Figure 4a, where the three processes are displayed and are named from the slowest to the fastest as α-, slower water- and faster water- relaxations (57). Two of these three relaxations have the same origin and characteristics as those discussed in Figure 2a: the slowest relaxation is the α-relaxation of the solute molecules, whereas the fast relaxation is due to the relaxation of water molecules. The slow water relaxation has different characteristics from the fast water relaxation. To observe this relaxation, a minimum number of water molecules need to be produced (this relaxation appears (57) at approximately 20 wt. % of water), and therefore, it is likely that there is some kind of connectivity between the water molecules. The time scale of the slow relaxation is strongly dependent on the solute molecule, and therefore, this water relaxation probably does not involve only the water motions but is more likely a collective and coupled motion of water molecules and surface groups of the solute molecules. In fact, in NMR studies (65) on one of these systems (ε-poly(lysine)), it is observed that the dynamics of polar peptide groups are involved in this relaxation, and only a small water fraction contributes to this process.

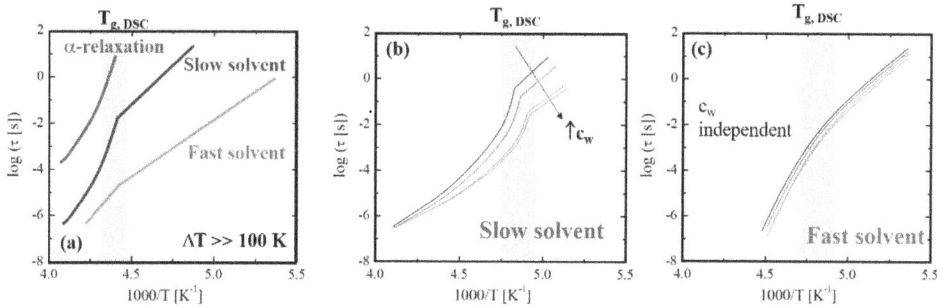

Figure 4. (a) Temperature dependence of the relaxation times of water solutions with $\Delta T_g >> 100$ K where apart from the α-relaxation, two water relaxations are observed (fast and slow solvent relaxations). (b) Concentration dependence of the relaxation times corresponding to the slow solvent relaxation. (c) Concentration dependence of the relaxation times corresponding to fast solvent relaxation. In (b) and (c), the same colors correspond to the same water concentration (c_w).

In both types of solutions (with one or two water relaxations), the experimental results have shown that the dynamics of fast water relaxation for different solutes and at high water contents are rather universal (*11*). Therefore, in the following section, we will mainly analyze the isotope effect on the dynamics of fast water relaxation.

Dynamics of Hydrophilic Solutions with Three Different Isotopes at Supercooled Temperatures

We now analyze the influence of three different isotopes (H_2O, $H_2{}^{18}O$ and D_2O) on the dynamics of water in solutions of 3PG (and 5EG) at supercooled temperatures. 3PG (see Figure 5) with the molecular formula of $C_9H_{20}O_4$ is a colorless liquid and is highly soluble in water, with two OH groups that can form hydrogen bonds with water molecules both as a donor and as an acceptor. 5EG ($C_{10}H_{22}O_6$) is an alcoholic compound with two –OH groups per molecule. Although the structure of 3PG with water has not been studied in the literature, propylene glycol has been analyzed by neutron diffraction. It was found that at a relatively high water content (40 wt%), hydration interactions of PG are dominant over PG self-self interactions in a solution (*66*), and a similar behavior is expected for 3PG. Moreover, previous dielectric studies (*59*) have shown that at high water contents ($c_w > 30$ wt%), water-water interactions prevail.

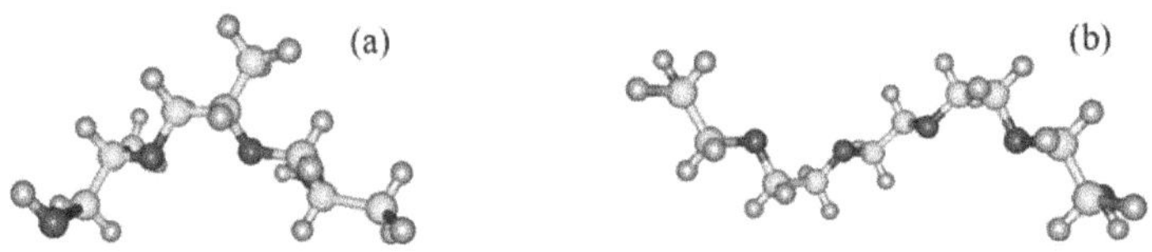

Figure 5. (a) Molecule of tripropylene glycol. (b) Molecule of pentaethylene glycol (n = 5).

Tables 1 and 2 show the samples that were analyzed, from the dried 3PG (and EG) to solutions with 50 wt% water. It is important to note that, to compare the behavior of the different solutions, the content in weight of H_2O, $H_2{}^{18}O$ and D_2O is slightly different to obtain solutions with the same number of solvent molecules per solute molecule (N).

Effect of Isotope Substitution on the Calorimetric Glass Transition Temperature

Figure 6a shows the heat flow versus temperature for 3PG solutions with $c_w = 35$ wt% using the three different isotopes. At a cooling rate of 10 K/min, it is possible to avoid crystallization upon cooling for all the solutions to obtain a glass phase at low temperatures. Therefore, there is no isotope effect in the freezing conditions of these samples. In the heating scans, we can observe the glass transition (T_g), which is very similar for 3PG-H_2O and 3PG-$H_2{}^{18}O$, whereas for 3PG-D_2O, T_g is shifted to higher temperatures. The T_g values of the samples were determined as the onset of the heat flow step, and the resulting T_g values are shown in Table 1 and Figure 6b for all the water contents.

Table 1. The glass transition temperature ($T_{g,DSC}$) of 3PG protonated and deuterated and 3PG solutions measured by DSC at 10 K/min; $\Delta T_{g,iso}$ ($\Delta T_{g,iso} = T_{g,H2O} - T_{g,D2O} = T_{g,H2^{18}O} - T_{g,D2O}$) indicates the difference between the glass transition of solutions with H_2O or $H_2{}^{18}O$ and D_2O. N indicates the number of water molecules per solvent molecule.

Sample	*Isotope*	*$T_{g,DSC}$ [K]*	*N*	*$\Delta T_{g,iso}$ [K]*
3PG	protonated	193.7	---	0.2 ± 0.2
3PG	deuterated	193.9	---	
3PG85-W15	H_2O	189.6	1.88	0.6 ± 0.1
	$H_2{}^{18}O$	189.6		
	D_2O	190.2		
3PG65-W35	H_2O	188.1	5.75	1.2 ± 0.2
	$H_2{}^{18}O$	188.2		
	D_2O	189.4		
3PG60-W40	H_2O	187.0	7.12	1.4 ± 0.2
	$H_2{}^{18}O$	187.1		
	D_2O	188.5		
3PG50-W50	H_2O	182.6	10.6	1.4 ± 0.3
	D_2O	184.0		

Table 2. Glass transition temperature ($T_{g,DSC}$) of 5EG measured by DSC at 10 K/min; $\Delta Tg,iso = T_{g,H2O} - T_{g,D2O}$ indicates the difference between the glass transition of solutions with H_2O and D_2O.

Sample	*Isotope*	*$T_{g,DSC}$ [K]*	*N*	*$\Delta Tg,iso$ [K]*
5EG	---	190.4	---	----
90-5EG-10-D_2O	D_2O	192.1	1.5	0.5 ± 0.1
90-5EG-10- H_2O	H_2O	191.6		
80-5EG-20-D_2O	D_2O	185.7	3.3	1.1 ± 0.2
80-5EG-20-H_2O	H_2O	184.6		
60-5EG-40-D_2O	D_2O	176.7	8.8	2.7 ± 0.2
60-5EG-40-H_2O	H_2O	174.0		

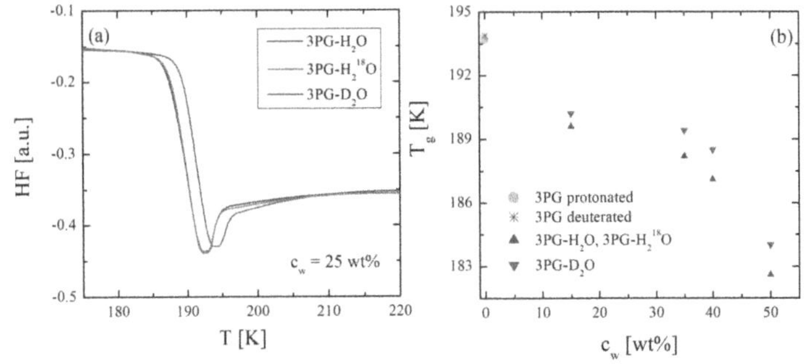

Figure 6. (a) DSC scans of a water solution of H_2O (c_w = 35 wt%), D_2O and $H_2{}^{18}O$ with 3PG at a rate of 10 K/min. The glass-transition temperatures were determined as the onset point. The Tg corresponding to the D_2O solution is higher than that of solutions with H_2O or $H_2{}^{18}O$. (b) The T_g values of the solutions with different water contents.

It is well known that the glass transition temperature of 3PG is affected by the amount of water present in the solutions (59) and, as a consequence, there is a monotonous drop in the T_g value with increasing water content. Water molecules act as a plasticizer for the solvent molecules, and consequently, the T_g of the solutions is reduced. However, here, we emphasize that this effect is the same with the substitution of H_2O by $H_2{}^{18}O$, whereas it is less pronounced for samples with D_2O.

We also notice (see Table 1) that when increasing the solvent contents, the difference between the T_g values of solutions with D_2O and H_2O (or $H_2{}^{18}O$) increases from 0.6 at c_w = 15 wt% to 1.4 degrees at c_w = 50 wt%. Similar results, such as those shown for 3PG solutions, are observed in Table 2 for 5EG solutions.

Below the T_g, the samples remain as a disordered liquid at the molecular scale but macroscopically appear as a solid. By increasing the temperature at the glass transition, both the rotational and translational motions of 3PG unfreeze. However, the T_g value of neat 3PG deuterated and protonated is not affected by the fact that the mass of the H atoms is doubled with the H → D substitution, as shown in Table 1. However, when 3PG is in solution and forms hydrogen bonds with solvent molecules, the T_g value of the 3PG-D_2O solutions increases relative to that of 3PG-H_2O and 3PG-$H_2{}^{18}O$ solutions. This increase in T_g becomes larger with increasing water contents. From these results, we can deduce that there is a stronger hydrogen-bonding network in 3PG/D_2O solutions than in 3PG/$H_2{}^{18}O$ and 3PG/H_2O solutions.

Effect of Isotope Substitution on the Dynamics at Supercooled Temperatures

In this section, we present the permittivity spectra of 3PG and 5EG solutions with different water contents using three different isotopes. As previously explained, in these solutions, we will find two dynamics (the α-relaxation and the fast solvent relaxation).

Figure 7 shows spectra of the real (ε′) and imaginary (ε″) parts of the dielectric permittivity at different temperatures (below and above the T_g) for aqueous 3PG-$H_2{}^{18}O$ solutions. As previously explained, the two relaxation processes are detected, which are displayed as being step-like in $\varepsilon'(\omega)$ and as peaks or shoulders in $\varepsilon''(\omega)$. These two dynamic processes are called α-relaxation and fast water relaxation, respectively, in agreement with data previously published for water solutions with hydrophilic solutes (59). The experimental data on H_2O and D_2O solutions show the same behavior. According to these results, the dielectric spectra were described by using a CC function to account for the fast water relaxation and an HN function for the α-relaxation. At higher temperatures, a conductivity term was also added. The shape and relaxation strengths (not shown) are highly similar for solutions with the three different isotopes.

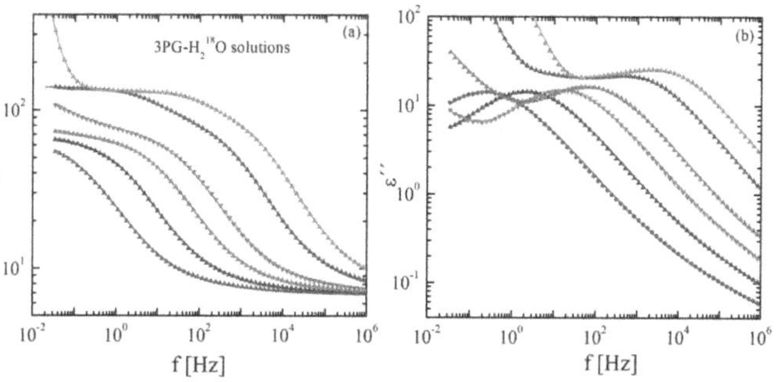

Figure 7. Dielectric spectra of the 3PG- $H_2{}^{18}O$ solution (c_w = 35 wt%) at the given temperatures (from left to right: 160, 170, 180, 186, 195, and 201 K). The lines are simultaneous fits of the real and imaginary parts using Havriliak-Negami and Cole-Cole functions for the α-relaxation and the fast water relaxation (see the text).

It is more interesting to make a comparison between the spectra, at the same water contents, using the different isotopes (see in Figure 8 the comparison of 3PG solutions with c_w = 35 wt%). The dielectric spectra are the same under the substitution of ^{16}O by ^{18}O at any temperature either below (180 K) or above (192 K) the T_g. In contrast, both α- and fast water relaxations become slower under the substitution of H →D. The temperature dependence of the relaxation times obtained from the fitting is shown in Figure 8, where we can observe the typical crossover in the fast solvent relaxation at the T_g. Under isotopic substitution, the solutions show the same behavior, and in the three cases, the temperature coincides with the corresponding calorimetric T_g (see Table 1).

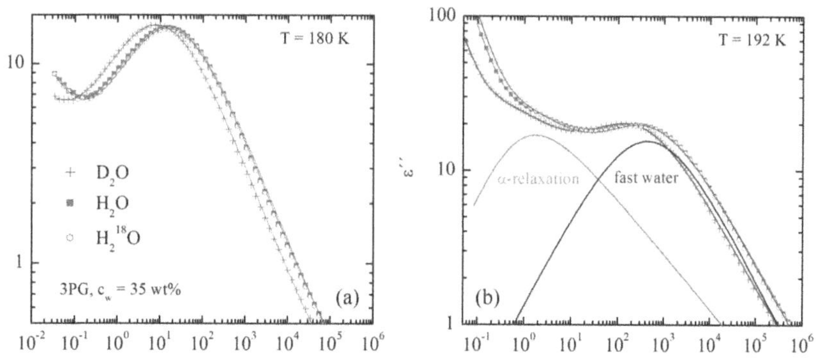

Figure 8. Comparison of the dielectric loss spectra of 3PG solutions with c_w = 35 wt% with H_2O, D_2O and $H_2{}^{18}O$ at two temperatures (180 K, below the T_g and 192 K, above the T_g). In (b), an example is shown of the fitting procedure. The data are described by two relaxation peaks caused by the structural α-relaxation and a secondary (fast water) relaxation.

The relaxation time corresponding to the α-relaxation of the three solutions in Figure 9 can be described by the Vogel-Fulcher-Tammann (VFT) law. In addition, the α-relaxation times corresponding to H_2O and $H_2{}^{18}O$ are the same as those of D_2O solutions by shifting the VFT curves in 1.5 K; i.e., we find the same relaxation time for the three isotopes on a temperature scale shifted by 1.5 K (see the inset in Figure 9), which represents a frequency shift of ~3 Hz. This value corresponds to the shift observed in the calorimetric response (see Table 1). In contrast, the relaxation times of the fast water relaxation appear to be much more sensitive to the H→D substitution. For this water content (35 wt%), the resultant shift is 3.4 K (which represents ~9 Hz in frequency), more than double that corresponding to the α-relaxation (1.5 K). For the fast water relaxation, below the crossover temperature, the activation energy is 0.55 eV for H_2O an $H_2{}^{18}O$ solutions, whereas is 0.54 eV for D_2O. This small difference between the activation energies is consistent with experiments on confined water (67).

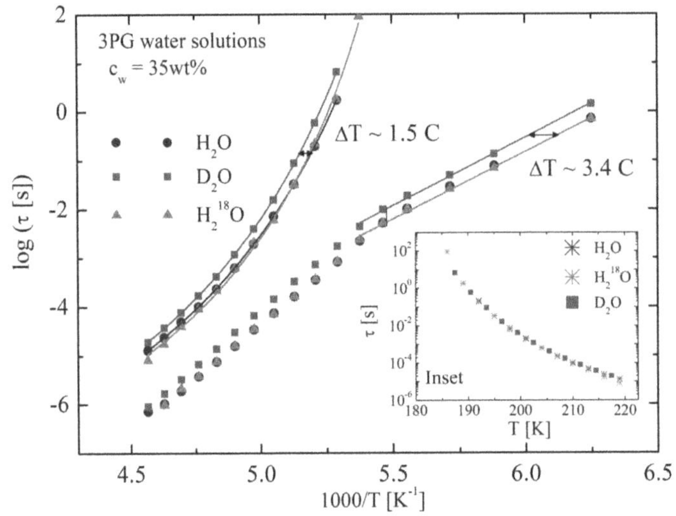

Figure 9. Dielectric relaxation times of 3PG aqueous solutions using the three different isotopes. The solid lines represent the Vogel-Fulcher-Tammann (VFT) and Arrhenius fitting for the α-relaxation and fast water relaxation, respectively. ***Inset:*** *temperature dependence of the α-relaxation times in H_2O, $H_2{}^{18}O$ and D_2O compared with shifted scale. The relaxation times of D_2O were shifted by 1.5 K.*

We now analyze the sensitivity of the shift of the relaxation times of the fast water relaxation for different water contents. At low water contents, fast water relaxation is more affected by the solute, but by increasing the water contents, water-water interactions are become more relevant. In fact, it has been shown that at the highest water concentration before crystallization occurs, the fast water relaxation in glassy solutions becomes similar to the nearly universal low-temperature relaxation of water in hard confinements (53). Figure 10 shows the temperature dependence of the relaxation times of the fast water relaxation for 3PG solutions at different water contents and 5EG at the highest water content. We can observe a monotonous increase in the shift from 1.2 to 5.3 K with increasing water content. Thus, when water-water interactions are more relevant, the difference between the relaxation times of solutions with H_2O (or $H_2{}^{18}O$) and D_2O increases. However, the times for H_2O or $H_2{}^{18}O$ remain the same for all water contents.

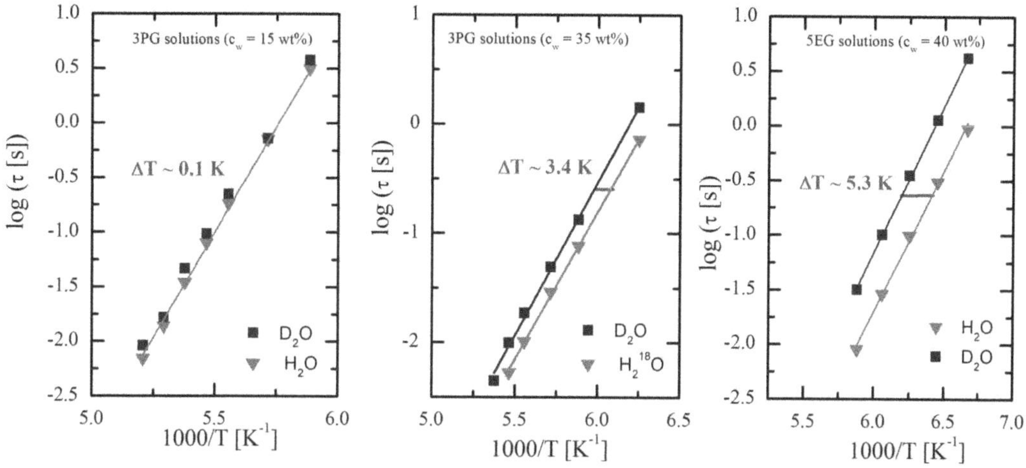

Figure 10. Fast water dielectric relaxation times of 3PG solutions (a) c_w = 15 wt%, (b) c_w = 35 wt% and 5EG solutions (c) c_w = 40 wt% using D_2O and H_2O. The solid lines represent the Arrhenius fitting for fast water relaxation.

Effect of Isotope Substitution on the Ratio of Slow Water Relaxation Times

As mentioned previously, the dynamics of D2O solutions becomes slower compared with solutions made with H_2O or $H_2{}^{18}O$. However, there is almost no change under the substitution of $^{16}O \rightarrow {}^{18}O$.

First, we analyze the substitution of H→D. The slow down could be associated with the increase in the total mass (related to translational motions) or with the change in the moment of inertia (related to rotational motions). Considering the classical mechanical theory, for pure rotational motions, an isotope effect proportional to $\sqrt{m_D/m_H} = \sqrt{2}$ ~ 1.4 is expected if we consider that the water molecule rotates around the O atom. For translational motions, $\sqrt{m_{D2O}/m_{H2O}} = \sqrt{m_{H218O}/m_{H2O}} = \sqrt{20/18}$ ~ 1.05 if we consider the diffusion of the whole molecule.

Figure 11 shows the ratio of τ_{D2O}/τ_{H2O} (which is the same considering $H_2{}^{18}O$ instead H_2O) for the different samples analyzed. The ratio of τ_{D2O}/τ_{H2O} increases with water content. For c_w = 15 wt%, the ratio is, on average, 1.25, whereas for c_w = 35 wt%, it increases to 1.9 and to ~ 4 for c_w = 40 wt%. Thus, for low water contents, the slow down of the dynamics can be explained using the trivial mass effect (H →D) (23). However, for higher water contents, when water-water interactions are relevant, the mass effects underestimate the obtained values. Therefore, we conclude that part of the acceleration of the dynamics arises mainly from NQEs at high water contents. Moreover, recent molecular dynamics simulations (68) compare the relaxation time of bulk water at room temperature using classical and quantum simulations. For bulk water at room temperature, ratios of τ_{D2O}/τ_{H2O} ~ 1.31 and 1.11 were found considering quantum and classical simulations, respectively. Here, we found even higher ratios, probably because at supercooled temperatures, NQEs are even larger. In fact, Gainaru et al (22) found ratios much larger for amorphous solid water-ASW (~6) and for LDA (~13) at low temperatures.

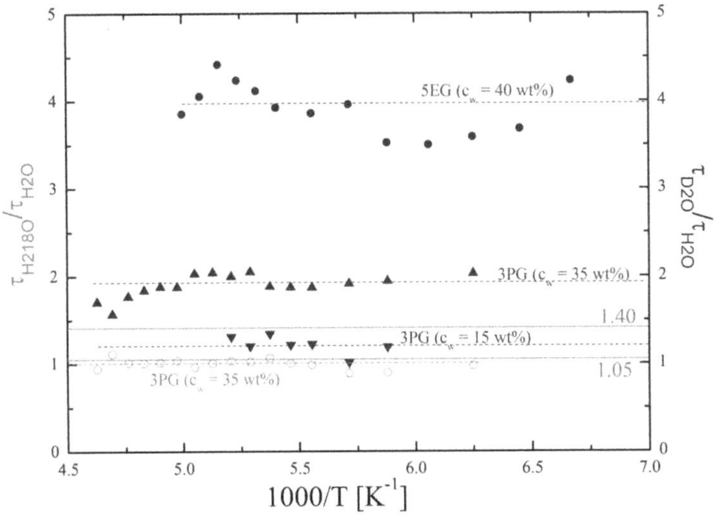

Figure 11. Ratio of τ_{D2O}/τ_{H2O} (solid data points) and τ_{H218O}/τ_{H2O} (open data points) for 3PG and 5EG solutions with different water contents. The lines indicate the values of 1.05 and 1.41 (see text), and the dashed lines indicate the average value of the time ratios.

On the other hand, the ratio of τ_{H218O}/τ_{H2O} (shown in Figure 11 in open data points) remains almost constant for both solutes and the different water contents, although the mass of $H_2{}^{18}O$ increases as much as that of D_2O. In addition, the moment of inertia (*26*) of $H_2{}^{18}O$ and H_2O is almost the same but is 40% smaller than that of D_2O. However, the times remain the same for both isotopes and faster than those of D_2O. Once again, this indicates that the mass and moment of inertia are not the only variables that determine the relaxation times of the fast water relaxation.

Conclusion

In this chapter, we presented in some detail how the dielectric relaxation of water in solutions is affected by two different isotope substitutions: H → D and $^{16}O \rightarrow {}^{18}O$. In particular, we studied the dynamics of two different hydrophilic solutes (3PG and 5EG) in solution with three isotopes ($H_2{}^{18}O$, H_2O and D_2O) in the supercooled temperature region (between 150 and 250 K), where water remains amorphous.

The DSC results indicate that the glass transition temperature of the solutions significantly changes under the substitution of H → D but changes less under $^{16}O \rightarrow {}^{18}O$. The variation in the Tg under the substitution of H → D increases with increasing water content, with a value of ~1.5 K at the highest water content analyzed (c_W = 50 wt%). For the substitution of $^{16}O \rightarrow {}^{18}O$, the glass transitions for both samples remain very similar and are independent of the water content, with a difference of 0.1 K. The same result is observed for the α-relaxation of the solutions when analyzed by BDS: the times of the α-relaxation of solutions with H_2O and $H_2{}^{18}O$ remain the same, and the relaxation times of solutions with D_2O become slower. However, for fast water relaxation, we find a more pronounced temperature shift (up to 4 K for solutions at high water contents) for H/D substitution and no difference for $H_2O/H_2{}^{18}O$ substitution. This result indicates that the translational mobility of the water molecules in solution freezes at temperatures below the glass transition of the solutions.

On the other hand, we analyzed the ratio of the relaxation times of the fast water relaxation among the three different solutions (τ_{D2O}/τ_{H2O}, τ_{D2O}/τ_{H218O} and τ_{H218O}/τ_{H2O}). We found that the ratio of τ_{D2O}/τ_{H2O} for c_w= 15 wt% is between the two classical limits (1.05 and 1.4 for pure translations and rotations, respectively). However, at high water contents, the ratio increases up to 4. These results indicate that when the water content is sufficiently high, the mass of the isotope and the moment of inertia are not the only factors that determine the isotope effect on the relaxation time of water in solutions. Instead, NQEs also govern the relaxation times of water in solutions.

Acknowledgments

The authors gratefully acknowledge CSIC (i-LINK+ program LINKB 20012) and Ministerio de Ciencia, Innovacion y Universidades code: PID2019-104650GB-C21.

References

1. Ebbinghaus, S.; Kim, S. J.; Heyden, M.; Yu, X.; Heugen, U.; Gruebele, M.; Leitner, D. M.; Havenith, M. An extended dynamical hydration shell around proteins. *Proc. Natl. Acad. Sci. U. S. A.* **2007**, *104* (52), 20749–20752.
2. Bellissent-Funel, M. C.; Hassanali, A.; Havenith, M.; Henchman, R.; Pohl, P.; Sterpone, F.; van der Spoel, D.; Xu, Y.; Garcia, A. E. Water Determines the Structure and Dynamics of Proteins. *Chem. Rev.* **2016**, *116* (13), 7673–7697.
3. Laage, D.; Elsaesser, T.; Hynes, J. T. Water Dynamics in the Hydration Shells of Biomolecules. *Chem. Rev.* **2017**, *117* (16), 10694–10725.
4. Debenedetti, P. G. Supercooled and glassy water. *J. Phys.: Condens. Matter* **2003**, *15* (45), R1669.
5. Franks, F. *Biophysics and Biochemistry at Low Temperatures*; Cambridge University Press: Cambridge, 1985.
6. Pettersson, L. G. M.; Henchman, R. H.; Nilsson, A. Water—The Most Anomalous Liquid. *Chem. Rev.* **2016**, *116* (13), 7459–7462.
7. Angell, C. A. Supercooled Water. *Annu. Rev. Phys. Chem.* **1983**, *34* (1), 593–630.
8. Mishima, O.; Stanley, H. E. The relationship between liquid, supercooled and glassy water. *Nature* **1998**, *396* (6709), 329–335.
9. Swenson, J. Possible relations between supercooled and glassy confined water and amorphous bulk ice. *Phys. Chem. Chem. Phys.* **2018**, *20* (48), 30095–30103.
10. Poole, P. H.; Sciortino, F.; Essmann, U.; Stanley, H. E. Phase behaviour of metastable water. *Nature* **1992**, *360* (6402), 324–328.
11. Cerveny, S.; Mallamace, F.; Swenson, J.; Vogel, M.; Xu, L. Confined Water as Model of Supercooled Water. *Chem. Rev.* **2016**, *116* (13), 7608–7625.
12. Ruiz, G. N.; Amann-Winkel, K.; Bove, L. E.; Corti, H. R.; Loerting, T. Calorimetric study of water's two glass transitions in the presence of LiCl. *Phys. Chem. Chem. Phys.* **2018**, *20* (9), 6401–6408.

13. Murata, K.-i.; Tanaka, H. Liquid–liquid transition without macroscopic phase separation in a water–glycerol mixture. *Nat. Mater.* **2012,** *11* (5), 436–443.
14. Soper, A. K.; Benmore, C. J. Quantum Differences between Heavy and Light Water. *Phys. Rev. Lett.* **2008**, *101* (6), 065502.
15. Kim, K. H.; Pathak, H.; Späh, A.; Perakis, F.; Mariedahl, D.; Sellberg, J. A.; Katayama, T.; Harada, Y.; Ogasawara, H.; Pettersson, L. G. M.; Nilsson, A. Temperature-Independent Nuclear Quantum Effects on the Structure of Water. *Phys. Rev. Lett.* **2017**, *119* (7), 075502.
16. RØnne, C.; Åstrand, P.-O.; Keiding, S. THz Spectroscopy of Liquid H2O and D2O. *Phys. Rev. Lett.* **1999**, *82* (14), 2888–2891.
17. Berger, A.; Ciardi, G.; Sidler, D.; Hamm, P.; Shalit, A. Impact of nuclear quantum effects on the structural inhomogeneity of liquid water. *Proc. Natl. Acad. Sci. U. S. A.* **2019**, *116* (7), 2458.
18. Kell, G. S. Effects of isotopic composition, temperature, pressure, and dissolved gases on the density of liquid water. *J. Phys. Chem. Ref. Data* **1977**, *6* (4), 1109–1131.
19. Baker, W. N. A New Comparison of the Viscosity of D2O with that of H2O. *J. Chem. Phys.* **1936**, *4* (4), 294–295.
20. Cho, C. H.; Urquidi, J.; Singh, S.; Robinson, G. W. Thermal Offset Viscosities of Liquid H2O, D2O, and T2O. *J. Phys. Chem. B* **1999**, *103* (11), 1991–1994.
21. Harris, K. R.; Woolf, L. A. Temperature and Volume Dependence of the Viscosity of Water and Heavy Water at Low Temperatures. *J. Chem. Eng. Data* **2004**, *49* (4), 1064–1069.
22. Gainaru, C.; Agapov, A. L.; Fuentes-Landete, V.; Amann-Winkel, K.; Nelson, H.; Köster, K. W.; Kolesnikov, A. I.; Novikov, V. N.; Richert, R.; Böhmer, R.; Loerting, T.; Sokolov, A. P. Anomalously large isotope effect in the glass transition of water. *Proc. Natl. Acad. Sci. U. S. A.* **2014**.
23. Ceriotti, M.; Fang, W.; Kusalik, P. G.; McKenzie, R. H.; Michaelides, A.; Morales, M. A.; Markland, T. E. Nuclear Quantum Effects in Water and Aqueous Systems: Experiment, Theory, and Current Challenges. *Chem. Rev.* **2016**, *116* (13), 7529–7550.
24. Shephard, J. J.; Salzmann, C. G. Molecular Reorientation Dynamics Govern the Glass Transitions of the Amorphous Ices. *J. Phys. Chem. Lett.* **2016**, *7* (12), 2281–2285.
25. Amann-Winkel, K.; Gainaru, C.; Handle, P. H.; Seidl, M.; Nelson, H.; Bohmer, R.; Loerting, T. Water's second glass transition. *Proc. Natl. Acad. Sci. U. S. A.* **2013**, *110* (44), 17720–17725.
26. Eisenberg, D.; Kauzmann, W. *The Structure and Properties of Water*; Oxford University Press: New York, 1969.
27. Marx, D.; Chandra, A.; Tuckerman, M. E. Aqueous Basic Solutions: Hydroxide Solvation, Structural Diffusion, and Comparison to the Hydrated Proton. *Chem. Rev.* **2010**, *110* (4), 2174–2216.
28. Schweizer, K. S.; Stillinger, F. H. High pressure phase transitions and hydrogen-bond symmetry in ice polymorphs. *J. Chem. Phys.* **1984**, *80* (3), 1230–1240.
29. Yen, F.; Gao, T. Dielectric Anomaly in Ice near 20 K: Evidence of Macroscopic Quantum Phenomena. *J. Phys. Chem. Lett.* **2015**, *6* (14), 2822–2825.
30. Fang, W.; Chen, J.; Feng, Y.; Li, X.-Z.; Michaelides, A. The quantum nature of hydrogen. *Int. Rev. Phys. Chem.* **2019**, *38* (1), 35–61.

31. Kremer, F.; Schönhals, A. *Broadband Dielectric Spectroscopy*; Springer: Berlin, 2003.
32. Cerveny, S.; Colmenero, J.; Alegria, A. Dynamics of confined water in different environments. *Eur. Phys. J.: Spec. Top.* **2007**, *141*, 49–52.
33. Cerveny, S.; Colmenero, J.; Alegria, A. Dielectric properties of water in amorphous mixtures of polymers and other glass forming materials. *J. Non-Cryst. Solids* **2007**, *353* (47–51), 4523–4527.
34. Sasaki, K.; Bandai, K.; Takatsuka, M.; Fujii, M.; Takagi, M.; Kita, R.; Yagihara, S.; Kimura, H.; Shinyashiki, N. Heterogeneous Solvent Dielectric Relaxation in Polymer Solutions of Water and Alcohols. *Front. Phys.* **2020**, *8*, 11.
35. Sato, T.; Buchner, R.; Fernandez, S.; Chiba, A.; Kunz, W. Dielectric relaxation spectroscopy of aqueous amino acid solutions: dynamics and interactions in aqueous glycine. *J. Mol. Liq.* **2005**, *117* (1–3), 93–98.
36. Cerveny, S.; Swenson, J. Dynamics of supercooled water in a biological model system of the amino acid l-lysine. *Phys. Chem. Chem. Phys.* **2014**, *16* (40), 22382–22390.
37. Saiz, L. M.; Cerveny, S. Dynamics of amorphous and partially crystallized proline solutions. *J. Non-Cryst. Solids* **2015**, *407*, 486–493.
38. Cerveny, S.; Combarro-Palacios, I.; Swenson, J. Evidence of Coupling between the Motions of Water and Peptides. *J. Phys. Chem. Lett.* **2016**, 7 (20), 4093–4098.
39. Jansson, H.; Bergman, R.; Swenson, J. Role of Solvent for the Dynamics and the Glass Transition of Proteins. *J. Phys. Chem. B* **2011**, *115*, 4099–4109.
40. Panagopoulou, A.; Kyritsis, A.; Shinyashiki, N.; Pissis, P. Protein and Water Dynamics in Bovine Serum Albumin-Water Mixtures over Wide Ranges of Composition. *J. Phys. Chem. B* **2012**, *116*, 4593–4602.
41. Wolf, M.; Gulich, R.; Lunkenheimer, P.; Loidl, A. Broadband dielectric spectroscopy on human blood. *Biochim. Biophys. Acta, Gen. Subj.* **2011**, *1810* (8), 727–740.
42. Levy, E.; Barshtein, G.; Livshits, L.; Ishai, P. B.; Feldman, Y. Dielectric Response of Cytoplasmic Water and Its Connection to the Vitality of Human Red Blood Cells: I. Glucose Concentration Influence. *J. Phys. Chem. B* **2016**, *120* (39), 10214–10220.
43. Li, Y.; Han, M.; Zhou, W.; Chen, Z. Radio-frequency dielectric relaxation behavior of selected vegetable tissues: Spectra analysis with logarithmic derivative method and simulation with double-shell model. *J. Food Eng.* **2020**, *277*, 109914.
44. Khaled, D. E.; Novas, N.; Gazquez, J. A.; Garcia, R. M.; Manzano-Agugliaro, F. Fruit and Vegetable Quality Assessment via Dielectric Sensing. *Sensors* **2015**, *15* (7), 15363–15397.
45. Pethig, R. Dielectric properties of body tissues. *Clinical Phys. Physiol. Measure.* **1987**, *8* (4A), 5–12.
46. Souli, M. P.; Klonos, P.; Fragopoulou, A. F.; Mavragani, I. V.; Pateras, I. S.; Kostomitsopoulos, N.; Margaritis, L. H.; Zoumpoulis, P.; Kaklamanis, L.; Kletsas, D.; Gorgoulis, V. G.; Kyritsis, A.; Pissis, P.; Georgakilas, A. G. Applying Broadband Dielectric Spectroscopy (BDS) for the Biophysical Characterization of Mammalian Tissues under a Variety of Cellular Stresses. *International Journal of Molecular Sciences* **2017**, *18* (4), 838.
47. Martins, M. L.; Bordallo, H. N.; Arrese-Igor, S.; Alegría, A.; Colmenero de Leon, J. Effect of Paclitaxel in the Water Dynamics of MCF-7 Breast Cancer Cells Revealed by Dielectric Spectroscopy. *ACS Omega* **2020**, *5* (30), 18602–18607.

48. Siwach, P.; Levy, E.; Livshits, L.; Feldman, Y.; Kaganovich, D. Water is a biomarker of changes in the cellular environment in live animals. *Sci. Rep.* **2020**, *10* (1), 8.
49. Cerveny, S.; Alegria, A.; Colmenero, J. Universal features of water dynamics in solutions of hydrophilic polymers, biopolymers, and small glass-forming materials. *Phys. Rev. E: Stat., Nonlinear, Soft Matter Phys.* **2008**, 77, 031803.
50. Singh, L. P.; Cerveny, S.; Alegria, A.; Colmenero, J. Broadband Dielectric Spectroscopic, Calorimetric, and FTIR-ATR Investigations of D-Arabinose Aqueous Solutions. *ChemPhysChem* **2011**, *12* (18), 3624–3633.
51. Swenson, J.; Cerveny, S. Dynamics of Deeply Supercooled Interfacial Water. *J. Phys.: Condens. Matter* **2015**, 27, 033102.
52. Immergut, E. H.; Mark, H. F., Principles of Plasticization. In *Plasticization and Plasticizer Processes*; American Chemical Society: 1965; Vol. 48, pp 1–26.
53. Sjostrom, J.; Swenson, J.; Bergman, R.; Kittaka, S. Investigating hydration dependence of dynamics of confined water: Monolayer, hydration water and Maxwell-Wagner processes. *J. Chem. Phys.* **2008**, *128* (15), 154503.
54. Fischer, J. K. H.; Sippel, P.; Denysenko, D.; Lunkenheimer, P.; Volkmer, D.; Loidl, A. Supercooled water confined in a metal-organic framework. *Commun. Phys.* **2020**, *3* (1), 95.
55. Fulcher, G. S. Analysis of recent measurements of the viscosity of glasses. *J. Am. Ceram. Soc.* **1925**, *8*, 339–355.
56. Vogel, H. Das Temperaturabhängigkeitsgesetz der Viskosität von Flüssigkeiten. *Phys. Z.* **1921**, *22*, 645–646.
57. Cerveny, S.; Swenson, J. Water dynamics in the hydration shells of biological and non-biological polymers. *J. Chem. Phys.* **2019**, *150* (23), 234904.
58. Shinyashiki, N.; Yamamoto, W.; Yokoyama, A.; Yoshinari, T.; Yagihara, S.; Kita, R.; Ngai, K. L.; Capaccioli, S. Glass Transitions in Aqueous Solutions of Protein (Bovine Serum Albumin). *J. Phys. Chem. B* **2009**, *113*, 14448–14456.
59. Cerveny, S.; Schwartz, G. A.; Alegria, A.; Bergman, R.; Swenson, J. Water dynamics in n-propylene glycol aqueous solutions. *J. Chem. Phys.* **2006**, *124* (19), 194501.
60. Cerveny, S.; Colmenero, J.; Alegría, A. Dielectric Investigation of the Low-Temperature Water Dynamics in the Poly(vinyl methyl ether)/H2O System. *Macromolecules* **2005**, *38* (16), 7056–7063.
61. Combarro Palacios, I.; Olsson, C.; Kamma-Lorger, C. S.; Swenson, J.; Cerveny, S. Motions of water and solutes—Slaving versus plasticization phenomena. *J. Chem. Phys.* **2019**, *150* (12), 124902.
62. Kripotou, S.; Zafeiris, K.; Culebras-Martínez, M.; Gallego Ferrer, G.; Kyritsis, A. Dynamics of hydration water in gelatin and hyaluronic acid hydrogels. *Eur. Phys. J. E* **2019**, *42* (8), 109.
63. Cerveny, S.; Alegria, A.; Colmenero, J. Broadband dielectric investigation on poly(vinyl pyrrolidone) and its water mixtures. *J. Chem. Phys.* **2008**, *128* (4), 044901.
64. Sasaki, K.; Matsui, Y.; Miyara, M.; Kita, R.; Shinyashiki, N.; Yagihara, S. Glass Transition and Dynamics of the Polymer and Water in the Poly(vinylpyrrolidone)–Water Mixtures Studied by Dielectric Relaxation Spectroscopy. *J. Phys. Chem. B* **2016**, *120* (27), 6882–6889.

65. Weigler, M.; Combarro-Palacios, I.; Cerveny, S.; Vogel, M. On the microscopic origins of relaxation processes in aqueous peptide solutions undergoing a glass transition. *J. Chem. Phys.* **2020**, *152* (23), 234503–234503.

66. Rhys, N. H.; Gillams, R. J.; Collins, L. E.; Callear, S. K.; Lawrence, M. J.; McLain, S. E. On the structure of an aqueous propylene glycol solution. *J. Chem. Phys.* **2016**, *145* (22), 224504.

67. Agapov, A.; Novikov, V. N.; Kisliuk, A.; Richert, R.; Sokolov, A. P. Role of quantum fluctuations in structural dynamics of liquids of light molecules. *J. Chem. Phys.* **2016**, *145* (23), 234507.

68. Wilkins, D. M.; Manolopoulos, D. E.; Pipolo, S.; Laage, D.; Hynes, J. T. Nuclear Quantum Effects in Water Reorientation and Hydrogen-Bond Dynamics. *J. Phys. Chem. Lett.* **2017**, *8* (12), 2602–2607.

Chapter 13

The Microwave Response of Water as the Measure of Interactions in a Complex Liquid

Yuri Feldman[1] and Paul Ben Ishai[2,*]

[1]Department of Applied Physics, The Hebrew University of Jerusalem, Givat Ram, Jerusalem 91904, Israel

[2]Department of Physics, Ariel University, Ariel 40700, Israel

*Email: paulbi@ariel.ac.il

Water is the universal solvent in nature. Does this imply, however, that its interaction with its environment is also a universal feature? While this question maybe too fundamental to be answered by one method only, we present evidence that in the broadening of dielectric spectra of water one can define universal features of dipolar interactions with different types of matrices. To explore this statement, we examine the dielectric spectrum of aqueous solutions of ions, monosaccharides, nucleotides, amino acids and finally proteins. In these systems the characteristic relaxation spectrum of bulk water can be described by the Cole-Cole (CC) function. The parameters of the CC function can be used to construct a new phenomenological phase space in which the evolution of the system can be traced as a trajectory, as a parameter such as the solute concentration is varied. The approach, first published in 2010 (Puzenko, A.; Ben Ishai, P.; Feldman, Y. Cole-Cole Broadening in Dielectric Relaxation and Strange Kinetics. Phys. Rev. Lett. 2010, 105), offers new insights into the dynamic of the relaxation and the organization of water in the solute hydration sphere.

Introduction

Water is of fundamental importance for all life and plays a critical role in many organic and non-organic systems. In biology an organism can function only if its macromolecules are hydrated and in chemistry the properties of many compounds can be sensitively determined by their water content. Bulk water possesses strongly polar hydrogen bonds. These are the root of much of the anomalous physical and chemical properties (*1*, *2*) of water. So much so that one can define it as a complex system (CS). Given its fundamental importance, it is surprising that there are still major questions in our understanding of how the molecular properties of water translate into its ability to solvate and interact with macromolecules. The mystery continues when one questions how in turn interfaces and the interactions therein affect water's structure and properties. One underlying premise is clear, that the hydrogen bond network, its fluctuations and the dynamics of its rearrangement, to a great extent determine the properties of the water containing soft matter materials (*3–6*). Inclusive to this description of water is the idea of the cluster. X-ray emission spectra (XES) (*7*), IR Spectroscopy (*8*), Raman Spectroscopy (*9*) and wide angle X-ray spectroscopy (*10*) reveals the traces of an underlying tetragonal structure in liquid water. IR spectroscopy (*11*) and Raman Spectroscopy (*9*) show that in liquid water up to 50% of all water molecules are fully coordinated, fully bonded, in any measurement window that one assumes. It is the dynamic rearrangement of these coordinated volumes that gives rise to an apparent dipole moment and dielectric response in the microwave frequency range (*12, 13*).

For frequencies up to 40 GHz the complex permittivity spectra of bulk water can be well described by the Debye function (*14–17*)

$$\varepsilon^*(\omega) = \varepsilon'(\omega) - i\varepsilon''(\omega) = \varepsilon_h + \frac{\Delta\varepsilon}{1 + i\omega\tau_D} \quad (1)$$

where ε' is the real component and ε'' the imaginary component of the complex dielectric permittivity, $\omega = 2\pi f$ is the cyclic frequency, τ_D is the relaxation time, i is the root of -1 and $\Delta\varepsilon = \varepsilon_s - \varepsilon_h$ is the dielectric strength, with ε_s and ε_h representing the low-frequency and high-frequency permittivity limits, respectively. eq. 1 holds for temperatures up to 70 °C and is shown in Figure 1 (*5*) for water at 25 °C. In order to derive the Debye equation, it is enough to consider a polarization process characterized by a single exponential dipole autocorrelation function. However, it would be a mistake to think that this necessarily implies that the relaxation is related to the simple rotational diffusion of one water molecule in viscous media (*16, 18*). Although using the viscosity of water one can predict with reasonable accuracy the hydrodynamic radius of the water molecule, continuous rotational diffusion (CRD) fails to describe the strong dielectric strength $\Delta\varepsilon$ of water (*16, 19, 20*). There are a number of competing models to describe the relaxation of bulk water (*21–24*). The current leader among them is the wait-and-switch relaxation model of water (*25–28*). The underlying theme to this model is that the reorientation of a water molecule happens only in the vicinity of a defect in the H-bond network in its vicinity. Until that moment the molecule 'waits', hence the name of the model. The wait induces a delay to the relaxation in comparison to the gas state relaxation time of 1ps (*29*). In short it is the stunted migration of defects through the lattice that is the cause of the cluster relaxation as a cumulative reaction of these polarization changes that leads to the validity of eq. 1 for bulk water's dielectric behaviour (*16, 17*).

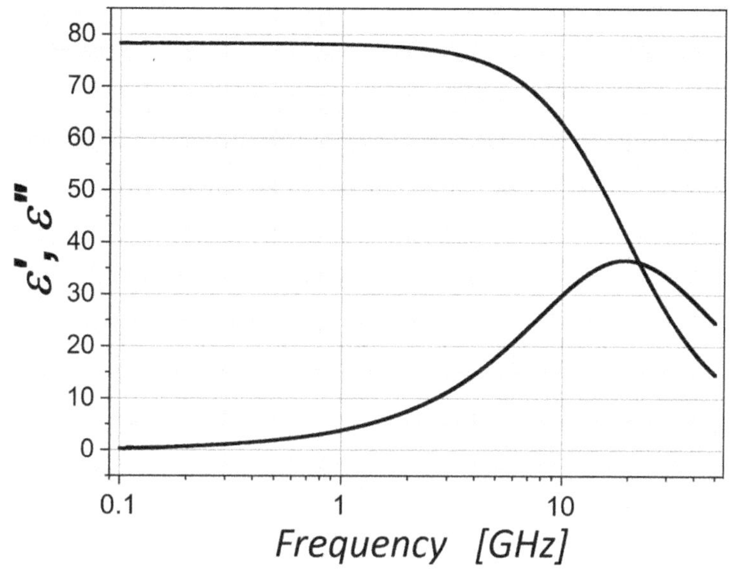

Figure 1. The real part ε'(ν) and the imaginary ε"(ν) of the water dielectric spectra at 25 °C (1021 frequency points). Reproduced with permission from reference (5). Copyright 2012 American Institute of Physics.

This balance of wait and switch is disturbed the moment another entity, charged or dipolar, is introduced. Once water has become a solvent, a symmetrical broadening of its dielectric loss peak occurs (*30*) (See Figure 2).

The modification of the water relaxation peak can be described by the phenomenological Cole-Cole (CC) function (*31*).

$$\varepsilon^*(\omega) = \varepsilon_h + \frac{\Delta\varepsilon}{1 + (i\omega\tau)^{\alpha}}\,, \tag{2}$$

where the parameters maintain the meanings defined for eq. 1. The exponent α ($0 < \alpha \leq 1$) is a measure of the symmetrical broadening. The polar or charged nature of the solute subtly effects the apparent dipole moment of the cluster, leading to inter-related changes in the parameters of eq. 2. By tracing this phenomenology one can gain a new insight into the dynamics of the water/solute interaction (*30, 32*). The broadening parameter, α, is a time fractal dimension, which is controlled by macroscopic physical quantities and reflects the rate of interactions of the dipole relaxation units with their surroundings (*30, 33, 34*). The basic features are illustrated in Figure 2.

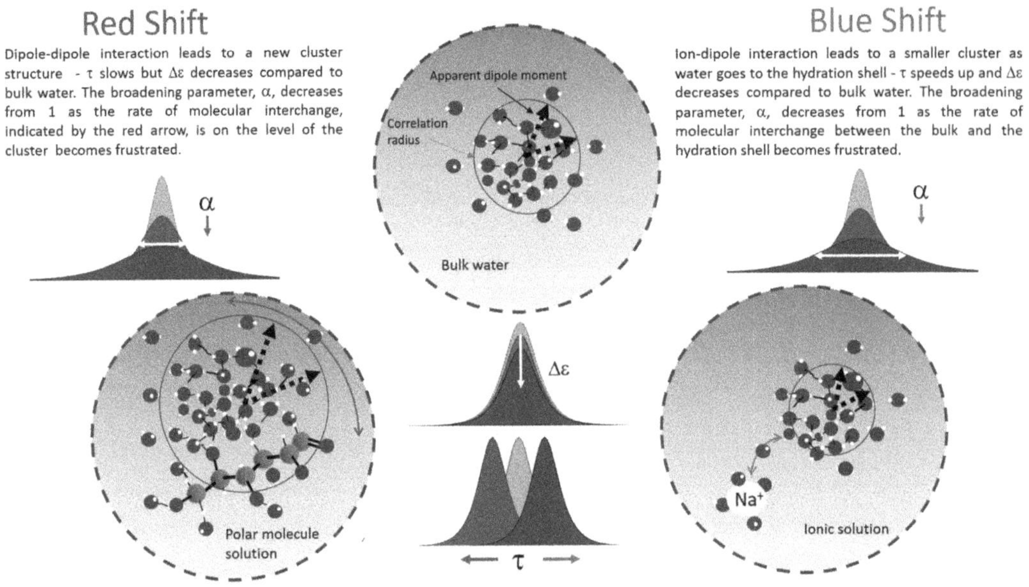

Figure 2. A schematic describing the variation in the behavior of the parameters, $\Delta\varepsilon$, τ and α, of the CC equation (eq. 2). For clarity, a single cluster is illustrated in each pane, along with a single solute entity. A dipole-dipole interaction leads to longer relaxation time, a "red shift", while an ion-dipole interaction leads to faster reorientation of the apparent dipole moment, a "blue shift". Concurrently in both cases the dielectric strength decreases, as water molecules shift to hydration shells. However, in the case of ion dipole interaction, shielding of the ion by the hydration shell leads to decoupling of the cluster–blue shift in the relaxation time. For the dipole-dipole interaction the hydration shell becomes a bridge to increase coupling– "red shift" in the relaxation time. In either case there is a broadening of the peak as α decreases from 1. α can be viewed as a fractal rate of interaction and is influenced by the nature of interchange between the water in the bulk and that constrained by the solute.

Theoretical Background

The thermodynamic state of the sample is subject to temperature, pressure, composition, structure and other thermodynamic variables. All of these factors will impinge on the parameters of the CC function of eq. 2. It therefore makes sense to trace the modification of these same parameters to gain an understanding of how the sample evolves. Furthermore, one can assume that these behaviors will be coupled as they represent the same basic thermodynamic state. By considering the plurality of length and time scales involved in any complex system, this coupling must be expressed through self-similarity. This idea is embodied in the formalism of the fractional Fokker-Planck equation (35–38) and the CC function as a solution to it. A direct result is that the exponent α expresses fractional time scaling involved in the dielectric relaxation process and this suggests a coupling between the broadening parameter and the relaxation time.

Implicit is the elemental relaxation event and its time interval, τ_c. In any observation time window one can enumerate the relaxation acts and reduce this to the average number density of N_τ over an appropriate mesoscopic volume. In particular, this must be true of the relaxation time. The relationship between the macroscopic relaxation time, the elementary time, τ_c and the number density, N_τ, will also reflect the self-similarity in the system and can be associated with the time fractal dimension α using the following scaling relationship: $N_\tau = (\tau/\tau_c)^\alpha$. Rearranging reveals the nature of α (*30*) as a time fractal dimension:

$$\alpha = \frac{\ln N_\tau}{ln(\tau/\tau_c)} \quad (3)$$

Moving to a dimensionless time interval $\xi = \tau/\tau_c$ the average number of discrete interactions N_τ, obeys a recursive fractal model:

$$N_\tau = N_0 \left(\frac{\tau}{\tau_c}\right)^A , \quad (4)$$

where $0 < A \leq 1$ and τ_c is a cutoff time that defines the time scaling (*30*). The factor N_0 is the average number of elementary relaxation acts happening during τ_c. Substituting eq. 4 into eq. 3 and using a new variable $x = ln\tau$, the behavior of α can be described by a hyperbolic equation (*30*).

$$\alpha = A + \frac{G}{x - x_0} \quad (5)$$

where, $x_0 = ln\,\tau_c$, $N_0 = exp(G)$ and $\tau_c = exp(x_0)$. In any two-component CS, where one component is a dipole subsystem and the other sub-system is distinct (dipolar or ionic) (*30*, *33*), the formulism of eq. 5 provides a framework to study the nature of the relaxation and any critical points that may be discovered. As a hyperbolic curve eq. 5 has two asymptotes; the constant A, representing the asymptotic value of the parameter α, and the asymptote $x_0 = ln\,\tau_c$, such that $\tau > \tau_c$ and $\tau < \tau_c$ (see Figure 3).

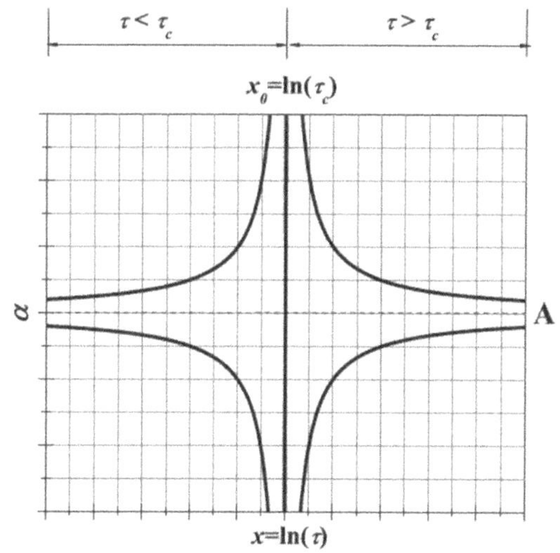

Figure 3. The functional behavior of eq. 5, illustrating its asymptotes and hyperbolic nature. Reproduced with permission from reference (30). Copyright 2010 American Institute of Physics.

In many cases, as the underlying thermodynamic parameter is varied, for instance temperature, a transition can be observed from one hyperbolic branch of $\alpha(ln\tau)$ to another (*6*). These critical points represent a change in the fundamental dynamic of the relaxation. The average number of interactions of a mesoscopic volume N_τ must perforce be coupled to the molecular structure within which the relaxation occurs. The parameters of the CC equation that are related to structural elements of the relaxation are $\Delta\varepsilon$, ε_s and ε_h. The static limit of the dielectric strength, ε_s, is related to structural aspects of the CS and its dipole moments. This was definitively described by Fröhlich for the temperature behavior of the static permittivity by (39)

$$B(T) = \Delta\varepsilon(T)\frac{2\varepsilon_s(T) + \varepsilon_\infty}{3\varepsilon_s(T)}T = \frac{1}{3\varepsilon_0 kV}\langle \boldsymbol{M}^2(T)\rangle, \tag{6}$$

where k is the Boltzmann constant, T is the absolute temperature, $\varepsilon_0 = 8.85 \times 10^{-12}$ F/m, V is a volume with N microscopic cells, each containing a dipole moment. The electric dipole moment of this volume is given by the vector sum of the individual dipole moments $\boldsymbol{M} = \sum_{i=1}^{N} \boldsymbol{m}_i$, with $\boldsymbol{m}_i$ as the average dipole moment of the i-th cell. The brackets in eq. 6 indicate a statistical averaging over all possible cell configurations. Theoretically, one can calculate the electric dipole moment of the mesoscopic volume in the case of a dipole-dipole interaction by considering that it must contain three contributions; 1) from the dipole moment of water clusters, $N_w\mu_w^2 g_w$, where N_w is the number density of clustered water molecules, μ_w is the dipole moment of water and g_w is the Kirkwood correlation factor between them; 2) similarly there is a contribution from the solute molecules $N_s\mu_s^2 g_s$, where the subscript refers to the solute population and, 3) $2N_sN_{ws}\mu_w\mu_s\langle\cos\theta_{ws}\rangle$, where N_{ws} is the number of water molecules in the hydration shell and θ_{ws} is the average angle between the bound water molecule dipoles and the dipole of the solute. While eq. 6 should give access to these parameters when the known dipole moments are taken into consideration, N_{ws} and θ_{ws} are in fact implicit. They can be condensed into a new parameter,

$$N_{cl} = N_{ws}\langle\cos\theta_{ws}\rangle, \tag{7}$$

which represents the effective number of water molecules interacting with the solute dipole.

As eq. 6 depends on one hand only on the macroscopic parameters of the CC function (eq. 2) it can be calculated independently of the microscopic structural details of the system. On the other hand, it relates to the factors governing the nature of dipole correlation at the microscopic level. Therefore, while explicitly isothermal, it will depend on the concentrations, the compositions and on any other thermodynamic parameter that relates to the size and structure of the appropriate mesoscale that governs the relaxation. This is reflected by defining a new 3D space, in which the actual set of parameters of a CC function represent a coordinate (30). The most convenient coordinate system is defined by $X = lnB$, $Y = ln\tau$ and $Z = \alpha$. The dynamics of the relaxation define a trajectory in this space. The following sections will show how this phenomenological trajectory approach can be used to investigate the interplay between the bulk and bound water populations in aqueous solutions.

Aqueous Solutions

Whenever water interacts with another polar system, one finds a broadening of its dielectric relaxation peak. This is true not just for aqueous solutions, but also for water interacting with a host matrix, like porous materials such as glasses and clays. In this case the dielectric relaxation arises from the interaction of the water dipoles organized into ice-like structures around hydrated centers

(*30, 32, 40*). However, in aqueous solutions the origin of the microwave CC broadening is defined by the dynamic rearrangements in the H-bond network in the vicinity of the solute molecules. The question of how the hydration shell forms and how it interacts with the bulk water around it is of paramount importance. By studying the effect of solute concentration on the trajectories through this new space one can shed light onto the dynamics therein involved. As the solute can be ionic or dipolar we present below examples of both non-ionic aqueous solutions (sugars (*4*), polar amino acids (*19*)) as well as for ionic ones (simple electrolytes (*5*), AMP/ATP (*41*), charged amino acids (*19*) and MetHb (*42*)). The results of an isothermal study in the frequency window of 500 MHz-50 GHz and varied by solute concentrations are presented in Figure 4. The samples were; Glucose–dipole nature, NaCl–ionic one, ATP–dipole with the strong negative charge, and Arginine-dipole with the weak positive charge.

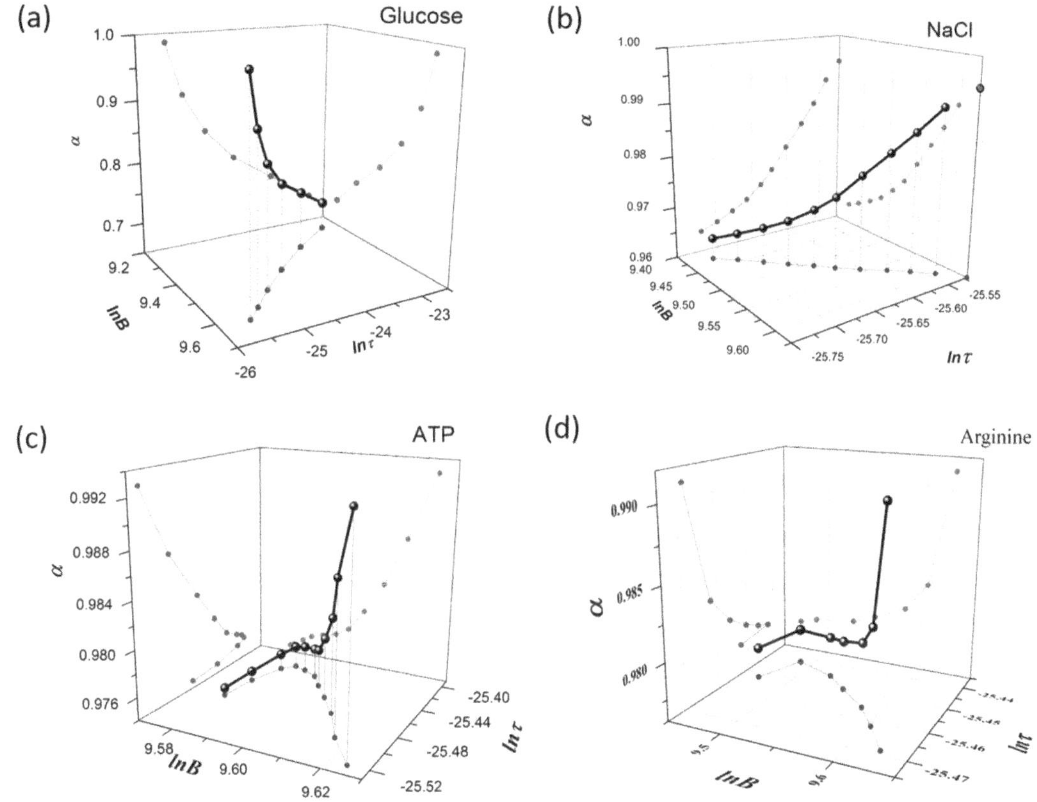

Figure 4. 3D Trajectories of CC relaxation processes of different aqueous solutions at 25 °C: (a) Glucose. Reproduced with permission from reference (4). Copyright 2012 American Institute of Physics; (b) NaCl. Reproduced with permission from reference (5). Copyright 2012 American Institute of Physics; (c) ATP. Reproduced with permission from reference (6). Copyright 2012 American Institute of Physics); (d) Arginine. Reproduced with permission from reference (19). Copyright 2014 American Institute of Physics.

The primary difference between the solutions is that of dipole-dipole interaction as opposed to ion-dipole interaction and this leads to two differing types of trajectory in the $\alpha(ln\tau)$ projection (See YZ projections in Figure 4a and 4b). In the case of a dipole solute (glucose) the dependence $\alpha(ln\tau)$ is in the first quadrant ($\tau > \tau_c, \alpha > A$). For non-ionic solutions, the curve $\alpha(ln\tau)$ (YZ projection) is placed in the second quadrant $\tau < \tau_c$; $\alpha > A$. In the case of complex organic molecules like nucleotides and amino acids, where both dipolar and ionic groups are present, both of these features appear in the trajectories, depending on solute concentration. Let us consider these different types consistently.

Non-Ionic Solutions

The microwave relaxation spectra of monosaccharide solutions have lower static permittivity, slower relaxation time and greater spectral broadening when compared to bulk water (*4*). Intuitively one would expect two separated relaxation peaks in the spectra as the solvent and solute relax independently of each other, rather than the noted single broadened and shifted peak (*4*). This conundrum of a growing relaxation time for the bulk water peak in these solutions has been extensively discussed in the literature (*4, 43*). The existence of H-bonded clusters (*44, 45*) posits an idea, by which increasing saccharide concentration leads to a new scale of cooperativity as the interaction of water clusters with glucose or fructose molecules (*46*) increases. Such an interaction could "slow down" the relaxation of water clusters with increasing saccharide concentration. Analyzing this tendency (See projection YZ in Figure 4a) in terms of the formulism derived from eq. 3 and expounded in eq. 5 provides the following fundamental parameters for the interaction of water molecules with glucose: $\tau_c \cong 1.3\ ps$, $N_{0\tau}=2.2$. Similar values were obtained for fructose (*4*):$\tau_c \cong 1.2\ ps$, $N_{0\tau}=2.8$. The fundamental relaxation time $\tau_c \sim 1\ ps$ agrees with the well-known reorientation time of a single water molecule in the bulk (*47, 48*). The number density associated with this, $N_{0\tau}$, implies that at the basic mesoscopic scale–the interaction of water clusters with a single saccharide molecule – only 2 to 3 relaxation events can occur. To understand this further one can analyze the structural information implicit in the B function (eq. 6). The righthand side of eq. 6 includes the mesoscopic dipole moment, ***M***, which itself consists of the averaged sum of the microscopic dipole moments of both water and saccharide molecules at any given concentration. The number of water molecules interacting with the one molecule of sugar at low concentration limit has been derived from eq. 6 as: $N_{cl} \sim 17$-18 (*4*), close to the value from molecular dynamic simulations of about 11 water molecules around glucose (*49*) for the immediate hydration shell.

Simple Ionic Solutions

The dielectric permittivity of ionic solutions has been studied intensively (*50–55*) and the mechanisms of relaxation are dominated by dipole/charge interactions. The formulism above can offer some new insight to the nature of these interactions. To demonstrate this, we studied isothermal diluted NaCl and KCl solutions driven by salt concentration.[5] In general, the dielectric strength decreases ($\Delta\varepsilon$), relaxation times become shorter (the 'blue shift' of Figure 2) and the spectra broaden with increasing ionic concentration. As water molecules become entrapped in hydration shells the mesoscopic water clusters also shrink (*5*). This process leads to markedly different trajectories compared to non-ionic solutions (compare Figures 4a and 4b). The YZ projection $\alpha\ (\ln\tau)$ is placed in the second quadrant of the plane ($\tau < \tau_c$; $\alpha > A$). Applying eq. 5 one finds $\tau_c \cong 8.7$ ps, $N_{0\tau} \approx 1$ for NaCl and $\tau_c \cong 8.5$ ps, $N_{0\tau} \approx 1$ for KCl solutions (*5*). The cutoff relaxation time τ_c corresponds to relaxation time of water cluster in bulk water ($c \rightarrow 0$) (*16, 47, 56*) and $N_{0\tau} \sim 1$ implies that the mesoscopic scale in this case is indeed the water cluster. This contrasts the behavior of the same parameters for monosaccharide solutions (*4*). To accentuate this difference one can consider the $B(c)$ function (eq. 6). For an ionic system, there are two basic dipolar entities depending on the ionic concentrations; shrinking water clusters and the increasing number of hydration shells around the ions (5). Together these constitute the Fröhlich mesoscopic cells. If the solutions are dilute, enough then there is no need to consider overlapping hydration shells. Under these conditions, one can derive the following expression for the fluctuating dipole moment induced by a single ion pair:

$$1 - \frac{B(c)}{B(0)} = \frac{c}{c_w g_w} N_{av} \quad (8)$$

where B(0) is the asymptote value for the Fröhlich function, c_w is the concentration of water clusters in the mesoscale volume defined by the ion pair, g_w is the Kirkwood factor for bulk water and N_{av} is the average number of water molecules perturbed by a single ion pair. The concept is illustrated in Figure 5 and the details of this derivation can be found in Ref (5). For ion pair Na^+, Cl^- this value decreases from about 33 molecules at concentration 0.1 mol/L to 24 molecules at concentration 1 mol/L. At the same concentrations the corresponding values for pair K^+, Cl^- are 40 and 22 molecules (5), respectively.

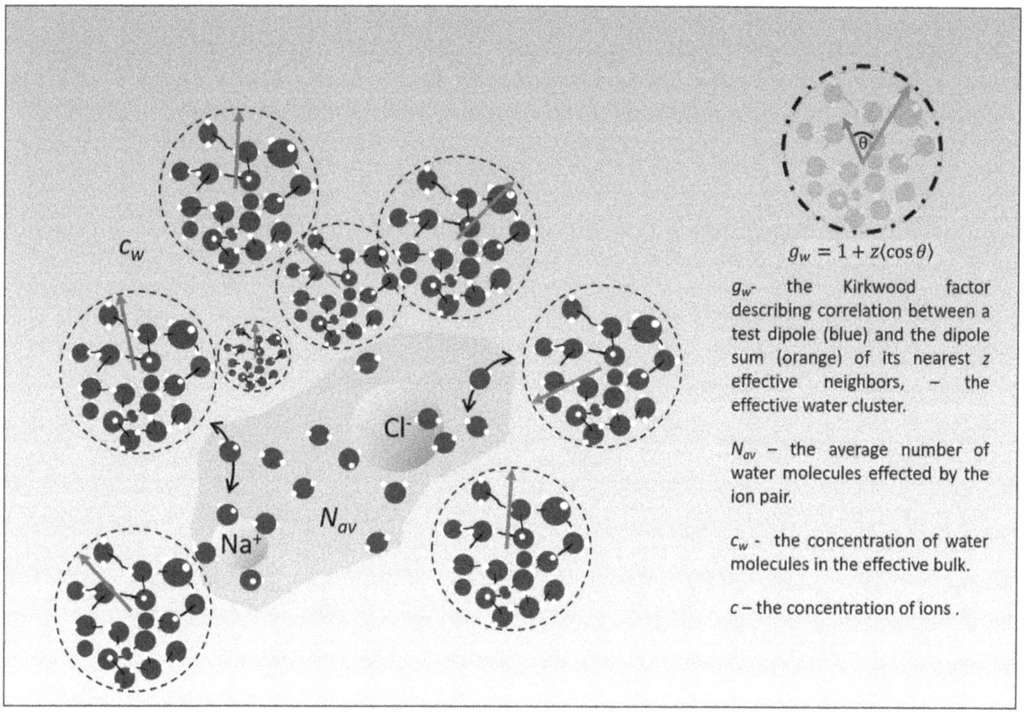

Figure 5. A schematic of the interplay between bulk water relaxation and the ion pair, illustrating the elements of Fröhlich B function, their interaction and how this influences the macroscopic fluctuating dipole moment of water structures in the solution. The black arrow indicates the exchange of a single water molecule between the hydration shell region and the water cluster, causing reorientation of the cluster dipole moment and a bulk water relaxation event. The elemental time for this to occur is bound by the relaxation time of the pure bulk water cluster, $\tau_c \cong 8.5$ ps, and in the mesoscopic region defining the cluster the exchange act requires only 1 water molecule, $N_{0\tau} \approx 1$.

Nucleotide and Amino Acid Solutions

A more complex situation is the aqueous solutions of nucleotides (AMP/ATP) and several amino acids (*6, 19*). The main feature here is the dual dipole/charge nature of their interfaces. Nucleotides have relatively large dipole moments (~45 D for AMP (*57, 58*) and~230 D for ATP (*59*)) and at low concentrations (lower than 3.5%) the behavior of $\alpha(ln\tau)$ for AMP/ATP solutions is similar to monosaccharides and amino acids solutions (*4, 19*) with $\alpha(ln\tau)$) lying in the first quadrant of Figure 3 ($\tau > \tau_c$; $\alpha > A$). However, by increasing their concentration the 3D trajectory in Figure 4c transforms to those of ionic solutions and the YZ projection of the trajectory places $\alpha(ln\tau)$) in the second quadrant of the plane ($\tau < \tau_c$; $\alpha > A$). Consequently, both AMP and ATP solutions are characterized by the two cutoff times τ_c, placing these extremes in separate quadrants of Figure 3. The low concentration (dipole-dipole like interactions) cutoff time, τ_c, is much longer than 1 ps, a marked difference from monosaccharide aqueous solutions (*4*). It is close to the relaxation time of the bulk solvent at the same temperature (~8.28 ps at 25°C (*60*)). The number of elemental acts, $N_{0\tau}$, in both cases of nucleotides and amino acids is still approximately 1 (*6, 19*).

In high concentrations (2nd quadrant, $\tau < \tau_c$, $\alpha > A$), the asymptotic value τ_c tends to a value typical for ionic solutions (*5*), but remains higher than for simple ionic solutions, as can be seen in Table 1.

Table 1. The Parameters of the Fit of $\alpha(\ln\tau)$ to eq. 3 for Solutions of AMP and ATP[a]

	Low concentrations $\tau > \tau_c$; $\alpha > A$			*High concentrations* $\tau < \tau_c$, $\alpha > A$		
	$N_{0\tau}$	τ_c *[ps]*	*A*	$N_{0\tau}$	τ_c *[ps]*	*A*
AMP	1	6.95	0.96	1	9.36	0.98
ATP	1	7.32	0.96	1	9.4	0.97

[a] Reproduced with permission from Reference (*6*) Copyright 2012, American Institute of Physics.

This indicates another level of cooperativity beyond the direct influence on water cluster relaxation. AMP/ATP and amino acid matrixes present simultaneously the dual nature of water–ion and water–dipole interaction. This impression is further strengthened, when studying the $lnB(ln\tau)$ (XY projection). The same kink points to a dramatic change in the dynamics of water relaxation, as the concentration of nucleotide is changed (*6*).

One can use the *B* function to calculate the effective number of water molecules interacting with the solute, N_{cl}, by following a similar procedure to that used to obtain eq. 7 for pure dipole–dipole interactions, only this time counting both ionic and dipolar contributions. The details are described in Ref (*6*). The results, Figure 6, demonstrate the decay of the effective number of water molecules bound to the AMP/ATP molecules as their concentration increases. Furthermore, ATP benefits from two additional phosphate groups compared to AMP and this significantly adds to the transfer of water molecules from the bulk to the hydration shell (*6*).

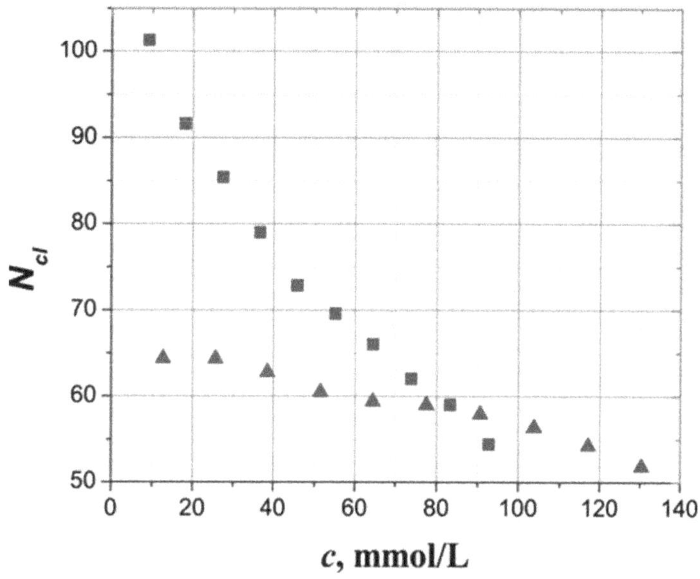

Figure 6. The effective number of water molecules in the shell around ATP (blue squares) and AMP (red triangles) molecule as a function of concentration at 25 °C. Reproduced with permission from reference (6). Copyright 2012 American Institute of Physics

Despite being zwitterionic in form the amino acids on the whole are dominated by dipole-dipole interactions in solution and their characteristic $\alpha(ln\tau)$ trace places them in the first quadrant of Figure 3. The total number of N_{cl} for most of them does not exceed few dozens of water molecules, except for Proline (~70 molecules) that has an unusual zwitterion structure (*19*). In contrast to AMP/ATP solutions, the role of charges in the case of amino acids is weak. However, in Figure 4d we present the trajectory of Arginine, as this amino acid breaks the mold by having a strong ionic contribution on its residue (*61*) and its trajectory is reminiscent of the nucleotides. Once again by using the B function one can elucidate the number N_{cl} of water dipoles perturbed by one amino acid molecule (*19*). The ability to decipher water's structure around the amino acids–the building blocks–and nucleotides opens a window to understanding nature's machinery itself, the protein.

Protein Aqueous Solutions

A range of different dipole-matrix systems and their interactions are present in the literature (*30, 32, 40, 62*), the most complex of which are proteins with their primary, secondary, tertiary and quaternary structures presenting varied layers of complexity (*63*). Their size and intricate surfaces present different areas of dipole-dipole and dipole-ion interactions with their aqueous environment. This competition can be analyzed using the approach outlined above.

We take Methemoglobin (MetHb) solutions (*42*) at different concentrations as a model to understand dynamics and structure in terms of $\alpha(ln\tau)$ and $B(c)$. MetHb has been intensively studied (*42, 64–66*). It is known that biological macromolecules perform their function in specific cellular environments (subcellular compartments or tissues); therefore, they are adapted to the biophysical characteristics of the corresponding environment, one of them being the characteristic pH (*67*). Dielectric measurements were carried out in phosphate-buffered saline (PBS) with pH=7.4 to maintain a valid physiological environment. As the control, MetHb solution at different concentrations was measured in double-distilled water (DDW). The 3-dimensional trajectories are presented in Figures 7a and 7b. In DDW the $\alpha(ln\tau)$ dependence for MetHb corresponds to the first quadrant ($\tau > \tau_c$; $\alpha > A$) (See Figure 7a), in PBS there is a transition to the second quadrant ($\tau < \tau_c$; $\alpha > A$) for low concentrations (See Figure 4b). The behavior in high concentrations is similar to what occurs in monosaccharide solutions (*4*), inferring that dipole-dipole interactions dominate. For low concentrations in PBS and in pure water, dipole-ion interactions prevail (Second quadrant. See Figures 7a and 7b). Consequently, for PBS the cutoff times τ_c correspond to different branches of eq. 5.

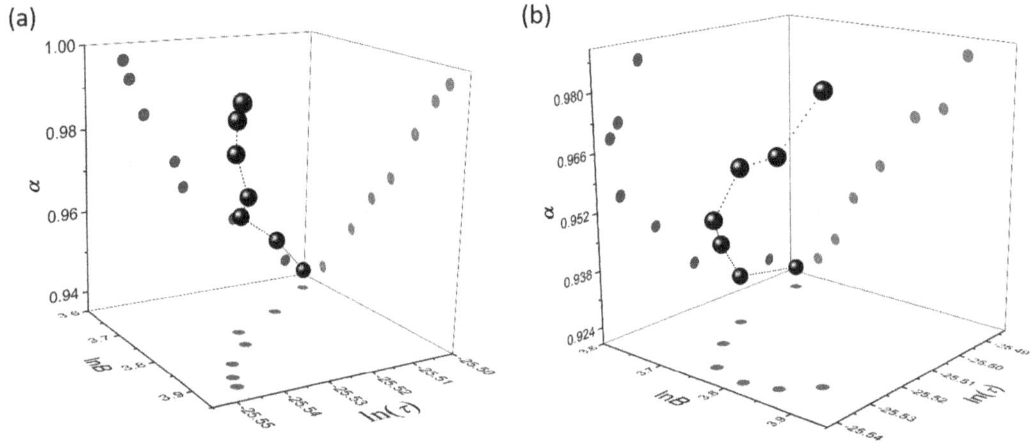

Figure 7. 3D trajectories of the CC relaxation process for MetHb solution in DDW (a) and MetHb in PBS (b) at 25 °C. Reproduced with permission from reference (65). Copyright 2020 American Institute of Physics.

For low concentrations in PBS, the asymptotic value τ_c =8.24 ps and $N_{0\tau} \cong 1$, typical for ionic solutions. In DDW $\tau_c \cong 7.85$ ps and $N_{0\tau} \cong 1$. In both cases, the asymptotic values of τ_c are slightly smaller than in the case of the relaxation of bulk water. The experimental relaxation time is associated with the minimal size of water clusters in bulk water (*16*). The implication is in MetHb solutions at low concentrations bulk waters clusters are smaller. However, their reorientation is still associated with a single relaxation event as $N_{0\tau} \cong 1$.

To gain further insight we must revaluate eq. 6 (39). For MetHb in DDW the total fluctuating dipole moment in any mesoscopic volume, *V*, of the sample is

$$M_{DDW} = \sum_{i=1}^{N_{free}} \mu_i + M_{HB}, \tag{9}$$

where $\boldsymbol{\mu}_i$ is the dipole moment of a water molecule and $\boldsymbol{M}_{HB}$ is the dipole moment of the Hb macromolecule surrounded by bound water. One can define the number of 'free' water molecules in this volume as the difference between the numbers of molecules in DDW for the same volume less the number in the hydration shell of MetHb. The function *B*, defined by eq. 6, establishes the relationship between the experimental values of the dielectric strength for the main water process in the Hb solution and the mean square $\langle \boldsymbol{M}^2 \rangle$ of the fluctuation dipole moment eq. 9 (*42*). Substituting the calculated value of $\langle \boldsymbol{M}^2 \rangle$ into eq. 6 we can determine the volume concentration of hydrated water molecules $n_b^{(DDW)}$ associated with the *Hb* globules. For MetHb we obtain

$$n_b^{(DDW)} = n_0 \frac{1 - \frac{B_{exp}}{B_{exp}^{(0)}}}{\left(1 + \frac{4}{g_w}\left(\frac{n_p}{n_p + n_n}\right)^2\right)} \tag{10}$$

Here, $n_0 = N_0/V$ is the volume concentration of the bulk water molecules (N_0 is number of the bulk water molecules in volume V) $n_p = N_{Hb}N_p/V$ and $n_n = N_{Hb}N_n/V$ are the concentrations of positive and negative charges assigned to MetHb macromolecules, respectively. In eq. 10 $B_{exp}^{(0)}$ and B_{exp} are the experimental functions for the bulk water and for the MetHb concentration respectively, and g_w is the Kirkwood factor for the bulk water (*16, 39*).

For the buffer solution concentrations of MetHb the situation is more complicated because water molecules are tied to the hydration shells of the PBS ions (Na^+, Cl^-, K^+, HPO_4^{2-}, $H_2PO_4^-$). This is done by introducing the term $\Delta\boldsymbol{M}$ to $\boldsymbol{M}_{DDW}$ in eq. 6:

$$\boldsymbol{M}_{PBS} = \boldsymbol{M}_{DDW} + \Delta\boldsymbol{M}, \tag{11}$$

or

$$\Delta\boldsymbol{M} = \sum_{i=1}^{N_{jp}}\sum_{k=1}^{N_{jw}^{+}} \boldsymbol{\mu}_{ik} + \sum_{i=1}^{N_{jn}}\sum_{k=1}^{N_{jw}^{-}} \boldsymbol{\mu}_{ik} \tag{12}$$

Here the index *i* runs of the number of ions with their total positive and negative numbers N_{jp} and N_{jn} in the volume *V*, respectively. The second summation index *k* is related to the number of water molecules bound to all positive and negative ions with the upper limits N_{jw}^{+} and N_{jw}^{-}, respectively. A number of assumptions are required:

a) One can ignore the interaction between the shells around ions with the same sign;
b) In PBS there are ions that include both phosphates and simple salts. We assume that $N_{jp} > N_{jn}$. In this case, the number of PBS buffer ions is the sum of N_{jn} of coupled ions and a surfeit $\Delta N_{jp} = N_{jp} - N_{jn}$ of positive ions. On the surface of the MetHb there are more negative charges than positive $(N_n > N_p)$, so we assume $\Delta N_n = N_n - N_p$ of MetHb are compensated by the rest of the positive ions ΔN_{jp} of PBS.

The calculation for the $\langle \boldsymbol{M}_{PBS}^2 \rangle$ in MetHb solution in PBS is presented elsewhere (*42*). The concentration of water molecules n_b associated with the MetHb macromolecules in the buffer is:

$$n_b^{(PBS)} = n_0 \frac{1 - \frac{B_{PBS}}{B_{PBS}^{(0)}} - \frac{n_{jwb}}{n_0}\left[1 - \frac{1}{g_w} + \frac{4}{g_w}\left(\frac{n_{jn}}{n_{jn} + n_{jp}}\right)^2\right]}{\left[1 + \frac{4}{g_w}\left(\frac{n_p}{n_p + n_n}\right)^2 + \frac{4}{g_w}\left(\frac{n_{jp}}{n_{Hb}\Delta N_n + n_{jp}}\right)^2\right]}, \tag{13}$$

where n_{jwb} is the concentration of water molecules in the hydration shells of solvent ions; n_{jn} and n_{jp} are the concentrations of the negative and positive ions of PBS, respectively; and n_{Hb} is the concentration of the MetHb (42). The ratio of the volume concentrations of water molecules $n_b^{(DDW)}$ and $n_b^{(PBS)}$ bound to MetHb in DDW and PBS solutions and normalized by bulk water is presented in Figure 8 as a function of MetHb concentration. In PBS solution the number of hydrated molecules in MetHb is higher than in DDW. The additional dipole contribution due to the PBS coupled ions modifies the concentration dependence of hydrated water.

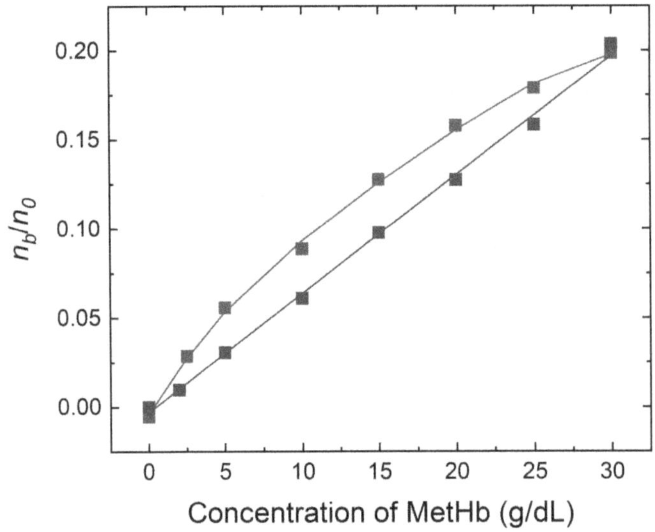

Figure 8. Volume concentrations $n_b^{(DDW)}$ of water molecules bound to MetHb in DDW solution and concentration $n_b^{(PBS)}$ water molecules bound by MetHb in PBS solution normalized to the volume concentration n_0 of water molecules in the pure bulk water. Reproduced with permission from reference (65). Copyright 2020 American Institute of Physics.

Conclusion

In this review we have tried to present a comprehensive consideration of the dielectric relaxation of water in aqueous solutions of the basic elements of biology: monosaccharides, ions, nucleotides, amino acids, and proteins. Even though the Cole-Cole relaxation is a universal feature, which reflects the interaction of water dipoles with solute molecules, the nature of the hydration shell is quite different in each case. Starting with simple ionic solutions, the hydration is determined by the dipole–charge interaction. The dominant feature for the monosaccharides is a dipole-dipole interaction. For amino acids and nucleotides, hydration could be either by the dipole–dipole interaction or by a mix involving a transition from a dipole–ion to a dipole–dipole. Finally, we come to proteins, macro-molecules that dwarf the size of the water dipole. Here, there are hydrophilic interactions of water molecules with regions of charges and dipoles, each localized on the rough surface of the macromolecule. In the case of a solution of MetHb in PBS, the hydration of the solvent ions and their interaction with charges on the surface of the protein are taken into account. These results have made it possible to unravel the mechanisms governing the role of water in protein hydration.

In summary we want to conclude that the concept of a dipole-matrix interaction and its manifestation in the Cole-Cole relaxation process unlocks new and intriguing insights into the physics of complex aqueous solutions. The 3D trajectory approach is a powerful technique for following the kinetics of a system as it evolves. This power is derived from the combination of the

Fröhlich B function, presenting structural dipole information, the relaxation time τ, representing the dynamics, and α the time fractal dimension of the relaxation. In particular, coupled with the Kirkwood-Froehlich theory of static permittivity, it allowed the determination of the number of water molecules involved with the solute molecule or ion as a function of the solute concentration.

References

1. Bagchi, B. From anomalies in neat liquid to structure, dynamics and function in the biological world. *Chem. Phys. Lett.* **2012**, *529*, 1–9.
2. Chaplin, M. A proposal for the structuring of water. *Biophys. Chem.* **2000**, *83*, 211–221.
3. Oleinikova, A.; Brovchenko, I.; Geiger, A.; Guillot, B. Percolation of water in aqueous solution and liquid-liquid immiscibility. *J. Chem. Phys. Eng. Data* **2002**, *117*, 3296–3304.
4. Levy, E.; Puzenko, A.; Kaatze, U.; Ben Ishai, P.; Feldman, Y. Dielectric spectra broadening as the signature of dipole-matrix interaction. I. Water in nonionic solutions. *J. Chem. Phys.* **2012**, *136*, 114502.
5. Levy, E.; Puzenko, A.; Kaatze, U.; Ben Ishai, P.; Feldman, Y. Dielectric spectra broadening as the signature of dipole-matrix interaction. II. Water in ionic solutions. *J. Chem. Phys.* **2012**, *136* (11), 114503.
6. Puzenko, A.; Levy, E.; Shendrik, A.; Talary, M. S.; Caduff, A.; Feldman, Y. Dielectric spectra broadening as a signature for dipole-matrix interaction. III. Water in adenosine monophosphateadenosine-5′-triphosphate solutions. *J. Chem. Phys.* **2012**, *137* (19), 194502 .
7. Huang, C.; Wikfeldt, K. T.; Tokushima, T.; Nordlund, D.; Harada, Y.; Bergmann, U.; Niebuhr, M.; Weiss, T. M.; Horikawa, Y.; Leetmaa, M.; Ljungberg, M. P.; Takahashi, O.; Lenz, A.; Ojamäe, L.; Lyubartsev, A. P.; Shin, S.; M. Pettersson, L. G.; Nilsson, A. The inhomogeneous structure of water at ambient conditions. *Proc. Natl. Acad. Sci. U. S. A.* **2009**, *106* (36), 15214–15218.
8. Maréchal, Y. The molecular structure of liquid water delivered by absorption spectroscopy in the whole IR region completed with thermodynamics data. *J. Mol. Struct.* **2011**, *1004*, 146–155.
9. Paolantoni, M.; Faginas Lago, N.; Albertí, M.; Laganà, A. Tetrahedral ordering in water: Raman profiles and their temperature dependence. *J. Phys. Chem. A* **2009**, *113*, 15100–15105.
10. Huang, C.; Wikfeldt, K. T.; Nordlund, D.; Bergmann, U.; McQueen, T.; Sellberg, J.; Pettersson, L. G. M.; Nilsson, A. Wide-angle X-ray diffraction and molecular dynamics study of medium-range order in ambient and hot water. *Phys. Chem. Chem. Phys.* **2011**, *13*, 19997–20007.
11. Brubach, J.-B.; Mermet, A.; Filabozzi, A.; Gerschel, A.; Roy, P. Signatures of the hydrogen bonding in the infrared bands of water. *J. Chem. Phys.* **2005**, *122*, 184509.
12. Ben Ishai, P.; Tripathi, S. R.; Kawasa, K.; Puzenko, A.; Feldman, Y. What is the primary mover of water dynamics? *Phys. Chem. Chem. Phys.* **2015**, *17* (23), 15428–15434.
13. Popov, I.; Ben Ishai, P.; Khamzin, A.; Feldman, Y. The mechanism of the dielectric relaxation in water. *Phys. Chem. Chem. Phys.* **2016**, *18*, 13941–13953.
14. Kaatze, U. Complex permittivity of water as a function of frequency and temperature. *J. Chem. Phys. Eng. Data* **1989**, *34*, 371–374.

15. Ellison, W. J.; Lamkaouchi, K.; Moreau, J. M. Water: a dielectric reference. *J. Mol. Liq.* **1996**, *68*, 171–279.
16. Popov, I.; Ben Ishai, P.; Khamzin, A.; Feldman, Y. The mechanism of the dielectric relaxation in water. *Phys. Chem. Chem. Phys.* **2016**, *18* (20), 13941–13953.
17. Kaatze, U. Dielectric Relaxation of Water. In *Dielectric relxaxtion in biologucal systems*; Raicu, V., Feldman, Y., Eds.; Oxford University Press: Oxford, 2015; pp 189−278.
18. Agmon, N. Tetrahedral displacement: the mechanism behind the Debye relaxation in water. *J. Phys. Chem.* **1996**, *100*, 1072–1080.
19. Levy, E.; Silvina, C.; Ermolina, I.; Puzenko, A.; Feldman, Y. Dielectric spectra broadening as a signature for dipole-matrix interaction. IV. Water in amino acids solutions. *J. Chem. Phys.* **2014**, *140*, 135104.
20. Ben Ishai, P.; Tripathi, S. R.; Kawase, K.; Puzenko, A.; Feldman, Y. What is the primary mover of water dynamics? *8th International Conference on Broadband Dielectric Spectroscopy and its Applications*, Wisla, Poland, 2014.
21. Buchner, R.; Barthel, J.; Stauber, J. The dielectric relaxation of water between 0 degrees C and 35 degrees C. *Chemical Physics Letters* **1999**, *306* (1−2), 57–63.
22. Ronne, C.; Astrand, P. O.; Keiding, S. R. THz spectroscopy of liquid H2O and D2O. *Phys. Rev. Lett.* **1999**, *82* (14), 2888–2891.
23. Yada, H.; Nagai, M.; Tanaka, K. Origin of the fast relaxation component of water and heavy water revealed by terahertz time-domain attenuated total reflection spectroscopy. *Chem. Phys. Lett.* **2008**, *464* (4−6), 166–170.
24. Zasetsky, A. Y. Dielectric Relaxation in Liquid Water: Two Fractions or Two Dynamics? *Phys. Rev. Lett.* **2011**, *107* (11), 117601 .
25. Singwi, K. S.; Sjolander, A. Diffusive Motions in Water and Cold Neutron Scattering. *Phys. Rev.* **1960**, *119* (3), 863–871.
26. Sciortino, F.; Geiger, A.; Stanley, H. E. Network Defects and Molecular Mobility in Liquid Water. *J. Chem. Phys.* **1992**, *96* (5), 3857–3865.
27. Sciortino, F.; Geiger, A.; Stanley, H. E. Effect of Defects on Molecular Mobility in Liquid Water. *Nature* **1991**, *354* (6350), 218–221.
28. Raicu, V.; Feldman, Y.; *Dielectric Relaxation in Biological Systems. Physical Principles, Methods, and Applications*; Oxford University Press: 2015.
29. Okada, K.; Yao, M.; Hiejima, Y.; Kohno, H.; Kajihara, Y. Dielectric relaxation of water and heavy water in the whole fluid phase. *J. Chem. Phys.* **1999**, *110* (6), 3026–3036.
30. Puzenko, A.; Ben Ishai, P.; Feldman, Y. Cole-Cole Broadening in Dielectric Relaxation and Strange Kinetics. *Phys. Rev. Lett.* **2010**, *105*, 037601.
31. Cole, K. S.; Cole, R. H. Dispersion and absorption in dielectrics. I. Alternating current characteristics. *J. Chem. Phys. Eng Data* **1941**, 9, 341–351.
32. Feldman, Y.; Puzenko, A.; Ben Ishai, P.; Gutina Greenbaum, A. The dielectric response of interfacial water - From the ordered structures to the single hydrated shell. *Colloid Polym. Sci.* **2014**, *292* (8), 1923–1932.
33. Feldman, Y.; Puzenko, A.; Ryabov, Y. Dielectric Relaxation Phenomena in Complex Materials. In *Fractals, Diffusion, and Relaxation in Disordered Complex Systems*; John Wiley & Sons, Inc.: 2006; pp 1−125.

34. Raicu, V.; Feldman, Y. *Dielectric Relaxation in Biological Systems: Physical Principles, Methods, and Applications*; Oxford University Press: Oxford, U.K., 2015; p 432.
35. Metzler, R.; Barkai, E.; Klafter, J. Anomalous Diffusion and Relaxation Close to Thermal Equilibrium: A Fractional Fokker-Planck Equation Approach. *Phys. Rev. Lett.* **1999**, *82* (18), 3563–3567.
36. Hilfer, R. Fractional time evolution. In *Applications of Fractional Calculus in Physics*; Hilfer, R., Ed.; World Scientific: Singapore, 2000; pp 87−130.
37. Coffey, W. T. Dielectric relaxation: an overview. *J. Mol. Liq.* **2004**, *114* (1–3), 5–25.
38. Coffey, W. T.; Kalmykov, Y. P.; Titov, S. V. Fractional rotational diffusion and anomalous dielectric relaxation in dipole systems. In *Advances in Chemical Physics*; Coffey, W. T., Kalmykov, Y. P. , Eds.; Wiley: New York, 2006; Vol. 133, p 285.
39. Fröhlich, H. *Theory of dielectrics; dielectric constant and dielectric loss*, 2nd ed.; Clarendon Press: Oxford, 1958; p 192.
40. Vasilyeva, M. A.; Gusev, Y. A.; Shtyrlin, V. G.; Greenbaum Gutina, A.; Puzenko, A.; Ishai, P. B.; Feldman, Y. Dielectric relaxation of water in clay minerals. *Clays Clay Minerals* **2014**, *62* (1), 62–73.
41. Puzenko, A.; Levy, E.; Shendrik, A.; Talary, M. S.; Caduff, A.; Feldman, Y. Dielectric spectra broadening as a signature for dipole-matrix interaction. III. Water in adenosine monophosphate/adenosine-5′-triphosphate solutions. *J. Chem. Phys.* **2012**, *137* (19), 194502.
42. Latypova, L.; Puzenko, A.; Levy, E.; Feldman, Y. Dielectric spectra broadening as a signature for dipole-matrix interactions. V. Water in protein solutions. *J. Chem. Phys.* **2020**, *153* (4), 045102.
43. Fuchs, K.; Kaatze, U. Molecular Dynamics of Carbohydrate Aqueous Solutions. Dielectric Relaxation as a Function of Glucose and Fructose Concentration. *J. Phys. Chem. B* **2001**, *105* (10), 2036–2042.
44. Pártay, L. B.; Jedlovszky, P.; Brovchenko, I.; Oleinikova, A. Formation of mesoscopic water networks in aqueous systems. *Phys. Chem. Chem. Phys.* **2007**, *9*, 1341–1346.
45. Brovchenko, I.; Geiger, A.; Oleinikova, A. Clustering of water molecules in aqueous solutions: Effect of water–solute interaction. *Phys. Chem. Chem. Phys.* **2004**, *6*, 1982–1987.
46. Suzuki, T. The hydration of glucose: the local configurations in sugar–water hydrogen bond. *Phys. Chem. Chem. Phys.* **2008**, *10*, 96–105.
47. Eisenberg, D.; Kauzmann, W. *The structure and Properties of Water*; Clarendon Press: Oxford, 1969.
48. Arkhipov, V. I. Hierarchy of dielectric relaxation times in water. *J. Non-Cryst Solids* **2002**, *305*, 127–135.
49. Köpf, M.; Corinth, C.; Haferkamp, O.; Nonnenmacher, T. F. Anomalous diffusion of water in biological tissues. *Biophys. J.* **1996**, *70* (6), 2950–8.
50. Noertemann, K.; Hilland, J.; Kaatze, U. Dielectric Properties of Aqueous NaCl Solutions at Microwave Frequencies. *J. Phys. Chem. A* **1997**, *101*, 6864–6869.
51. Buchner, R.; Hefter, G. T.; May, P. M. Dielectric Relaxation of Aqueous NaCl Solutions. *J. Phys. Chem. A* **1999**, *103* (1), 1–9.
52. Chen, T.; Hefter, G.; Buchner, R. Dielectric spectroscopy of aqueous solutions of KCl and CsCl. *J. Phys. Chem. A* **2003**, *107*, 4025–4031.

53. Gulich, R.; Köhler, M.; Lunkenheimer, P.; Loidl, A. Dielectric spectroscopy on aqueous electrolytic solutions. *Radiat. Environ. Biophys.* **2009**, *48*, 107–114.
54. Peyman, A.; Gabriel, C.; Grant, E. H. Complex permittivity of sodium chloride solutions at microwave frequencies. *Bioelectromagnetics* **2007**, *28*, 264–274.
55. Miyazaki, T.; Mogami, G.; Wazawa, T.; Kodama, T.; Suzuki, M. Measurement of the dielectric relaxation property of water-ion loose complex in aqueous solutions of salt at low concentrations. *J. Phys. Chem. A* **2008**, *112*, 10801–10806.
56. Kaatze, U. The dielectric properties of water in its different states of interaction. *J. Solut. Chem.* **1997**, *26* (11), 1049–1112.
57. Livshits, L.; Caduff, A.; Talary, M. S.; Lutz, H. U.; Hayashi, Y.; Puzenko, A.; Shendrik, A.; Feldman, Y. The Role of GLUT1 in the Sugar-Induced Dielectric Response of Human Erythrocytes. *J. Phys. Chem. B* **2009**, *113* (7), 2212–2220.
58. Meyer-Almes, F. J.; Porschke, D. J. The Cyclic AMP Receptor Promoter DNA Complex: A Comparison of Crystal and Solution Structure by Quantitative Molecular Electrooptics. *Mol. Biol.* **1997**, *269*, 842–850.
59. Lampinen, M. J.; Noponen, T. Electric dipole theory and thermodynamics of actomyosin molecular motor in muscle contraction. *Theor Biol* **2005**, *236*, 397–421.
60. Ellison, W. J. Permittivity of Pure Water, at Standard Atmospheric Pressure, over the Frequency Range 0–25THz and the Temperature Range. *J. Phys. Chem. Ref. Data* **2007**, *36* (1), 1–18.
61. Lodish, H.; Baltimore, D.; Berk, A.; Zipursky, S. L.; Matsudaira, P.; Darnell, J. *Molecular Cell Biology*, 3rd ed.; Scientific American Books Inc.: New York, 1995.
62. Caduff, A.; Ben Ishai, P.; Feldman, Y. Continuous noninvasive glucose monitoring; water as a relevant marker of glucose uptake in vivo. *Biophys. Rev.* **2019**, *8* (11), 1–19.
63. Whitford, D. *Proteins: Structure and Function*; John Wiley & Sons: 2013.
64. Latypova, L.; Barshtein, G.; Puzenko, A.; Poluektov, Y.; Anashkina, A.; Petrushanko, I.; Fenk, S.; Bogdanova, A.; Feldman, Y. Oxygenation state of hemoglobin defines dynamics of water molecules in its vicinity. *J. Chem. Phys.* **2020**, *153*, 135101.
65. Takashima, S.; Schwan, H. P. Dielectric Dispersion of Crystalline Powders of Amino Acids, Peptides, and Proteins. *J. Phys. Chem. B* **1965**, *69* (12), 4176–4182.
66. Chevion, M.; Ilan, Y. A.; Samuni, A.; Navok, T.; Czapski, G. Quaternary structure of methemoglobin. Pulse radiolysis study of the binding of oxygen to the valence hybrid. *J. Biol. Chem.* **1979**, *254*, 6370–6374.
67. Talley, K.; Alexov, E. On the pH-optimum of activity and stability of proteins. *Proteins* **2010**, *78*, 2699–2706.

Editor's Biography

William Henry Hunter Woodward

William Henry Hunter Woodward joined Dow Chemical Core R&D in 2011 and has contributed to several projects involving the electrical, thermal, and rheological characterization of polymers. Working in Dow Core R&D, he has enhanced Dow's capabilities in Broadband Dielectric Spectroscopy and applied it in innovative ways for such industries as Oil, Gas and Mining, Building Materials, Food and Pharmaceuticals, Electronic Materials, and Packaging and Specialty Plastics. Hunter has co-authored 12 patent applications and 30 journal articles. He received his B.S. in Chemistry from Dickinson College and his Ph.D. in Chemistry from Penn State.

Indexes

Author Index

Subject Index

P

W